Graduate Texts in Mathematics 160

BOOKS OF RELATED INTEREST BY SERGE LANG

Linear Algebra, Third Edition
1987, ISBN 96412-6

Undergraduate Algebra, Second Edition
1990, ISBN 97279-X

Complex Analysis, Third Edition
1993, ISBN 97886-0

Real and Functional Analysis, Third Edition
1993, ISBN 94001-4

Algebraic Number Theory, Second Edition
1994, ISBN 94225-4

Introduction to Complex Hyperbolic Spaces
1987, ISBN 96447-9

OTHER BOOKS BY LANG PUBLISHED BY
SPRINGER-VERLAG

Introduction to Arakelov Theory • Riemann-Roch Algebra (with William Fulton) • Complex Multiplication • Introduction to Modular Forms • Modular Units (with Daniel Kubert) • Fundamentals of Diophantine Geometry • Elliptic Functions • Number Theory III • Cyclotomic Fields I and II • $SL_2(R)$ • Abelian Varieties • Introduction to Algebraic and Abelian Functions • Undergraduate Analysis • Elliptic Curves: Diophantine Analysis • Introduction to Linear Algebra • Calculus of Several Variables • First Course in Calculus • Basic Mathematics • Geometry: A High School Course (with Gene Murrow) • Math! Encounters with High School Students • The Beauty of Doing Mathematics • THE FILE

Serge Lang

Differential and Riemannian Manifolds

With 20 Illustrations

Springer-Verlag
New York Berlin Heidelberg London Paris
Tokyo Hong Kong Barcelona Budapest

Serge Lang
Department of Mathematics
Yale University
New Haven, CT 06520
USA

Library of Congress Cataloging-in-Publication Data
Lang, Serge, 1927–
    Differential and Riemannian manifolds / Serge Lang.
        p.   cm. — (Graduate texts in mathematics ; 160)
    Includes bibliographical references (p.   -   ) and index.
    ISBN 0-387-94338-2 (acid-free)
    1. Differentiable manifolds. 2. Riemannian manifolds.   I. Title.
II. Series.
QA614.3.L34   1995b
516.3'6—dc20   95-1594

This is the third edition of *Differential Manifolds*, originally published by Addison-Wesley in 1962.

Printed on acid-free paper.

Production coordinated by Brian Howe and managed by Terry Kornak; manufacturing supervised by Jeffrey Taub.
Typeset by Asco Trade Typesetting Ltd., Hong Kong.
Printed and bound by Braun-Brumfield, Ann Arbor, MI.
Printed in the United States of America.

9 8 7 6 5 4 3 2 1

ISBN 0-387-94338-2 Springer-Verlag New York Berlin Heidelberg
ISBN 3-540-94338-2 Springer-Verlag Berlin Heidelberg New York

Preface

This is the third version of a book on differential manifolds. The first version appeared in 1962, and was written at the very beginning of a period of great expansion of the subject. At the time, I found no satisfactory book for the foundations of the subject, for multiple reasons. I expanded the book in 1971, and I expand it still further today. Specifically, I have added three chapters on Riemannian and pseudo Riemannian geometry, that is, covariant derivatives, curvature, and some applications up to the Hopf–Rinow and Hadamard–Cartan theorems, as well as some calculus of variations and applications to volume forms. I have rewritten the sections on sprays, and I have given more examples of the use of Stokes' theorem. I have also given many more references to the literature, all of this to broaden the perspective of the book, which I hope can be used among things for a general course leading into many directions. The present book still meets the old needs, but fulfills new ones.

At the most basic level, the book gives an introduction to the basic concepts which are used in differential topology, differential geometry, and differential equations. In differential topology, one studies for instance homotopy classes of maps and the possibility of finding suitable differentiable maps in them (immersions, embeddings, isomorphisms, etc.). One may also use differentiable structures on topological manifolds to determine the topological structure of the manifold (for example, à la Smale [Sm 67]). In differential geometry, one puts an additional structure on the differentiable manifold (a vector field, a spray, a 2-form, a Riemannian metric, ad lib.) and studies properties connected especially with these objects. Formally, one may say that one studies properties invariant under the group of differentiable automorphisms which preserve the additional structure. In differential equations, one studies vector fields and their integral curves, singular points, stable and unstable manifolds, etc. A certain number of concepts are essential for all three, and are so basic and elementary that it is worthwhile to collect them together so that more advanced expositions can be given without having to start from the very beginnings.

It is possible to lay down *at no extra cost* the foundations (and much more beyond) for manifolds modeled on Banach or Hilbert spaces rather than finite dimensional spaces. In fact, it turns out that the exposition gains considerably from the systematic elimination of the indiscriminate use of local coordinates $x_1, \ldots, x_n$ and $dx_1, \ldots, dx_n$. These are replaced by what they stand for, namely isomorphisms of open subsets of the manifold on open subsets of Banach spaces (local charts), and a local analysis of the situation which is more powerful and equally easy to use formally. In most cases, the finite dimensional proof extends at once to an invariant infinite dimensional proof. Furthermore, in studying differential forms, one needs to know only the definition of multilinear continuous maps. An abuse of multilinear algebra in standard treatises arises from an unnecessary double dualization and an abusive use of the tensor product.

I don't propose, of course, to do away with local coordinates. They are useful for computations, and are also especially useful when integrating differential forms, because the $dx_1 \wedge \cdots \wedge dx_n$ corresponds to the $dx_1 \cdots dx_n$ of Lebesgue measure, in oriented charts. Thus we often give the local coordinate formulation for such applications. Much of the literature is still covered by local coordinates, and I therefore hope that the neophyte will thus be helped in getting acquainted with the literature. I also hope to convince the expert that nothing is lost, and much is gained, by expressing one's geometric thoughts without hiding them under an irrelevant formalism.

It is profitable to deal with infinite dimensional manifolds, modeled on a Banach space in general, a self-dual Banach space for pseudo Riemannian geometry, and a Hilbert space for Riemannian geometry. In the standard pseudo Riemannian and Riemannian theory, readers will note that the differential theory works in these infinite dimensional cases, with the Hopf–Rinow theorem as the single exception, but not the Cartan–Hadamard theorem and its corollaries. Only when one comes to dealing with volumes and integration does finite dimensionality play a major role. Even if via the physicists with their Feynman integration one eventually develops a coherent analogous theory in the infinite dimensional case, there will still be something special about the finite dimensional case.

One major function of finding proofs valid in the infinite dimensional case is to provide proofs which are especially natural and simple in the finite dimensional case. Even for those who want to deal only with finite dimensional manifolds, I urge them to consider the proofs given in this book. In many cases, proofs based on coordinate free local representations in charts are clearer than proofs which are replete with the claws of a rather unpleasant prying insect such as Γ_{jkl}^{i}. Indeed, the bilinear map associated with a spray (which is the quadratic map corresponding to a symmetric connection) satisfies quite a nice local formalism in charts. I think the local representation of the curvature tensor as in Proposition 1.2 of Chapter IX shows the efficiency of this formalism and its superiority over local coordinates. Readers may also find it instructive to compare the proof of Proposition 2.6 of Chapter IX concerning the rate of growth of Jacobi fields with more classical ones involving coordinates as in [He 78], pp. 71–73.

Of course, there are also direct applications of the infinite dimensional case. Some of them are to the calculus of variations and to physics, for instance as in Abraham–Marsden [AbM 78]. It may also happen that one does not need formally the infinite dimensional setting, but that it is useful to keep in mind to motivate the methods and approach taken in various directions. For instance, by the device of using curves, one can reduce what is a priori an infinite dimensional question to ordinary calculus in finite dimensional space, as in the standard variation formulas given in Chapter IX, §4.

Similarly, the proper domain for the geodesic part of Morse theory is the loop space (or the space of certain paths), viewed as an infinite dimensional manifold, but a substantial part of the theory can be developed without formally introducing this manifold. The reduction to the finite dimensional case is of course a very interesting aspect of the situation, from which one can deduce deep results concerning the finite dimensional manifold itself, but it stops short of a complete analysis of the loop space. (Cf. Boot [Bo 60], Milnor [Mi 63].) This was already mentioned in the first version of the book, and since then, the papers of Palais [Pa 63] and Smale [Sm 64] appeared, carrying out the program. They determined the appropriate condition in the infinite dimensional case under which this theory works.

In addition, given two finite dimensional manifolds X, Y it is fruitful to give the set of differentiable maps from X to Y an infinite dimensional manifold structure, as was started by Eells [Ee 58], [Ee 59], [Ee 61], and [Ee 66]. By so doing, one transcends the purely formal translation of finite dimensional results getting essentially new ones, which would in turn affect the finite dimensional case.

Foundations for the geometry of manifolds of mappings are given in Abraham's notes of Smale's lectures [Ab 60] and Palais's monograph [Pa 68].

For more recent applications to critical point theory and submanifold geometry, see [PaT 88].

One especially interesting case of Banach manifolds occurs in the theory of Teichmuller spaces, which, as shown by Bers, can be embedded as submanifolds of a complex Banach space. Cf. [Ga 87], [Vi 73].

In the direction of differential equations, the extension of the stable and unstable manifold theorem to the Banach case, already mentioned as a possibility in the earlier version of this book, was proved quite elegantly by Irwin [Ir 70], following the idea of Pugh and Robbin for dealing with local flows using the implicit mapping theorem in Banach spaces. I have included the Pugh–Robbin proof, but refer to Irwin's paper for the stable manifold theorem which belongs at the very beginning of the theory of ordinary differential equations. The Pugh–Robbin proof can also be adjusted to hold for vector fields of class H^p (Sobolev spaces), of importance in partial differential equations, as shown by Ebin and Marsden [EbM 70].

It is a standard remark that the C^∞-functions on an open subset of a euclidean space do not form a Banach space. They form a Fréchet space (denumerably many norms instead of one). On the other hand, the implicit function theorem and the local existence theorem for differential equations are not true in the more general case. In order to recover similar results, a much more sophisticated theory is needed, which is only beginning to be developed. (Cf. Nash's paper on Riemannian metrics [Na 56], and subsequent contributions of Schwartz [Sc 60] and Moser [Mo 61].) In particular, some additional structure must be added (smoothing operators). Cf. also my Bourbaki seminar talk on the subject [La 61]. This goes beyond the scope of this book, and presents an active topic for research.

I have emphasized differential aspects of differential manifolds rather than topological ones. I am especially interested in laying down basic material which may lead to various types of applications which have arisen since the sixties, vastly expanding the perspective on differential geometry and analysis. For instance, I expect the marvelous book [BGV 92] to be only the first of many to present the accumulated vision from the seventies and eighties, after the work of Atiyah, Bismut, Bott, Gilkey, McKean, Patodi, Singer, and many others.

New Haven, 1994   SERGE LANG

Acknowledgments

I have greatly profited from several sources in writing this book. These sources include some from the 1960s, and some more recent ones.

First, I originally profited from Dieudonné's *Foundations of Modern Analysis*, which started to emphasize the Banach point of view.

Second, I originally profited from Bourbaki's *Fascicule de résultats* [Bou 69] for the foundations of differentiable manifolds. This provides a good guide as to what should be included. I have not followed it entirely, as I have omitted some topics and added others, but on the whole, I found it quite useful. I have put the emphasis on the differentiable point of view, as distinguished from the analytic. However, to offset this a little, I included two analytic applications of Stokes' formula, the Cauchy theorem in several variables, and the residue theorem.

Third, Milnor's notes [Mi 58], [Mi 59], [Mi 61] proved invaluable. They were of course directed toward differential topology, but of necessity had to cover ad hoc the foundations of differentiable manifolds (or, at least, part of them). In particular, I have used his treatment of the operations on vector bundles (Chapter III, §4) and his elegant exposition of the uniqueness of tubular neighborhoods (Chapter IV, §6, and Chapter VII, §4).

Fourth, I am very much indebted to Palais for collaborating on Chapter IV, and giving me his exposition of sprays (Chapter IV, §3). As he showed me, these can be used to construct tubular neighborhoods. Palais also showed me how one can recover sprays and geodesics on a Riemannian manifold by making direct use of the fundamental 2-form and the metric (Chapter VII, §7). This is a considerable improvement on past expositions.

Finally, in the direction of differential geometry, I found Berger–Gauduchon–Mazet [BGM 71] extremely valuable, especially in the way they lead to the study of the Laplacian and the heat equation. This book has been very influential, for instance for [GHL 87/93], which I have also found useful.

Contents

Preface v

CHAPTER I

Differential Calculus 1

§1. Categories 2
§2. Topological Vector Spaces 3
§3. Derivatives and Composition of Maps 6
§4. Integration and Taylor's Formula 10
§5. The Inverse Mapping Theorem 13

CHAPTER II

Manifolds 20

§1. Atlases, Charts, Morphisms 20
§2. Submanifolds, Immersions, Submersions 23
§3. Partitions of Unity 31
§4. Manifolds with Boundary 36

CHAPTER III

Vector Bundles 40

§1. Definition, Pull Backs 40
§2. The Tangent Bundle 48
§3. Exact Sequences of Bundles 49
§4. Operations on Vector Bundles 55
§5. Splitting of Vector Bundles 60

CHAPTER IV

Vector Fields and Differential Equations 64

§1. Existence Theorem for Differential Equations 65
§2. Vector Fields, Curves, and Flows 86

§3. Sprays 94
§4. The Flow of a Spray and the Exponential Map 103
§5. Existence of Tubular Neighborhoods 108
§6. Uniqueness of Tubular Neighborhoods 110

CHAPTER V

Operations on Vector Fields and Differential Forms 114

§1. Vector Fields, Differential Operators, Brackets 114
§2. Lie Derivative 120
§3. Exterior Derivative 122
§4. The Poincaré Lemma 135
§5. Contractions and Lie Derivative 137
§6. Vector Fields and 1-Forms Under Self Duality 141
§7. The Canonical 2-Form 146
§8. Darboux's Theorem 148

CHAPTER VI

The Theorem of Frobenius 153

§1. Statement of the Theorem 153
§2. Differential Equations Depending on a Parameter 158
§3. Proof of the Theorem 159
§4. The Global Formulation 160
§5. Lie Groups and Subgroups 163

CHAPTER VII

Metrics 169

§1. Definition and Functoriality 169
§2. The Hilbert Group 173
§3. Reduction to the Hilbert Group 176
§4. Hilbertian Tubular Neighborhoods 179
§5. The Morse–Palais Lemma 182
§6. The Riemannian Distance 184
§7. The Canonical Spray 188

CHAPTER VIII

Covariant Derivatives and Geodesics 191

§1. Basic Properties 191
§2. Sprays and Covariant Derivatives 194
§3. Derivative Along a Curve and Parallelism 199
§4. The Metric Derivative 203
§5. More Local Results on the Exponential Map 209
§6. Riemannian Geodesic Length and Completeness 216

CHAPTER IX

Curvature 225

§1. The Riemann Tensor 225
§2. Jacobi Lifts 233

§3. Application of Jacobi Lifts to $d\exp_x$ 240
§4. The Index Form, Variations, and the Second Variation Formula 249
§5. Taylor Expansions 257

CHAPTER X

Volume Forms 261

§1. The Riemannian Volume Form 261
§2. Covariant Derivatives 264
§3. The Jacobian Determinant of the Exponential Map 268
§4. The Hodge Star on Forms 273
§5. Hodge Decomposition of Differential Forms 279

CHAPTER XI

Integration of Differential Forms 284

§1. Sets of Measure 0 284
§2. Change of Variables Formula 288
§3. Orientation 297
§4. The Measure Associated with a Differential Form 299

CHAPTER XII

Stokes' Theorem 307

§1. Stokes' Theorem for a Rectangular Simplex 307
§2. Stokes' Theorem on a Manifold 310
§3. Stokes' Theorem with Singularities 314

CHAPTER XIII

Applications of Stokes' Theorem 321

§1. The Maximal de Rham Cohomology 321
§2. Moser's Theorem 328
§3. The Divergence Theorem 329
§4. The Adjoint of d for Higher Degree Forms 333
§5. Cauchy's Theorem 335
§6. The Residue Theorem 339

APPENDIX

The Spectral Theorem 343

§1. Hilbert Space 343
§2. Functionals and Operators 344
§3. Hermitian Operators 347

Bibliography 355

Index 361

CHAPTER I

Differential Calculus

We shall recall briefly the notion of derivative and some of its useful properties. As mentioned in the foreword, Chapter VIII of Dieudonné's book or my book on real analysis [La 93] give a self-contained and complete treatment for Banach spaces. We summarize certain facts concerning their properties as topological vector spaces, and then we summarize differential calculus. *The reader can actually skip this chapter* and start immediately with Chapter II if the reader is accustomed to thinking about the derivative of a map as a linear transformation. (In the finite dimensional case, when bases have been selected, the entries in the matrix of this transformation are the partial derivatives of the map.) We have repeated the proofs for the more important theorems, for the ease of the reader.

It is convenient to use throughout the language of categories. The notion of category and morphism (whose definitions we recall in §1) is designed to abstract what is common to certain collections of objects and maps between them. For instance, topological vector spaces and continuous linear maps, open subsets of Banach spaces and differentiable maps, differentiable manifolds and differentiable maps, vector bundles and vector bundle maps, topological spaces and continuous maps, sets and just plain maps. In an arbitrary category, maps are called morphisms, and in fact the category of differentiable manifolds is of such importance in this book that from Chapter II on, we use the word morphism synonymously with differentiable map (or *p*-times differentiable map, to be precise). All other morphisms in other categories will be qualified by a prefix to indicate the category to which they belong.

I, §1. CATEGORIES

A **category** is a collection of objects $\{X, Y, \ldots\}$ such that for two objects X, Y we have a set $\text{Mor}(X, Y)$ and for three objects X, Y, Z a mapping (composition law)

$$\text{Mor}(X, Y) \times \text{Mor}(Y, Z) \to \text{Mor}(X, Z)$$

satisfying the following axioms:

> **CAT 1.** *Two sets $\text{Mor}(X, Y)$ and $\text{Mor}(X', Y')$ are disjoint unless $X = X'$ and $Y = Y'$, in which case they are equal.*

> **CAT 2.** *Each $\text{Mor}(X, X)$ has an element id_X which acts as a left and right identity under the composition law.*

> **CAT 3.** *The composition law is associative.*

The elements of $\text{Mor}(X, Y)$ are called **morphisms**, and we write frequently $f: X \to Y$ for such a morphism. The composition of two morphisms f, g is written fg or $f \circ g$.

A **functor** $\lambda: \mathfrak{A} \to \mathfrak{A}'$ from a category $\mathfrak{A}$ into a category $\mathfrak{A}'$ is a map which associates with each object X in $\mathfrak{A}$ an object $\lambda(X)$ in $\mathfrak{A}'$, and with each morphism $f: X \to Y$ a morphism $\lambda(f): \lambda(X) \to \lambda(Y)$ in $\mathfrak{A}'$ such that, whenever f and g are morphisms in $\mathfrak{A}$ which can be composed, then $\lambda(fg) = \lambda(f)\lambda(g)$ and $\lambda(\text{id}_X) = \text{id}_{\lambda(X)}$ for all X. This is in fact a covariant functor, and a contravariant functor is defined by reversing the arrows (so that we have $\lambda(f): \lambda(Y) \to \lambda(X)$ and $\lambda(fg) = \lambda(g)\lambda(f)$).

In a similar way, one defines functors of many variables, which may be covariant in some variables and contravariant in others. We shall meet such functors when we discuss multilinear maps, differential forms, etc.

The functors of the same variance from one category $\mathfrak{A}$ to another $\mathfrak{A}'$ form themselves the objects of a category $\text{Fun}(\mathfrak{A}, \mathfrak{A}')$. Its morphisms will sometimes be called **natural transformations** instead of functor morphisms. They are defined as follows. If λ, μ are two functors from $\mathfrak{A}$ to $\mathfrak{A}'$ (say covariant), then a natural transformation $t: \lambda \to \mu$ consists of a collection of morphisms

$$t_X: \lambda(X) \to \mu(X)$$

as X ranges over $\mathfrak{A}$, which makes the following diagram commutative for any morphism $f: X \to Y$ in $\mathfrak{A}$:

$$
\begin{array}{ccc}
\lambda(X) & \xrightarrow{\ t_X\ } & \mu(X) \\
\lambda(f) \downarrow & & \downarrow \mu(f) \\
\lambda(Y) & \xrightarrow[\ t_Y\ ]{} & \mu(Y)
\end{array}
$$

In any category $\mathfrak{A}$, we say that a morphism $f\colon X \to Y$ is an **isomorphism** if there exists a morphism $g\colon Y \to X$ such that fg and gf are the identities. For instance, an isomorphism in the category of topological spaces is called a topological isomorphism, or a homeomorphism. In general, we describe the category to which an isomorphism belongs by means of a suitable prefix. In the category of sets, a set-isomorphism is also called a bijection.

If $f\colon X \to Y$ is a morphism, then a **section** of f is defined to be a morphism $g\colon Y \to X$ such that $f \circ g = \mathrm{id}_Y$.

I, §2. TOPOLOGICAL VECTOR SPACES

The proofs of all statements in this section, including the Hahn–Banach theorem and the closed graph theorem, can be found in [La 93].

A **topological vector space** E (over the reals $\mathbf{R}$) is a vector space with a topology such that the operations of addition and scalar multiplication are continuous. It will be convenient to assume also, as part of the definition, that the space is **Hausdorff**, and **locally convex**. By this we mean that every neighborhood of 0 contains an open neighborhood U of 0 such that, if x, y are in U and $0 \leq t \leq 1$, then $tx + (1 - t)y$ also lies in U.

The topological vector spaces form a category, denoted by TVS, if we let the morphisms be the continuous linear maps (by linear we mean throughout $\mathbf{R}$-linear). The set of continuous linear maps of one topological vector space E into F is denoted by $L(E, F)$. The continuous r-multilinear maps

$$\psi\colon \mathbf{E} \times \cdots \times \mathbf{E} \to \mathbf{F}$$

of E into F will be denoted by $L^r(\mathbf{E}, \mathbf{F})$. Those which are symmetric (resp. alternating) will be denoted by $L_s^r(\mathbf{E}, \mathbf{F})$ or $L_{\mathrm{sym}}^r(\mathbf{E}, \mathbf{F})$ (resp. $L_a^r(\mathbf{E}, \mathbf{F})$). The isomorphisms in the category TVS are called **toplinear** isomorphisms, and we write $\mathrm{Lis}(\mathbf{E}, \mathbf{F})$ and $\mathrm{Laut}(\mathbf{E})$ for the toplinear isomorphisms of E onto F and the toplinear automorphisms of E.

We find it convenient to denote by $L(\mathbf{E})$, $L^r(\mathbf{E})$, $L_s^r(\mathbf{E})$, and $L_s^r(\mathbf{E})$ the continuous linear maps of E into $\mathbf{R}$ (resp. the continuous, r-multilinear, symmetric, alternating maps of E into $\mathbf{R}$). Following classical terminology, it is also convenient to call such maps into $\mathbf{R}$ **forms** (of the corresponding type). If $\mathbf{E}_1, \ldots, \mathbf{E}_r$ and F are topological vector spaces, then we denote by $L(\mathbf{E}_1, \ldots, \mathbf{E}_r; F)$ the continuous multilinear maps of the product $\mathbf{E}_1 \times \cdots \times \mathbf{E}_r$ into F. We let:

$\mathrm{End}(\mathbf{E}) = L(\mathbf{E}, \mathbf{E})$,

$\mathrm{Laut}(\mathbf{E}) =$ elements of $\mathrm{End}(\mathbf{E})$ which are invertible in $\mathrm{End}(\mathbf{E})$.

The most important type of topological vector space for us is the **Banachable space** (a TVS which is complete, and whose topology can be defined by a norm). We should say **Banach** space when we want to put the norm into the structure. There are of course many norms which can be used to make a Banachable space into a Banach space, but in practice, one allows the abuse of language which consists in saying Banach space for Banachable space (unless it is absolutely necessary to keep the distinction).

For this book, we assume from now on that all our topological vector spaces are Banach spaces. We shall occasionally make some comments to indicate where it might be possible to generalize certain results to more general spaces. We denote our Banach spaces by **E, F, ...** .

The next two propositions give two aspects of what is known as the **closed graph theorem.**.

Proposition 2.1. *Every continuous bijective linear map of **E** onto **F** is a toplinear isomorphism.*

Proposition 2.2. *If **E** is a Banach space, and $\mathbf{F}_1$, $\mathbf{F}_2$ are two closed subspaces which are complementary (i.e. $\mathbf{E} = \mathbf{F}_1 + \mathbf{F}_2$ and $\mathbf{F}_1 \cap \mathbf{F}_2 = 0$), then the map of $\mathbf{F}_1 \times \mathbf{F}_2$ onto **E** given by the sum is a toplinear isomorphism.*

We shall frequently encounter a situation as in Proposition 2.2, and if **F** is a closed subspace of **E** such that there exists a closed complement $\mathbf{F}_1$ such that **E** is toplinearly isomorphic to the product of **F** and $\mathbf{F}_1$ under the natural mapping, then we shall say that **F splits** in **E**.

Next, we state a weak form of the Hahn–Banach theorem.

Proposition 2.3. *Let **E** be a Banach space and $x \neq 0$ an element of **E**. Then there exists a continuous linear map λ of **E** into **R** such that $\lambda(x) \neq 0$.*

One constructs λ by Zorn's lemma, supposing that λ is defined on some subspace, and having a bounded norm. One then extends λ to the subspace generated by one additional element, without increasing the norm.

In particular, every finite dimensional subspace of **E** splits if **E** is complete. More trivially, we observe that a finite codimensional closed subspace also splits.

We now come to the problem of putting a topology on $L(\mathbf{E}, \mathbf{F})$. Let **E, F** be Banach spaces, and let

$$A: \mathbf{E} \to \mathbf{F}$$

be a continuous linear map (also called a bounded linear map). We can then define the **norm** of A to be the greatest lower bound of all numbers K such that

$$|Ax| \leqq K|x|$$

for all $x \in \mathbf{E}$. This norm makes $L(\mathbf{E}, \mathbf{F})$ into a Banach space.

In a similar way, we define the topology of $L(\mathbf{E}_1, \ldots, \mathbf{E}_r; \mathbf{F})$, which is a Banach space if we define the norm of a multilinear continuous map

$$A: \mathbf{E}_1 \times \cdots \times \mathbf{E}_r \to \mathbf{F}$$

by the greatest lower bound of all numbers K such that

$$|A(x_1, \ldots, x_r)| \leqq K|x_1| \cdots |x_r|.$$

We have:

Proposition 2.4. *If* $\mathbf{E}_1, \ldots, \mathbf{E}_r, \mathbf{F}$ *are Banach spaces, then the canonical map*

$$L\big(\mathbf{E}_1, L(\mathbf{E}_2, \ldots, L(\mathbf{E}_r, \mathbf{F}), \ldots)\big) \to L^r(\mathbf{E}_1, \ldots, \mathbf{E}_r; \mathbf{F})$$

from the repeated continuous linear maps to the continuous multilinear maps is a toplinear isomorphism, which is norm-preserving, i.e. a Banach-isomorphism.

The preceding propositions could be generalized to a wider class of topological vector spaces. The following one exhibits a property peculiar to Banach spaces.

Proposition 2.5. *Let* $\mathbf{E}$, $\mathbf{F}$ *be two Banach spaces. Then the set of toplinear isomorphisms* $\mathrm{Lis}(\mathbf{E}, \mathbf{F})$ *is open in* $L(\mathbf{E}, \mathbf{F})$.

The proof is in fact quite simple. If $\mathrm{Lis}(\mathbf{E}, \mathbf{F})$ is not empty, one is immediately reduced to proving that $\mathrm{Laut}(\mathbf{E})$ is open in $L(\mathbf{E}, \mathbf{E})$. We then remark that if $u \in L(\mathbf{E}, \mathbf{E})$, and $|u| < 1$, then the series

$$1 + u + u^2 + \cdots$$

converges. Given any toplinear automorphism w of $\mathbf{E}$, we can find an open neighborhood by translating the open unit ball multiplicatively from 1 to w.

Again in Banach spaces, we have:

Proposition 2.6. *If* $\mathbf{E}$, $\mathbf{F}$, $\mathbf{G}$ *are Banach spaces, then the bilinear maps*

$$L(\mathbf{E}, \mathbf{F}) \times L(\mathbf{F}, \mathbf{G}) \to L(\mathbf{E}, \mathbf{G}),$$

$$L(\mathbf{E}, \mathbf{F}) \times \mathbf{E} \to \mathbf{F},$$

*obtained by composition of mappings are continuous, and similarly for
multilinear maps.*

Remark. The preceding proposition is false for more general spaces
than Banach spaces, say Fréchet spaces. In that case, one might hope
that the following may be true. Let U be open in a Fréchet space and
let

$$f: U \to L(\mathbf{E}, \mathbf{F}),$$

$$g: U \to L(\mathbf{F}, \mathbf{G}),$$

be continuous. Let γ be the composition of maps. Then $\gamma(f, g)$ is continuous. The same type of question arises later, with differentiable maps
instead, and it is of course essential to know the answer to deal with the
composition of differentiable maps.

I, §3. DERIVATIVES AND COMPOSITION OF MAPS

A real valued function of a real variable, defined on some neighborhood
of 0 is said to be $o(t)$ if

$$\lim_{t \to 0} o(t)/t = 0.$$

Let $\mathbf{E}$, $\mathbf{F}$ be two topological vector spaces, and φ a mapping of a
neighborhood of 0 in $\mathbf{E}$ into $\mathbf{F}$. We say that φ is **tangent to** 0 if, given a
neighborhood W of 0 in $\mathbf{F}$, there exists a neighborhood V of 0 in $\mathbf{E}$ such
that

$$\varphi(tV) \subset o(t)W$$

for some function $o(t)$. If both $\mathbf{E}$, $\mathbf{F}$ are normed, then this amounts to
the usual condition

$$|\varphi(x)| \leq |x| \psi(x)$$

with $\lim \psi(x) = 0$ as $|x| \to 0$.

Let $\mathbf{E}$, $\mathbf{F}$ be two topological vector spaces and U open in $\mathbf{E}$. Let
$f: U \to \mathbf{F}$ be a continuous map. We shall say that f is **differentiable** at a
point $x_0 \in U$ if there exists a continuous linear map λ of $\mathbf{E}$ into $\mathbf{F}$ such
that, if we let

$$f(x_0 + y) = f(x_0) + \lambda y + \varphi(y)$$

for small y, then φ is tangent to 0. It then follows trivially that λ is
uniquely determined, and we say that it is the **derivative** of f at x_0. We
denote the derivative by $Df(x_0)$ or $f'(x_0)$. It is an element of $L(\mathbf{E}, \mathbf{F})$. If

f is differentiable at every point of U, then f' is a map

$$f': U \to L(\mathbf{E}, \mathbf{F}).$$

It is easy to verify the chain rule.

Proposition 3.1. *If* $f: U \to V$ *is differentiable at* x_0, *if* $g: V \to W$ *is differentiable at* $f(x_0)$, *then* $g \circ f$ *is differentiable at* x_0, *and*

$$(g \circ f)'(x_0) = g'(f(x_0)) \circ f'(x_0).$$

Proof. We leave it as a simple (and classical) exercise.

The rest of this section is devoted to the statements of the differential calculus. All topological vector spaces are assumed to be Banach spaces (i.e. Banachable). Then $L(\mathbf{E}, \mathbf{F})$ is also a Banach space, if $\mathbf{E}$ and $\mathbf{F}$ are Banach spaces.

Let U be open in $\mathbf{E}$ and let $f: U \to \mathbf{F}$ be differentiable at each point of U. If f' is continuous, then we say that f is **of class** C^1. We define maps of class C^p ($p \geq 1$) inductively. The p-th derivative $D^p f$ is defined as $D(D^{p-1}f)$ and is itself a map of U into

$$L(\mathbf{E}, L(\mathbf{E}, \ldots, L(\mathbf{E}, \mathbf{F}) \cdots))$$

which can be identified with $L^p(\mathbf{E}, \mathbf{F})$ by Proposition 2.4. A map f is said to be **of class** C^p if its kth derivative $D^k f$ exists for $1 \leq k \leq p$, and is continuous.

Remark. *Let* f *be of class* C^p, *on an open set* U *containing the origin. Suppose that* f *is locally homogeneous of degree* p *near* 0, *that is*

$$f(tx) = t^p f(x)$$

for all t *and* x *sufficiently small. Then for all sufficiently small* x *we have*

$$f(x) = \frac{1}{p!} D^p f(0) x^{(p)},$$

where $x^{(p)} = (x, x, \ldots, x)$, p *times.*

This is easily seen by differentiating p times the two expressions for $f(tx)$, and then setting $t = 0$. The differentiation is a trivial application of the chain rule.

Proposition 3.2. *Let* U, V *be open in Banach spaces. If* $f: U \to V$ *and* $g: V \to \mathbf{F}$ *are of class* C^p, *then so is* $g \circ f$.

From Proposition 3.2, we can view open subsets of Banach spaces as the objects of a category, whose morphisms are the continuous maps of class C^p. These will be called C^p**-morphisms**. We say that f is of class C^∞ if it is of class C^p for all integers $p \geq 1$. From now on, p is an integer ≥ 0 or ∞ (C^0 maps being the continuous maps). In practice, we omit the prefix C^p if the p remains fixed. Thus by **morphism**, throughout the rest of this book, we mean C^p-morphism with $p \leq \infty$. We shall use the word morphism also for C^p-morphisms of manifolds (to be defined in the next chapter), but *morphisms in any other category will always be prefixed so as to indicate the category to which they belong* (for instance bundle morphism, continuous linear morphism, etc.).

Proposition 3.3. *Let* U *be open in the Banach space* $\mathbf{E}$, *and let* $f: U \to \mathbf{F}$ *be a* C^p-morphism. *Then* $D^p f$ (*viewed as an element of* $L^p(\mathbf{E}, \mathbf{F})$) *is symmetric*.

Proposition 3.4. *Let* U *be open in* $\mathbf{E}$, *and let* $f_i: U \to \mathbf{F}_i$ ($i = 1, \ldots, n$) *be continuous maps into spaces* $\mathbf{F}_i$. *Let* $f = (f_1, \ldots, f_n)$ *be the map of* U *into the product of the* $\mathbf{F}_i$. *Then* f *is of class* C^p *if and only if each* f_i *is of class* C^p, *and in that case*

$$D^p f = (D^p f_1, \ldots, D^p f_n).$$

Let U, V be open in spaces $\mathbf{E}_1$, $\mathbf{E}_2$ and let

$$f: U \times V \to \mathbf{F}$$

be a continuous map into a Banach space. We can introduce the notion of partial derivative in the usual manner. If (x, y) is in $U \times V$ and we keep y fixed, then as a function of the first variable, we have the derivative as defined previously. This derivative will be denoted by $D_1 f(x, y)$. Thus

$$D_1 f: U \times V \to L(\mathbf{E}_1, \mathbf{F})$$

is a map of $U \times V$ into $L(\mathbf{E}_1, \mathbf{F})$. We call it the **partial deriative** with respect to the first variable. Similarly, we have $D_2 f$, and we could take n factors instead of 2. The total derivative and the partials are then related as follows.

Proposition 3.5. *Let* $U_1, \ldots, U_n$ *be open in the spaces* $\mathbf{E}_1, \ldots, \mathbf{E}_n$ *and let* $f: U_1 \times \cdots \times U_n \to \mathbf{F}$ *be a continuous map. Then* f *is of class* C^p *if and only if each partial derivative* $D_i f: U_1 \times \cdots U_n \to L(\mathbf{E}_i, \mathbf{F})$ *exists and is* *of class* C^{p-1}. *If that is the case, then for* $x = (x_1, \ldots, x_n)$ *and*

$$v = (v_1, \ldots, v_n) \in \mathbf{E}_1 \times \cdots \times \mathbf{E}_n,$$

we have

$$Df(x) \cdot (v_1, \ldots, v_n) = \sum D_i f(x) \cdot v_i.$$

The next four propositions are concerned with continuous linear and multilinear maps.

Proposition 3.6. *Let* **E**, **F** *be Banach spaces and* $f: \mathbf{E} \to \mathbf{F}$ *a continuous linear map. Then for each* $x \in \mathbf{E}$ *we have*

$$f'(x) = f.$$

Proposition 3.7. *Let* **E**, **F**, **G** *be Banach spaces, and* U *open in* **E**. *Let* $f: U \to \mathbf{F}$ *be of class* C^p *and* $g: \mathbf{F} \to \mathbf{G}$ *continuous and linear. Then* $g \circ f$ *is of class* C^p *and*

$$D^p(g \circ f) = g \circ D^p f.$$

Proposition 3.8. *If* $\mathbf{E}_1, \ldots, \mathbf{E}_r$ *and* **F** *are Banach spaces and*

$$f: \mathbf{E}_1 \times \cdots \times \mathbf{E}_r \to \mathbf{F}$$

a continuous multilinear map, then f *is of class* C^∞, *and its* $(r + 1)$-*st derivative is* 0. *If* $r = 2$, *then* Df *is computed according to the usual rule for derivative of a product (first times the derivative of the second plus derivative of the first times the second).*

Proposition 3.9. *Let* **E**, **F** *be Banach spaces which are toplinearly isomorphic. If* $u: \mathbf{E} \to \mathbf{F}$ *is a toplinear isomorphism, we denote its inverse by* u^{-1}. *Then the map*

$$u \mapsto u^{-1}$$

from $\mathrm{Lis}(\mathbf{E}, \mathbf{F})$ *to* $\mathrm{Lis}(\mathbf{F}, \mathbf{E})$ *is a* C^∞-*isomorphism. Its derivative at a point* u_0 *is the linear map of* $L(\mathbf{E}, \mathbf{F})$ *into* $L(\mathbf{F}, \mathbf{E})$ *given by the formula*

$$v \mapsto u_0^{-1} v u_0^{-1}.$$

Finally, we come to some statements which are of use in the theory of vector bundles.

Proposition 3.10. *Let* U *be open in the Banach space* **E** *and let* **F**, **G** *be Banach spaces.*

(i)   If $f: U \to L(\mathbf{E}, \mathbf{F})$ is a C^p-morphism, then the map of $U \times \mathbf{E}$ into $\mathbf{F}$
      given by
$$(x, v) \mapsto f(x)v$$
      is a morphism.

(ii) If $f: U \to L(\mathbf{E}, \mathbf{F})$ and $g: U \to L(\mathbf{F}, \mathbf{G})$ are morphisms, then so is
      $\gamma(f, g)$ (γ being the composition).

(iii) If $f: U \to \mathbf{R}$ and $g: U \to L(\mathbf{E}, \mathbf{F})$ are morphisms, so is fg (the value
      of fg at x is $f(x)g(x)$, ordinary multiplication by scalars).

(iv) If $f, g: U \to L(\mathbf{E}, \mathbf{F})$ are morphisms, so is $f + g$.

This proposition concludes our summary of results assumed without
proof.

I, §4. INTEGRATION AND TAYLOR'S FORMULA

Let $\mathbf{E}$ be a Banach space. Let I denote a real, closed interval, say
$a \leqq t \leqq b$. A **step mapping**
$$f: I \to \mathbf{E}$$

is a mapping such that there exists a finite number of disjoint subintervals $I_1, \dots, I_n$ covering I such that on each interval I_j, the mapping
has constant value, say v_j. We do not require the intervals I_j to be
closed. They may be open, closed, or half-closed.

Given a sequence of mappings f_n from I into $\mathbf{E}$, we say that it
converges uniformly if, given a neighborhood W of 0 into $\mathbf{E}$, there exists
an integer n_0 such that, for all n, $m > n_0$ and all $t \in I$, the difference
$f_n(t) - f_m(t)$ lies in W. The sequence f_n then converges to a mapping f of
I into $\mathbf{E}$.

A **ruled** mapping is a uniform limit of step mappings. We leave to the
reader the proof that every continuous mapping is ruled.

If f is a step mapping as above, we define its integral
$$\int_a^b f = \int_a^b f(t)\, dt = \sum \mu(I_j)v_j,$$

where $\mu(I_j)$ is the length of the interval I_j (its measure in the standard
Lebesgue measure). This integral is independent of the choice of intervals
I_j on which f is constant.

If f is ruled and $f = \lim f_n$ (lim being the uniform limit), then the
sequence
$$\int_a^b f_n$$

converges in $\mathbf{E}$ to an element of $\mathbf{E}$ independent of the particular sequence

f_n used to approach f uniformly. We denote this limit by

$$\int_a^b f = \int_a^b f(t)\, dt$$

and call it the **integral** of f. The integral is linear in f, and satisfies the usual rules concerning changes of intervals. (If $b < a$ then we define $\int_a^b$ to be minus the integral from b to a.)

As an immediate consequence of the definition, we get:

Proposition 4.1. *Let* $\lambda: E \to R$ *be a continuous linear map and let* $f: I \to E$ *be ruled. Then* $\lambda f = \lambda \circ f$ *is ruled, and*

$$\lambda \int_a^b f(t)\, dt = \int_a^b \lambda f(t)\, dt.$$

Proof. If f_n is a sequence of step functions converging uniformly to f, then λf_n is ruled and converges uniformly to λf. Our formula follows at once.

Taylor's Formula. *Let* E, F *be Banach spaces. Let* U *be open in* E. *Let* x, y *be two points of* U *such that the segment* $x + ty$ *lies in* U *for* $0 \leq t \leq 1$. *Let*

$$f: U \to F$$

be a C^p-*morphism, and denote by* $y^{(p)}$ *the "vector"* $(y, \ldots, y)$ p *times. Then the function* $D^p f(x + ty) \cdot y^{(p)}$ *is continuous in* t, *and we have*

$$f(x + y) = f(x) + \frac{Df(x)y}{1!} + \cdots + \frac{D^{p-1}f(x)y^{(p-1)}}{(p-1)!}$$

$$+ \int_0^1 \frac{(1-t)^{p-1}}{(p-1)!} D^p f(x + ty)y^{(p)}\, dt.$$

Proof. By the Hahn–Banach theorem, it suffices to show that both sides give the same thing when we apply a functional λ (continuous linear map into R). This follows at once from Proposition 3.7 and 4.1, together with the known result when $F = R$. In this case, the proof proceeds by induction on p, and integration by parts, starting from

$$f(x + y) - f(x) = \int_0^1 Df(x + ty)y\, dt.$$

The next two corollaries are known as the **mean value theorem**.

Corollary 4.2. *Let* E, F *be two Banach spaces,* U *open in* E, *and* x, z *two distinct points of* U *such that the segment* $x + t(z - x)$ $(0 \leq t \leq 1)$ *lies in* U. *Let* $f : U \to F$ *be continuous and of class* C^1. *Then*

$$|f(z) - f(x)| \leq |z - x| \sup |f'(\xi)|,$$

the sup being taken over ξ *in the segment.*

Proof. This comes from the usual estimations of the integral. Indeed, for any continuous map $g : I \to F$ we have the estimate

$$\left| \int_a^b g(t)\, dt \right| \leq K(b - a)$$

if K is a bound for g on I, and $a \leq b$. This estimate is obvious for step functions, and therefore follows at once for continuous functions.

Another version of the mean value theorem is frequently used.

Corollary 4.3. *Let the hypotheses be as in Corollary* 4.2. *Let* x_0 *be a point on the segment between* x *and* z. *Then*

$$|f(z) - f(x) - f'(x_0)(z - x)| \leq |z - x| \sup |f'(\xi) - f'(x_0)|,$$

the sup taken over all ξ *on the segment.*

Proof. We apply Corollary 4.2 to the map

$$g(x) = f(x) - f'(x_0)x.$$

Finally, let us make some comments on the estimate of the remainder term in Taylor's formula. We have assumed that $D^p f$ is continuous. Therefore, $D^p f(x + ty)$ can be written

$$D^p f(x + ty) = D^p f(x) + \psi(y, t),$$

where ψ depends on y, t (and x of course), and for fixed x, we have

$$\lim |\psi(y, t)| = 0$$

as $|y| \to 0$. Thus we obtain:

Corollary 4.4. *Let* E, F *be two Banach spaces,* U *open in* E, *and* x *a point of* U. *Let* $f : U \to F$ *be of class* C^p, $p \geq 1$. *Then for all* y *such* *that the segment $x + ty$ lies in U $(0 \leq t \leq 1)$, we have*

$$f(x + y) = f(x) + \frac{Df(x)y}{1!} + \cdots + \frac{D^p f(x)y^{(p)}}{p!} + \theta(y)$$

with an error term $\theta(y)$ satisfying

$$\lim_{y \to 0} \theta(y)/|y|^p = 0.$$

I, §5. THE INVERSE MAPPING THEOREM

The inverse function theorem and the existence theorem for differential equations (of Chapter IV) are based on the next result.

Lemma 5.1 (Contraction Lemma or Shrinking Lemma). *Let M be a complete metric space, with distance function d, and let $f: M \to M$ be a mapping of M into itself. Assume that there is a constant K, $0 < K < 1$, such that, for any two points x, y in M, we have*

$$d(f(x), f(y)) \leq K\, d(x, y).$$

Then f has a unique fixed point (a point such that $f(x) = x$). Given any point x_0 in M, the fixed point is equal to the limit of $f^n(x_0)$ (iteration of f repeated n times) as n tends to infinity.

Proof. This is a trivial exercise in the convergence of the geometric series, which we leave to the reader.

Theorem 5.2. *Let E, F be Banach spaces, U an open subset of E, and let $f: U \to F$ a C^p-morphism with $p \geq 1$. Assume that for some point $x_0 \in U$, the derivative $f'(x_0): E \to F$ is a toplinear isomorphism. Then f is a local C^p-isomorphism at x_0.*

(By a **local C^p-isomorphism** at x_0, we mean that there exists an open neighborhood V of x_0 such that the restriction of f to V establishes a C^p-isomorphism between V and an open subset of E.)

Proof. Since a toplinear isomorphism is a C^∞-isomorphism, we may assume without loss of generality that $E = F$ and $f'(x_0)$ is the identity (simply by considering $f'(x_0)^{-1} \circ f$ instead of f). After translations, we may also assume that $x_0 = 0$ and $f(x_0) = 0$.

We let $g(x) = x - f(x)$. Then $g'(x_0) = 0$ and by continuity there exists $r > 0$ such that, if $|x| < 2r$, we have

$$|g'(x)| < \tfrac{1}{2}.$$

From the mean value theorem, we see that $|g(x)| \leq \frac{1}{2}|x|$ and hence g maps the closed ball of radius r, $\bar{B}_r(0)$ into $\bar{B}_{r/2}(0)$.

We contend: Given $y \in \bar{B}_{r/2}(0)$, there exists a unique element $x \in \bar{B}_r(0)$ such that $f(x) = y$. We prove this by considering the map

$$g_y(x) = y + x - f(x).$$

If $|y| \leq r/2$ and $|x| \leq r$, then $|g_y(x)| \leq r$ and hence g_y may be viewed as a mapping of the complete metric space $\bar{B}_r(0)$ into itself. The bound of $\frac{1}{2}$ on the derivative together with the mean value theorem shows that g_y is a contracting map, i.e. that

$$|g_y(x_1) - g_y(x_2)| = |g(x_1) - g(x_2)| \leq \frac{1}{2}|x_1 - x_2|$$

for x_1, $x_2 \in \bar{B}_r(0)$. By the contraction lemma, it follows that g_y has a unique fixed point. But the fixed point of g_y is precisely the solution of the equation $f(x) = y$. This proves our contention.

We obtain a local inverse $\varphi = f^{-1}$. This inverse is continuous, because

$$|x_1 - x_2| \leq |f(x_1) - f(x_2)| + |g(x_1) - g(x_2)|$$

and hence

$$|x_1 - x_2| \leq 2|f(x_1) - f(x_2)|.$$

Furthermore φ is differentiable in $B_{r/2}(0)$. Indeed, let $y_1 = f(x_1)$ and $y_2 = f(x_2)$ with y_1, $y_2 \in B_{r/2}(0)$ and x_1, $x_2 \in \bar{B}_r(0)$. Then

$$|\varphi(y_1) - \varphi(y_2) - f'(x_2)^{-1}(y_1 - y_2)| = |x_1 - x_2 - f'(x_2)^{-1}(f(x_1) - f(x_2))|.$$

We operate on the expression inside the norm sign with the identity

$$\text{id} = f'(x_2)^{-1}f'(x_2).$$

Estimating and using the continuity of f', we see that for some constant A, the preceding expression is bounded by

$$A|f'(x_2)(x_1 - x_2) - f(x_1) + f(x_2)|.$$

From the differentiability of f, we conclude that this expression is $o(x_1 - x_2)$ which is also $o(y_1 - y_2)$ in view of the continuity of φ proved above. This proves that φ is differentiable and also that its derivative is what it should be, namely

$$\varphi'(y) = f'(\varphi(y))^{-1},$$

for $y \in B_{r/2}(0)$. Since the mappings φ, f', "inverse" are continuous, it follows that φ' is continuous and thus that φ is of class C^1. Since taking inverses is C^∞ and f' is C^{p-1}, it follows inductively that φ is C^p, as was to be shown.

Note that this last argument also proves:

Proposition 5.3. *If $f: U \to V$ is a homeomorphism and is of class C^p with $p \geqq 1$, and if f is a C^1-isomorphism, then f is a C^p-isomorphism.*

In some applications it is necessary to know that if the derivative of a map is close to the identity, then the image of a ball contains a ball of only slightly smaller radius. The precise statement follows. In this book, it will be used only in the proof of the change of variables formula, and therefore may be omitted until the reader needs it.

Lemma 5.4. *Let U be open in $\mathbf{E}$, and let $f: U \to \mathbf{E}$ be of class C^1. Assume that $f(0) = 0$, $f'(0) = I$. Let $r > 0$ and assume that $\bar{B}_r(0) \subset U$. Let $0 < s < 1$, and assume that*

$$|f'(z) - f'(x)| \leqq s$$

for all x, $z \in \bar{B}_r(0)$. If $y \in \mathbf{E}$ and $|y| \leqq (1 - s)r$, then there exists a unique $x \in \bar{B}_r(0)$ such that $f(x) = y$.

Proof. The map g_y given by $g_y(x) = x - f(x) + y$ is defined for $|x| \leqq r$ and $|y| \leqq (1 - s)r$, and maps $\bar{B}_r(0)$ into itself because, from the estimate

$$|f(x) - x| = |f(x) - f(0) - f'(0)x| \leqq |x| \sup |f'(z) - f'(0)| \leqq sr,$$

we obtain

$$|g_y(x)| \leqq sr + (1 - s)r = r.$$

Furthermore, g_y is a shrinking map because, from the mean value theorem, we get

$$|g_y(x_1) - g_y(x_2)| = |x_1 - x_2 - (f(x_1) - f(x_2))|$$
$$= |x_1 - x_2 - f'(0)(x_1 - x_2) + \delta(x_1, x_2)|$$
$$= |\delta(x_1, x_2)|,$$

where

$$|\delta(x_1, x_2)| \leqq |x_1 - x_2| \sup |f'(z) - f'(0)| \leqq s|x_1 - x_2|.$$

Hence g_y has a unique fixed point $x \in \bar{B}_r(0)$ which is such that $f(x) = y$. This proves the lemma.

We shall now prove some useful corollaries, which will be used in dealing with immersions and submersions later. *We assume that morphism means C^p-morphism with $p \geq 1$.*

Corollary 5.5. *Let U be an open subset of $\mathbf{E}$, and $f: U \to \mathbf{F}_1 \times \mathbf{F}_2$ a morphism of U into a product of Banach spaces. Let $x_0 \in U$, suppose that $f(x_0) = (0, 0)$ and that $f'(x_0)$ induces a toplinear isomorphism of $\mathbf{E}$ and $\mathbf{F}_1 = \mathbf{F}_1 \times 0$. Then there exists a local isomorphism g of $\mathbf{F}_1 \times \mathbf{F}_2$ at $(0, 0)$ such that*

$$g \circ f: U \to \mathbf{F}_1 \times \mathbf{F}_2$$

maps an open subset U_1 of U into $\mathbf{F}_1 \times 0$ and induces a local isomorphism of U_1 at x_0 on an open neighborhood of 0 in $\mathbf{F}_1$.

Proof. We may assume without loss of generality that $\mathbf{F}_1 = \mathbf{E}$ (identify by means of $f'(x_0)$) and $x_0 = 0$. We define

$$\varphi: U \times \mathbf{F}_2 \to \mathbf{F}_1 \times \mathbf{F}_2$$

by the formula

$$\varphi(x, y_2) = f(x) + (0, y_2)$$

for $x \in U$ and $y_2 \in \mathbf{F}_2$. Then $\varphi(x, 0) = f(x)$, and

$$\varphi'(0, 0) = f'(0) + (0, \mathrm{id}_2).$$

Since $f'(0)$ is assumed to be a toplinear isomorphism onto $\mathbf{F}_1 \times 0$, it follows that $\varphi'(0, 0)$ is also a toplinear isomorphism. Hence by the theorem, it has a local inverse, say g, which obviously satisfies our requirements.

Corollary 5.6. *Let $\mathbf{E}, \mathbf{F}$ be Banach spaces, U open in $\mathbf{E}$, and $f: U \to \mathbf{F}$ a C^p-morphism with $p \geq 1$. Let $x_0 \in U$. Suppose that $f(x_0) = 0$ and $f'(x_0)$ gives a toplinear isomorphism of $\mathbf{E}$ on a closed subspace of $\mathbf{F}$ which splits. Then there exists a local isomorphism $g: \mathbf{F} \to \mathbf{F}_1 \times \mathbf{F}_2$ at 0 and an open subset U_1 of U containing x_0 such that the composite map $g \circ f$ induces an isomorphism of U_1 onto an open subset of $\mathbf{F}_1$.*

Considering the splitting assumption, this is a reformulation of Corollary 5.5.

It is convenient to define the notion of splitting for injections. If $\mathbf{E}, \mathbf{F}$ are topological vector spaces, and $\lambda: \mathbf{E} \to \mathbf{F}$ is a continuous linear map, which is injective, then we shall say that λ **splits** if there exists a toplinear isomorphism $\alpha: \mathbf{F} \to \mathbf{F}_1 \times \mathbf{F}_2$ such that $\alpha \circ \lambda$ induces a toplinear isomorphism of $\mathbf{E}$ onto $\mathbf{F}_1 = \mathbf{F}_1 \times 0$. In our corollary, we could have rephrased our assumption by saying that $f'(x_0)$ is a splitting injection.

For the next corollary, dual to the preceding one, we introduce the notion of a **local projection**. Given a product of two open sets of Banach spaces $V_1 \times V_2$ and a morphism $f: V_1 \times V_2 \to \mathbf{F}$, we say that f is a **projection** (on the first factor) if f can be factored

$$V_1 \times V_2 \to V_1 \to \mathbf{F}$$

into an ordinary projection and an isomorphism of V_1 onto an open subset of $\mathbf{F}$. We say that f is a local projection at (a_1, a_2) if there exists an open neighborhood $U_1 \times U_2$ of (a_1, a_2) such that the restriction of f to this neighborhood is a projection.

Corollary 5.7. *Let U be an open subset of a product of Banach spaces $\mathbf{E}_1 \times \mathbf{E}_2$ and (a_1, a_2) a point of U. Let $f: U \to \mathbf{F}$ be a morphism into a Banach space, say $f(a_1, a_2) = 0$, and assume that the partial derivative*

$$D_2 f(a_1, a_2): \mathbf{E}_2 \to \mathbf{F}$$

is a toplinear isomorphism. Then there exists a local isomorphism h of a product $V_1 \times V_2$ onto an open neighborhood of (a_1, a_2) contained in U such that the composite map

$$V_1 \times V_2 \xrightarrow{\ h\ } U \xrightarrow{\ f\ } \mathbf{F}$$

is a projection (on the second factor).

Proof. We may assume $(a_1, a_2) = (0, 0)$ and $\mathbf{E}_2 = \mathbf{F}$. We define

$$\varphi: \mathbf{E}_1 \times \mathbf{E}_2 \to \mathbf{E}_1 \times \mathbf{E}_2$$

by

$$\varphi(x_1, x_2) = (x_1, f(x_1, x_2))$$

locally at (a_1, a_2). Then φ' is represented by the matrix

$$\begin{pmatrix} \mathrm{id}_1 & O \\ D_1 f & D_2 f \end{pmatrix}$$

and is therefore a toplinear isomorphism at (a_1, a_2). By the theorem, it has a local inverse h which clearly satisfies our requirements.

Corollary 5.8. *Let U be an open subset of a Banach space $\mathbf{E}$ and $f: U \to \mathbf{F}$ a morphism into a Banach space $\mathbf{F}$. Let $x_0 \in U$ and assume that $f'(x_0)$ is surjective, and that its kernel splits. Then there exists an open subset U' of U containing x_0 and an isomorphism*

$$h: V_1 \times V_2 \to U'$$

such that the composite map $f \circ h$ is a projection

$$V_1 \times V_2 \to V_1 \to \mathbf{F}.$$

Proof. Again this is essentially a reformulation of the corollary, taking into account the splitting assumption.

Theorem 5.9 (The Implicit Mapping Theorem). *Let U, V be open sets in Banach spaces $\mathbf{E}$, $\mathbf{F}$ respectively, and let*

$$f: U \times V \to G$$

be a C^p mapping. Let $(a, b) \in U \times V$, and assume that

$$D_2 f(a, b): \mathbf{F} \to G$$

is a toplinear isomorphism. Let $f(a, b) = 0$. Then there exists a continuous map $g: U_0 \to V$ defined on an open neighborhood U_0 of a such that $g(a) = b$ and such that

$$f(x, g(x)) = 0$$

for all $x \in U_0$. If U_0 is taken to be a sufficiently small ball, then g is uniquely determined, and is also of class C^p.

Proof. Let $\lambda = D_2 f(a, b)$. Replacing f by $\lambda^{-1} \circ f$ we may assume without loss of generality that $D_2 f(a, b)$ is the identity. Consider the map

$$\varphi: U \times V \to \mathbf{E} \times \mathbf{F}$$

given by

$$\varphi(x, y) = (x, f(x, y)).$$

Then the derivative of φ at (a, b) is immediately computed to be represented by the matrix

$$D\varphi(a, b) = \begin{pmatrix} \mathrm{id}_{\mathbf{E}} & 0 \\ D_1 f(a, b) & D_2 f(a, b) \end{pmatrix} = \begin{pmatrix} \mathrm{id}_{\mathbf{E}} & 0 \\ D_1 f(a, b) & \mathrm{id}_{\mathbf{F}} \end{pmatrix}$$

whence φ is locally invertible at (a, b) since the inverse of $D\varphi(a, b)$ exists and is the matrix

$$\begin{pmatrix} \mathrm{id}_{\mathbf{E}} & 0 \\ -D_1 f(a, b) & \mathrm{id}_{\mathbf{F}} \end{pmatrix}.$$

We denote the local inverse of φ by ψ. We can write

$$\psi(x, z) = (x, h(x, z))$$

where h is some mapping of class C^p. We define

$$g(x) = h(x, 0).$$

Then certainly g is of class C^p and

$$(x, f(x, g(x))) = \varphi(x, g(x)) = \varphi(x, h(x, 0)) = \varphi(\psi(x, 0)) = (x, 0).$$

This proves the existence of a C^p map g satisfying our requirements.

Now for the uniqueness, suppose that g_0 is a continuous map defined near a such that $g_0(a) = b$ and $f(x, g_0(x)) = c$ for all x near a. Then $g_0(x)$ is near b for such x, and hence

$$\varphi(x, g_0(x)) = (x, 0).$$

Since φ is invertible near (a, b) it follows that there is a unique point (x, y) near (a, b) such that $\varphi(x, y) = (x, 0)$. Let U_0 be a small ball on which g is defined. If g_0 is also defined on U_0, then the above argument shows that g and g_0 coincide on some smaller neighborhood of a. Let $x \in U_0$ and let $v = x - a$. Consider the set of those numbers t with $0 \leq t \leq 1$ such that $g(a + tv) = g_0(a + tv)$. This set is not empty. Let s be its least upper bound. By continuity, we have $g(a + sv) = g_0(a + sv)$. If $s < 1$, we can apply the existence and that part of the uniqueness just proved to show that g and g_0 are in fact equal in a neighborhood of $a + sv$. Hence $s = 1$, and our uniqueness statement is proved, as well as the theorem.

Note. The particular value $f(a, b) = 0$ in the preceding theorem is irrelevant. If $f(a, b) = c$ for some $c \neq 0$, then the above proof goes through replacing 0 by c everywhere.

CHAPTER II

Manifolds

Starting with open subsets of Banach spaces, one can glue them together with C^p-isomorphisms. The result is called a manifold. We begin by giving the formal definition. We then make manifolds into a category, and discuss special types of morphisms. We define the tangent space at each point, and apply the criteria following the inverse function theorem to get a local splitting of a manifold when the tangent space splits at a point.

We shall wait until the next chapter to give a manifold structure to the union of all the tangent spaces.

II, §1. ATLASES, CHARTS, MORPHISMS

Let X be a set. An **atlas of class** C^p $(p \geq 0)$ on X is a collection of pairs (U_i, φ_i) (i ranging in some indexing set), satisfying the following conditions:

AT 1. *Each U_i is a subset of X and the U_i cover X.*

AT 2. *Each φ_i is a bijection of U_i onto an open subset $\varphi_i U_i$ of some Banach space $\mathbf{E}_i$ and for any i, j, $\varphi_i(U_i \cap U_j)$ is open in $\mathbf{E}_i$.*

AT 3. *The map*

$$\varphi_j \varphi_i^{-1} \colon \varphi_i(U_i \cap U_j) \to \varphi_j(U_i \cap U_j)$$

is a C^p-isomorphism for each pair of indices i, j.

It is a trivial exercise in point set topology to prove that one can give X a topology in a unique way such that each U_i is open, and the φ_i

are topological isomorphisms. We see no reason to assume that X is Hausdorff. If we wanted X to be Hausdorff, we would have to place a separation condition on the covering. This plays no role in the formal development in Chapters II and III. It is to be understood, however, that any construction which we perform (like products, tangent bundles, etc.) would yield Hausdorff spaces if we start with Hausdorff spaces.

Each pair (U_i, φ_i) will be called a **chart** of the atlas. If a point x of X lies in U_i, then we say that (U_i, φ_i) is a **chart at** x.

In condition **AT 2**, we did not require that the vector spaces be the same for all indices i, or even that they be toplinearly isomorphic. If they are all equal to the same space **E**, then we say that the atlas is an **E**-atlas. If two charts (U_i, φ_i) and (U_j, φ_j) are such that U_i and U_j have a non-empty intersection, and if $p \geqq 1$, then taking the derivative of $\varphi_j \varphi_i^{-1}$ we see that $\mathbf{E}_i$ and $\mathbf{E}_j$ are toplinearly isomorphic. Furthermore, the set of points $x \in X$ for which there exists a chart (U_i, φ_i) at x such that $\mathbf{E}_i$ is toplinearly isomorphic to a given space **E** is both open and closed. Consequently, on each connected component of X, we could assume that we have an **E**-atlas for some fixed **E**.

Suppose that we are given an open subset U of X and a topological isomorphism $\varphi: U \to U'$ onto an open subset of some Banach space **E**. We shall say that (U, φ) is **compatible** with the atlas $\{(U_i, \varphi_i)\}$ if each map $\varphi_i \varphi^{-1}$ (defined on a suitable intersection as in **AT 3**) is a C^p-isomorphism. Two atlases are said to be **compatible** if each chart of one is compatible with the other atlas. One verifies immediately that the relation of compatibility between atlases is an equivalence relation. An equivalence class of atlases of class C^p on X is said to define a structure of C^p-**manifold** on X. If all the vector spaces $\mathbf{E}_i$ in some atlas are toplinearly isomorphic, then we can always find an equivalent atlas for which they are all equal, say to the vector space **E**. We then say that X is an **E-manifold** or that X is **modeled** on **E**.

If $\mathbf{E} = \mathbf{R}^n$ for some fixed n, then we say that the manifold is n-**dimensional**. In this case, a chart

$$\varphi: U \to \mathbf{R}^n$$

is given by n coordinate functions $\varphi_1, \ldots, \varphi_n$. If P denotes a point of U, these functions are often written

$$x_1(P), \ldots, x_n(P),$$

or simply $x_1, \ldots, x_n$. They are called **local coordinates** on the manifold.

If the integer p (which may also be ∞) is fixed throughout a discussion, we also say that X is a manifold.

The collection of C^p-manifolds will be denoted by Man^p. If we look only at those modeled on spaces in a category $\mathfrak{A}$ then we write Man$^p(\mathfrak{A})$. Those modeled on a fixed **E** will be denoted by Man$^p(\mathbf{E})$. We shall make these into categories by defining morphisms below.

Let X be a manifold, and U an open subset of X. Then it is possible, in the obvious way, to induce a manifold structure on U, by taking as charts the intersections

$$(U_i \cap U, \varphi_i|(U_i \cap U)).$$

If X is a topological space, covered by open subsets V_j, and if we are given on each V_j a manifold structure such that for each pair j, j' the induced structure on $V_j \cap V_{j'}$ coincides, then it is clear that we can give to X a unique manifold structure inducing the given ones on each V_j.

Example. Let X be the real line, and for each open interval U_i, let φ_i be the function $\varphi_i(t) = t^3$. Then the $\varphi_j \varphi_i^{-1}$ are all equal to the identity, and thus we have defined a C^∞-manifold structure on **R**!

If X, Y are two manifolds, then one can give the product $X \times Y$ a manifold structure in the obvious way. If $\{(U_i, \varphi_i)\}$ and $\{(V_j, \psi_j)\}$ are atlases for X, Y respectively, then

$$\{(U_i \times V_j, \varphi_i \times \psi_j)\}$$

is an atlas for the product, and the product of compatible atlases gives rise to compatible atlases, so that we do get a well-defined product structure.

Let X, Y be two manifolds. Let $f: X \to Y$ be a map. We shall say that f is a C^p-**morphism** if, given $x \in X$, there exists a chart (U, φ) at x and a chart (V, ψ) at $f(x)$ such that $f(U) \subset V$, and the map

$$\psi \circ f \circ \varphi^{-1} : \varphi U \to \psi V$$

is a C^p-morphism in the sense of Chapter I, §3. One sees then immediately that this same condition holds for any choice of charts (U, φ) at x and (V, ψ) at $f(x)$ such that $f(U) \subset V$.

It is clear that the composite of two C^p-morphisms is itself a C^p-morphism (because it is true for open subsets of vector spaces). The C^p-manifolds and C^p-morphisms form a category. The notion of isomorphism is therefore defined, and we observe that in our example of the real line, the map $t \mapsto t^3$ gives an isomorphism between the funny differentiable structure and the usual one.

If $f: X \to Y$ is a morphism, and (U, φ) is a chart at a point $x \in X$, while (V, ψ) is a chart at $f(x)$, then we shall also denote by

$$f_{V, U} : \varphi U \to \psi V$$

the map $\psi f \varphi^{-1}$.

It is also convenient to have a local terminology. Let U be an open set (of a manifold or a Banach space) containing a point x_0. By a **local isomorphism** at x_0 we mean an isomorphism

$$f: U_1 \to V$$

from some open set U_1 containing x_0 (and contained in U) to an open set V (in some manifold or some Banach space). Thus a local isomorphism is essentially a change of chart, locally near a given point.

Manifolds of maps. Even starting with a finite dimensional manifold, the set of maps satisfying various smoothness conditions forms an infinite dimensional manifold. This story started with Eells [Ee 58], [Ee 59], [Ee 61]. Palais and Smale used such manifolds of maps in their Morse theory [Pa 63], [Ab 62], [Sm 64]. For a brief discussion of subsequent developments, see [Mar 74], p. 67, referring to [Eb 70], [Ee 66], [El 67], [Kr 72], [Le 67], [Om 70], and [Pa 68]. Two kinds of maps have played a role: the C^p maps of course, with various values of p, but also maps satisfying Sobolev conditions, and usually denoted by H^s. The latter form Hilbert manifolds (definition to be given later).

II, §2. SUBMANIFOLDS, IMMERSIONS, SUBMERSIONS

Let X be a topological space, and Y a subset of X. We say that Y is **locally closed** in X if every point $y \in Y$ has an open neighborhood U in X such that $Y \cap U$ is closed in U. One verifies easily that a locally closed subset is the intersection of an open set and a closed set. For instance, any open subset of X is locally closed, and any open interval is locally closed in the plane.

Let X be a manifold (of class C^p with $p \geq 0$). Let Y be a subset of X and assume that for each point $y \in Y$ there exists a chart (V, ψ) at y such that ψ gives an isomorphism of V with a product $V_1 \times V_2$ where V_1 is open in some space $\mathbf{E}_1$ and V_2 is open in some space $\mathbf{E}_2$, and such that

$$\psi(Y \cap V) = V_1 \times a_2$$

for some point $a_2 \in V_2$ (which we could take to be 0). Then it is clear that Y is locally closed in X. Furthermore, the map ψ induces a bijection

$$\psi_1: Y \cap V \to V_1.$$

The collection of pairs $(Y \cap V, \psi_1)$ obtained in the above manner constitutes an atlas for Y, of class C^p. The verification of this assertion, whose formal details we leave to the reader, depends on the following obvious fact.

Lemma 2.1. *Let* U_1, U_2, V_1, V_2 *be open subsets of Banach spaces, and* $g: U_1 \times U_2 \to V_1 \times V_2$ *a* C^p*-morphism. Let* $a_2 \in U_2$ *and* $b_2 \in V_2$ *and assume that* g *maps* $U_1 \times a_2$ *into* $V_1 \times b_2$*. Then the induced map*

$$g_1: U_1 \to V_1$$

is also a morphism.

Indeed, it is obtained as a composite map

$$U_1 \to U_1 \times U_2 \to V_1 \times V_2 \to V_1,$$

the first map being an inclusion and the third a projection.

We have therefore defined a C^p-structure on Y which will be called a **submanifold** of X. This structure satisfies a universal mapping property, which characterizes it, namely:

Given any map $f: Z \to X$ *from a manifold* Z *into* X *such that* $f(Z)$ *is contained in* Y*. Let* $f_Y: Z \to Y$ *be the induced map. Then* f *is a morphism if and only if* f_Y *is a morphism.*

The proof of this assertion depends on Lemma 2.1, and is trivial.

Finally, we note that the inclusion of Y into X is a morphism.

If Y is also a closed subspace of X, then we say that it is a **closed submanifold**.

Suppose that X is finite dimensional of dimension n, and that Y is a submanifold of dimension r. Then from the definition we see that the local product structure in a neighborhood of a point of Y can be expressed in terms of local coordinates as follows. Each point P of Y has an open neighborhood U in X with local coordinates $(x_1, \ldots, x_n)$ such that the points of Y in U are precisely those whose last $n - r$ coordinates are 0, that is, those points having coordinates of type

$$(x_1, \ldots, x_r, 0, \ldots, 0).$$

Let $f: Z \to X$ be a morphism, and let $z \in Z$. We shall say that f is an **immersion** at z if there exists an open neighborhood Z_1 of z in Z such that the restriction of f to Z_1 induces an isomorphism of Z_1 onto a submanifold of X. We say that f is an **immersion** if it is an immersion at every point.

Note that there exist injective immersions which are not isomorphisms onto submanifolds, as given by the following example:

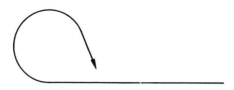

(The arrow means that the line approaches itself without touching.) An immersion which does give an isomorphism onto a submanifold is called an **embedding**, and it is called a **closed embedding** if this submanifold is closed.

A morphism $f: X \to Y$ will be called a **submersion** at a point $x \in X$ if there exists a chart (U, φ) at x and a chart (V, ψ) at $f(x)$ such that φ gives an isomorphism of U on a products $U_1 \times U_2$ (U_1 and U_2 open in some Banach spaces), and such that the map

$$\psi f \varphi^{-1} = f_{V,U}: U_1 \times U_2 \to V$$

is a projection. One sees then that the image of a submersion is an open subset (a submersion is in fact an open mapping). We say that f is a **submersion** if it is a submersion at every point.

For manifolds modelled on Banach spaces, we have the usual criterion for immersions and submersions in terms of the derivative.

Proposition 2.2. *Let X, Y be manifolds of class C^p ($p \geq 1$) modeled on Banach spaces. Let $f: X \to Y$ be a C^p-morphism. Let $x \in X$. Then:*

(i) *f is an immersion at x if and only if there exists a chart (U, φ) at x and (V, ψ) at $f(x)$ such that $f'_{V,U}(\varphi x)$ is injective and splits.*

(ii) *f is a submersion at x if and only if there exists a chart (U, φ) at x and (V, ψ) at $f(x)$ such that $f'_{V,U}(\varphi x)$ is surjective and its kernel splits.*

Proof. This is an immediate consequence of Corollaries 5.4 and 5.6 of the inverse mapping theorem.

The conditions expressed in (i) and (ii) depend only on the derivative, and if they hold for one choice of charts (U, φ) and (V, ψ) respectively, then they hold for every choice of such charts. It is therefore convenient to introduce a terminology in order to deal with such properties.

Let X be a manifold of class C^p ($p \geq 1$). Let x be a point of X. We consider triples (U, φ, v) where (U, φ) is a chart at x and v is an element of the vector space in which φU lies. We say that two such triples (U, φ, v) and (V, ψ, w) are **equivalent** if the derivative of $\psi \varphi^{-1}$ at φx maps v on w. The formula reads:

$$(\psi \varphi^{-1})'(\varphi x)v = w$$

(obviously an equivalence relation by the chain rule). An equivalence class of such triples is called a **tangent vector** of X at x. The set of such tangent vectors is called the **tangent space** of X at x and is denoted by $T_x(X)$. Each chart (U, φ) determines a bijection of $T_x(X)$ on a Banach space, namely the equivalence class of (U, φ, v) corresponds to the vector v. By means of such a bijection it is possible to transport to $T_x(X)$ the structure of topological vector space given by the chart, and it is immediate that this structure is independent of the chart selected.

If U, V are open in Banach spaces, then to every morphism of class C^p $(p \geq 1)$ we can associate its derivative $Df(x)$. If now $f: X \to Y$ is a morphism of one manifold into another, and x a point of X, then by means of charts we can interpret the derivative of f on each chart at x as a mapping

$$T_x f: T_x(X) \to T_{f(x)}(Y).$$

Indeed, this map $T_x f$ is the unique linear map having the following property. If (U, φ) is a chart at x and (V, ψ) is a chart at $f(x)$ such that $f(U) \subset V$ and $\bar{v}$ is a tangent vector at x represented by v in the chart (U, φ), then

$$T_x f(\bar{v})$$

is the tangent vector at $f(x)$ represented by $Df_{V,U}(x)v$. The representation of $T_x f$ on the spaces of charts can be given in the form of a diagram

$$
\begin{array}{ccc}
T_x(X) & \longrightarrow & \mathbf{E} \\
{\scriptstyle T_x f}\downarrow & & \downarrow{\scriptstyle f'_{V,U}(x)} \\
T_{f(x)}(Y) & \longrightarrow & \mathbf{F}
\end{array}
$$

The map $T_x f$ is obviously continuous and linear for the structure of topological vector space which we have placed on $T_x(X)$ and $T_{f(x)}(Y)$.

As a matter of notation, we shall sometimes write $f_{*,x}$ instead of $T_x f$.

The operation T satisfies an obvious functorial property, namely, if $f: X \to Y$ and $g: Y \to Z$ are morphisms, then

$$T_x(g \circ f) = T_{f(x)}(g) \circ T_x(f),$$

$$T_x(\mathrm{id}) = \mathrm{id}.$$

We may reformulate Proposition 2.2:

Proposition 2.3. *Let X, Y be manifolds of class C^p $(p \geq 1)$ modelled on Banach spaces. Let $f: X \to Y$ be a C^p-morphism. Let $x \in X$. Then:*

(i) *f is an immersion at x if and only if the map $T_x f$ is injective and splits.*

(ii) *f is a submersion at x if and only if the map $T_x f$ is surjective and its kernel splits.*

Note. If X, Y are finite dimensional, then the condition that $T_x f$ splits is superfluous. Every subspace of a finite dimensional vector space splits.

Example. Let **E** be a (real) Hilbert space, and let $\langle x, y \rangle \in \mathbf{R}$ be its inner product. Then the square of the norm $f(x) = \langle x, x \rangle$ is obviously of class C^∞. The derivative $f'(x)$ is given by the formula

$$f'(x)y = 2\langle x, y \rangle$$

and for any given $x \neq 0$, it follows that the derivative $f'(x)$ is surjective. Furthermore, its kernel is the orthogonal complement of the subspace generated by x, and hence splits. Consequently the unit sphere in Hilbert space is a submanifold.

If W is a submanifold of a manifold Y of class C^p ($p \geqq 1$), then the inclusion

$$i: W \rightarrow Y$$

induces a map

$$T_w i: T_w(W) \rightarrow T_w(Y)$$

which is in fact an injection. From the definition of a submanifold, one sees immediately that the image of $T_w i$ splits. It will be convenient to identify $T_w(W)$ in $T_w(Y)$ if no confusion can result.

A morphism $f: X \rightarrow Y$ will be said to be **transversal** over the submanifold W of Y if the following condition is satisfied.

Let $x \in X$ be such that $f(x) \in W$. Let (V, ψ) be a chart at $f(x)$ such that $\psi: V \rightarrow V_1 \times V_2$ is an isomorphism on a product, with

$$\psi(f(x)) = (0, 0) \qquad \text{and} \qquad \psi(W \cap V) = V_1 \times 0.$$

Then there exists an open neighborhood U of x such that the composite map

$$U \xrightarrow{\ f\ } V \xrightarrow{\ \psi\ } V_1 \times V_2 \xrightarrow{\ \text{pr}\ } V_2$$

is a submersion.

In particular, if f is transversal over W, then $f^{-1}(W)$ is a submanifold of X, because the inverse image of 0 by our local composite map

$$\text{pr} \circ \psi \circ f$$

is equal to the inverse image of $W \cap V$ by ψ.

As with immersions and submersions, we have a characterization of transversal maps in terms of tangent spaces.

Proposition 2.4. *Let X, Y be manifolds of class C^p ($p \geqq 1$) modeled on Banach spaces. Let $f: X \rightarrow Y$ be a C^p-morphism, and W a submanifold* *of Y. The map f is transversal over W if and only if for each $x \in X$ such that $f(x)$ lies in W, the composite map*

$$T_x(X) \xrightarrow{T_x f} T_w(Y) \to T_w(Y)/T_w(W)$$

with $w = f(x)$ is surjective and its kernel splits.

Proof. If f is transversal over W, then for each point $x \in X$ such that $f(x)$ lies in W, we choose charts as in the definition, and reduce the question to one of maps of open subsets of Banach spaces. In that case, the conclusion concerning the tangent spaces follows at once from the assumed direct product decompositions. Conversely, assume our condition on the tangent map. The question being local, we can assume that $Y = V_1 \times V_2$ is a product of open sets in Banach spaces such that $W = V_1 \times 0$, and we can also assume that $X = U$ is open in some Banach space, $x = 0$. Then we let $g \colon U \to V_2$ be the map $\pi \circ f$ where π is the projection, and note that our assumption means that $g'(0)$ is surjective and its kernel splits. Furthermore, $g^{-1}(0) = f^{-1}(W)$. We can then use Corollary 5.7 of the inverse mapping theorem to conclude the proof.

Remark. In the statement of our proposition, we observe that the surjectivity of the composite map is equivalent to the fact that $T_w(Y)$ is equal to the sum of the image of $T_x f$ and $T_w(W)$, that is

$$T_w(Y) = \mathrm{Im}(T_x f) + \mathrm{Im}(T_x i),$$

where $i \colon W \to Y$ is the inclusion. In the finite dimensional case, the other condition is therefore redundant.

If $\mathbf{E}$ is a Banach space, then the diagonal Δ in $\mathbf{E} \times \mathbf{E}$ is a closed subspace and splits: Either factor $\mathbf{E} \times 0$ or $0 \times \mathbf{E}$ is a closed complement. Consequently, the diagonal is a closed submanifold of $\mathbf{E} \times \mathbf{E}$. If X is any manifold of class C^p, $p \geq 1$, then the diagonal is therefore also a submanifold. (It is closed of course if and only if X is Hausdorff.)

Let $f \colon X \to Z$ and $g \colon Y \to Z$ be two C^p-morphisms, $p \geq 1$. We say that they are **transversal** if the morphism

$$f \times g \colon X \times Y \to Z \times Z$$

is transversal over the diagonal. We remark right away that the surjectivity of the map in Proposition 2.4 can be expressed in two ways. Given two points $x \in X$ and $y \in Y$ such that $f(x) = g(y) = z$, the condition

$$\mathrm{Im}(T_x f) + \mathrm{Im}(T_y g) = T_z(Z)$$

is equivalent to the condition

$$\operatorname{Im}(T_{(x,y)}(f \times g)) + T_{(z,z)}(\Delta) = T_{(z,z)}(Z \times Z).$$

Thus in the finite dimensional case, we could take it as definition of transversality.

We use transversality as a sufficient condition under which the fiber product of two morphisms exists. We recall that in any category, the **fiber product** of two morphisms $f: X \to Z$ and $g: Y \to Z$ over Z consists of an object P and two morphisms

$$g_1: P \to X \qquad \text{and} \qquad g_2: P \to Y$$

such that $f \circ g_1 = g \circ g_2$, and satisfying the universal mapping property: Given an object S and two morphisms $u_1: S \to X$ and $u_2: S \to Y$ such that $fu_1 = gu_2$, there exists a unique morphism $u: S \to P$ making the following diagram commutative:

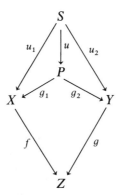

The triple (P, g_1, g_2) is uniquely determined, up to a unique isomorphism (in the obvious sense), and P is also denoted by $X \times_Z Y$.

One can view the fiber product unsymmetrically. Given two morphisms f, g as in the following diagram:

$$\begin{array}{ccc} & & Y \\ & & \downarrow g \\ X & \xrightarrow{\quad f \quad} & Z \end{array}$$

assume that their fiber product exists, so that we can fill in the diagram:

$$\begin{array}{ccc} X \times_Z Y & \longrightarrow & Y \\ g_1 \downarrow & & \downarrow g \\ X & \longrightarrow & Z \end{array}$$

We say that g_1 is the **pull back** of g by f, and also write it as $f^*(g)$. Similarly, we write $X \times_Z Y$ as $f^*(Y)$.

In our category of manifolds, we shall deal only with cases when the fiber product can be taken to be the set-theoretic fiber product on which a manifold structure has been defined. (The set-theoretic fiber product is the set of pairs of points projecting on the same point.) This determines the fiber product uniquely, and not only up to a unique isomorphism.

Proposition 2.5. *Let* $f: X \to Z$ *and* $g: Y \to Z$ *be two* C^p*-morphisms with* $p \geq 1$. *If they are transversal, then*

$$(f \times g)^{-1}(\Delta_Z),$$

together with the natural morphisms into X *and* Y *(obtained from the projections), is a fiber product of* f *and* g *over* Z.

Proof. Obvious.

To construct a fiber product, it suffices to do it locally. Indeed, let $f: X \to Z$ and $g: Y \to Z$ be two morphisms. Let $\{V_i\}$ be an open covering of Z, and let

$$f_i: f^{-1}(V_i) \to V_i \qquad \text{and} \qquad g_i: g^{-1}(V_i) \to V_i$$

be the restrictions of f and g to the respective inverse images of V_i. Let $P = (f \times g)^{-1}(\Delta_Z)$. Then P consists of the points (x, y) with $x \in X$ and $y \in Y$ such that $f(x) = g(y)$. We view P as a subspace of $X \times Y$ (i.e. with the topology induced by that of $X \times Y$). Similarly, we construct P_i with f_i and g_i. Then P_i is open in P. The projections on the first and second factors give natural maps of P_i into $f^{-1}(V_i)$ and $g^{-1}(V_i)$, and of P into X and Y.

Proposition 2.6. *Assume that each* P_i *admits a manifold structure (compatible with its topology) such that these maps are morphisms, making* P_i *into a fiber product of* f_i *and* g_i. *Then* P, *with its natural projections, is a fiber product of* f *and* g.

To prove the above assertion, we observe that the P_i form a covering of P. Furthermore, the manifold structure on $P_i \cap P_j$ induced by that of P_i or P_j must be the same, because it is the unique fiber product structure over $V_i \cap V_j$, for the maps f_{ij} and g_{ij} (defined on $f^{-1}(V_i \cap V_j)$ and $g^{-1}(V_i \cap V_j)$ respectively). Thus we can give P a manifold structure, in such a way that the two projections into X and Y are morphisms, and make P into a fiber product of f and g.

We shall apply the preceding discussion to vector bundles in the next chapter, and the following local criterion will be useful.

Proposition 2.7. *Let* $f: X \to Z$ *be a morphism, and* $g: Z \times W \to Z$ *be the projection on the first factor. Then* f, g *have a fiber product, namely the product* $X \times W$ *together with the morphisms of the following diagram:*

$$
\begin{array}{ccc}
X \times W & \xrightarrow{f \times \mathrm{id}} & Z \times W \\
\text{pr}_1 \downarrow & & \downarrow \text{pr}_1 \\
X & \xrightarrow{\quad f \quad} & Z
\end{array}
$$

II, §3. PARTITIONS OF UNITY

Let X be a manifold of class C^p. A **function** on X will be a morphism of X into **R**, of class C^p, unless otherwise specified. The C^p functions form a ring $\mathfrak{F}^p(X)$. The **support** of a function f is the closure of the set of points x such that $f(x) \neq 0$.

Let X be a topological space. A covering of X is **locally finite** if every point has a neighborhood which intersects only finitely many elements of the covering. A **refinement** of a covering of X is a second covering, each element of which is contained in an element of the first covering. A topological space is **paracompact** if it is Hausdorff, and every open covering has a locally finite open refinement.

Proposition 3.1. *If* X *is a paracompact space, and if* $\{U_i\}$ *is an open covering, then there exists a locally finite open covering* $\{V_i\}$ *such that* $V_i \subset U_i$ *for each i.*

Proof. Let $\{V_k\}$ be a locally finite open refinement of $\{U_i\}$. For each k there is an index $i(k)$ such that $V_k \subset U_{i(k)}$. We let W_i be the union of those V_k such that $i(k) = i$. Then the W_i form a locally finite open covering, because any neighborhood of a point which meets infinitely many W_i must also meet infinitely many V_k.

Proposition 3.2. *If* X *is paracompact, then* X *is normal. If, furthermore,* $\{U_i\}$ *is a locally finite open covering of* X, *then there exists a locally finite open covering* $\{V_i\}$ *such that* $\bar{V}_i \subset U_i$.

Proof. We refer the reader to Bourbaki [Bou 68].

Observe that Proposition 3.1 shows that the insistence that the indexing set of a refinement be a given one can easily be achieved.

A **partition of unity** (of class C^p) on a manifold X consists of an open covering $\{U_i\}$ of X and a family of functions

$$\psi_i \colon X \to \mathbf{R}$$

satisfying the following conditions:

PU 1. *For all $x \in X$ we have $\psi_i(x) \geq 0$.*

PU 2. *The support of ψ_i is contained in U_i.*

PU 3. *The covering is locally finite.*

PU 4. *For each point $x \in X$ we have*

$$\sum \psi_i(x) = 1.$$

(The sum is taken over all i, but is in fact finite for any given point x in view of **PU 3**.)

We sometimes say that $\{(U_i, \psi_i)\}$ is a partition of unity.

A manifold X will be said to **admit partitions of unity** if it is paracompact, and if, given a locally finite open covering $\{U_i\}$, there exists a partition of unity $\{\psi_i\}$ such that the support of ψ_i is contained in U_i.

If $\{U_i\}$ is a covering of X, then we say that a covering $\{V_k\}$ is subordinated to $\{U_i\}$ if each V_k is contained in some U_i.

It is desirable to give sufficient conditions on a manifold in order to insure the existence of partitions of unity. There is no difficulty with the topological aspects of this problem. It is known that a metric space is paracompact (cf. Bourbaki [Bou 68], [Ke 55]), and on a paracompact space, one knows how to construct continuous partitions of unity (loc. cit.). However, in the case of infinite dimensional manifolds, certain difficulties arise to construct differentiable ones, and it is known that a Banach space itself may not admit partitions of unity (say of class C^∞). The construction of differentiable partitions of unity depends on the construction of a differentiable norm. Readers will find examples, theorems, and counterexamples in [BoF 65], [BoF 66], and [Re 64]. In the finite dimensional case, the existence will follow from the next theorem.

If $\mathbf{E}$ is a Banach space, we denote by $B_r(a)$ the open ball of radius r and center a, and by $\bar{B}_r(a)$ the closed ball of radius r and center a. If $a = 0$, then we write B_r and $\bar{B}_r$ respectively. Two open balls (of finite radius) are obviously C^∞-isomorphic. If X is a manifold and (V, φ) is a chart at a point $x \in X$, then we say that (V, φ) (or simply V) is a ball of radius r if φV is a ball of radius r in the Banach space.

Theorem 3.3. *Let X be a manifold which is locally compact, Hausdorff, and whose topology has a countable base. Given an open covering of X,* *then there exists an atlas* $\{(V_k, \varphi_k)\}$ *such that the covering* $\{V_k\}$ *is locally finite and subordinated to the given covering, such that* $\varphi_k V_k$ *is the open ball* B_3, *and such that the open sets* $W_k = \varphi_k^{-1}(B_1)$ *cover* X.

Proof. Let U_1, U_2, ... be a basis for the open sets of X such that each $\overline{U}_i$ is compact. We construct inductively a sequence A_1, A_2, ... of compact sets whose union is X, such that A_i is contained in the interior of A_{i+1}. We let $A_1 = \overline{U}_1$. Suppose we have constructed A_i. We let j be the smallest integer such that A_i is contained in $U_1 \cup \cdots \cup U_j$. We let A_{i+1} be the closed and compact set

$$\overline{U}_1 \cup \cdots \cup \overline{U}_j \cup \overline{U}_{i+1}.$$

For each point $x \in X$ we can find an arbitrarily small chart (V_x, φ_x) at x such that $\varphi_x V_x$ is the ball of radius 3 (so that each V_x is contained in some element of U). We let $W_x = \varphi_x^{-1}(B_1)$ be the ball of radius 1 in this chart. We can cover the set

$$A_{i+1} - \text{Int}(A_i)$$

(intuitively the closed annulus) by a finite number of these balls of radius 1, say $W_1, \ldots, W_n$, such that, at the same time, each one of $V_1, \ldots, V_n$ is contained in the open set $\text{Int}(A_{i+2}) - A_{i-1}$ (intuitively, the open annulus of the next bigger size). We let $\mathfrak{B}_i$ denote the collection $V_1, \ldots, V_n$ and let $\mathfrak{B}$ be composed of the union of the $\mathfrak{B}_i$. Then $\mathfrak{B}$ is locally finite, and we are done.

Corollary 3.4. *Let* X *be a manifold which is locally compact Hausdorff, and whose topology has a countable base. Then* X *admits partitions of unity.*

Proof. Let $\{(V_k, \varphi_k)\}$ be as in the theorem, and $W_k = \varphi_k^{-1}(B_1)$. We can find a function ψ_k of class C^p such that $0 \leq \psi_k \leq 1$, such that $\psi_k(x) = 1$ for $x \in W_k$ and $\psi_k(x) = 0$ for $x \notin V_k$. (The proof is recalled below.) We now let

$$\psi = \sum \psi_k$$

(a sum which is finite at each point), and we let $\gamma_k = \psi_k/\psi$. Then $\{(V_k, \gamma_k)\}$ is the desired partition of unity.

We now recall the argument giving the function ψ_k. First, given two real numbers r, s with $0 \leq r < s$, the function defined by

$$\exp\left(\frac{-1}{(t - r)(s - t)}\right)$$

in the open interval $r < t < s$ and 0 outside the interval determines a bell-shaped C^∞-function from $\mathbf{R}$ into $\mathbf{R}$. Its integral from minus infinity to t, divided by the area under the bell yields a function which lies strictly between 0 and 1 in the interval $r < t < s$, is equal to 0 for $t \leq r$ and is equal to 1 for $t \geq s$. (The function is even monotone increasing.)

We can therefore find a real valued function of a real variable, say $\eta(t)$, such that $\eta(t) = 1$ for $|t| < 1$ and $\eta(t) = 0$ for $|t| \geq 1 + \delta$ with small δ, and such that $0 \leq \eta \leq 1$. If $\mathbf{E}$ is a Hilbert space, then $\eta(|x|^2) = \psi(x)$ gives us a function which is equal to 1 on the ball of radius 1 and 0 outside the ball of radius $1 + \delta$. This function can then be transported to the manifold by any given chart whose image is the ball of radius 3.

In a similar way, one would construct a function which is > 0 on a given ball and $= 0$ outside this ball.

Partitions of unity constitute the only known means of gluing together local mappings (into objects having an addition, namely vector bundles, discussed in the next chapter). It is therefore important, in both the Banach and Hilbert cases, to determine conditions under which they exist. In the Banach case, there is the added difficulty that the argument just given to get a local function which is 1 on B_1 and 0 outside B_2 fails if one cannot find a differentiable function of the norm, or of an equivalent norm used to define the Banachable structure.

Even though it is not known whether Theorem 3.3 extends to Hilbert manifolds, it is still possible to construct partitions of unity in that case. As Eells pointed out to me, Dieudonné's method of proof showing that separable metric space is paracompact can be applied for that purpose (this is Lemma 3.5 below), and I am indebted to him for the following exposition.

We need some lemmas. We use the notation ^cA for the complement of a set A.

Let M be a metric space with distance function d. We can then speak of open and closed balls. For instance $\bar{B}_a(x)$ denotes the closed ball of radius a with center x. It consists of all points y with $d(y, x) \leq a$. An open subset V of M will be said to be **scalloped** if there exist open balls $U, U_1, \ldots, U_m$ in M such that

$$V = U \cap {}^c\bar{U}_1 \cap \cdots \cap {}^c\bar{U}_m.$$

A covering $\{V_i\}$ of a subset W of M is said to be locally finite (with respect to W) if every point $x \in W$ has a neighborhood which meets only a finite number of elements of the covering.

Lemma 3.5. *Let M be a metric space and $\{U_i\}$ $(i = 1, 2, \ldots)$ a countable covering of a subset W by open balls. Then there exists a locally finite open covering $\{V_i\}$ $(i = 1, 2, \ldots)$ of W such that $V_i \subset U_i$ for all i, and such that V_i is scalloped for all i.*

Proof. We define V_i inductively as follows. Each U_i is a ball, say $B_{a_i}(x_i)$. Let $V_1 = U_1$. Having defined V_{i-1}, let

$$r_{1i} = a_1 - \frac{1}{i}, \quad \ldots, \quad r_{i-1,i} = a_{i-1} - \frac{1}{i}$$

and let

$$V_i = U_i \cap {}^c\overline{B}_{r_{1i}}(x_1) \cap \cdots \cap {}^c\overline{B}_{r_{i-1,i}}(x_{i-1}),$$

it being understood that a ball of negative radius is empty. Then each V_i is scalloped, and is contained in U_i. We contend that the V_i cover W. Indeed, let x be an element of W. Let j be the smallest index such that $x \in U_j$. Then $x \in V_j$, for otherwise, x would be in the complement of V_j which is equal to the union of ${}^cU_j$ and the balls

$$\overline{B}_{r_{1,j}}(x_1) \cup \cdots \cup \overline{B}_{r_{j-1,j}}(x_{j-1}).$$

Hence x would lie in some U_i with $i < j$, contradiction.

There remains to be shown that our covering $\{V_i\}$ is locally finite. Let $x \in W$. Then x lies in some U_n. Let s be such a small number > 0 that the ball $B_s(x)$ is contained in U_n. Let $t = s/2$. For all i sufficiently large, the ball $B_t(x)$ is contained in $\overline{B}_{a_n - 1/i}(x_n) = \overline{B}_{r_{ni}}(x_n)$ and therefore this ball does not meet V_i. We have found a neighborhood of x which meets only a finite number of members of our covering, which is consequently locally finite (with respect to W).

Lemma 3.6. *Let U be an open ball in Hilbert space $\mathbf{E}$ and let*

$$V = U \cap {}^c\overline{U}_1 \cap \cdots \cap {}^c\overline{U}_m$$

be a scalloped open subset. Then there exists a C^∞-function $\omega: \mathbf{E} \to \mathbf{R}$ such that $\omega(x) > 0$ if $x \in V$ and $\omega(x) = 0$ otherwise.

Proof. For each U_i let $\varphi_i: \mathbf{E} \to \mathbf{R}$ be a function such that

$$0 \leq \varphi_i(x) < 1 \quad \text{if} \quad x \in {}^c\overline{U}_i,$$
$$\varphi_i(x) = 1 \quad \text{if} \quad x \in \overline{U}_i.$$

Let $\varphi(x)$ be a function such that $\varphi(x) > 0$ on U and $\varphi(x) = 0$ outside U. Let

$$\omega(x) = \varphi(x) \prod (1 - \varphi_i(x)).$$

Then $\omega(x)$ satisfies our requirements.

Theorem 3.7. *Let A_1, A_2 be non-void, closed, disjoint subsets of a separable Hilbert space* **E**. *Then there exists a C^∞-function $\psi: \mathbf{E} \to \mathbf{R}$ such that $\psi(x) = 0$ if $x \in A_1$ and $\psi(x) = 1$ if $x \in A_2$, and $0 \leqq \psi(x) \leqq 1$ for all x.*

Proof. By Lindelöf's theorem, we can find a countable collection of open balls $\{U_i\}$ $(i = 1, 2, \ldots)$ covering A_2 and such that each U_i is contained in the complement of A_1. Let W be the union of the U_i. We find a locally finite refinement $\{V_i\}$ as in Lemma 3.5. Using Lemma 3.6, we find a function ω_i which is > 0 on V_i and 0 outside V_i. Let $\omega = \sum \omega_i$ (the sum is finite at each point of W). Then $\omega(x) > 0$ if $x \in A_2$, and $\omega(x) = 0$ if $x \in A_1$.

Let U be the open neighborhood of A_2 on which ω is > 0. Then A_2 and ${}^c U$ are disjoint closed sets, and we can apply the above construction to obtain a function $\sigma: \mathbf{E} \to \mathbf{R}$ which is > 0 on ${}^c U$ and $= 0$ on A_2. We let $\psi = \omega/(\sigma + \omega)$. Then ψ satisfies our requirements.

Corollary 3.8. *Let X be a paracompact manifold of class C^p, modeled on a separable Hilbert space* **E**. *Then X admits partitions of unity (of class C^p).*

Proof. It is trivially verified that an open ball of finite radius in **E** is C^∞-isomorphic to **E**. (We reproduce the formula in Chapter VII.) Given any point $x \in X$, and a neighborhood N of x, we can therefore always find a chart (G, γ) at x such that $\gamma G = \mathbf{E}$, and $G \subset N$. Hence, given an open covering of X, we can find an atlas $\{(G_\alpha, \gamma_\alpha)\}$ subordinated to the given covering, such that $\gamma_\alpha G_\alpha = \mathbf{E}$. By paracompactness, we can find a refinement $\{U_i\}$ of the covering $\{G_\alpha\}$ which is locally finite. Each U_i is contained in some $G_{\alpha(i)}$ and we let φ_i be the restriction of $\gamma_{\alpha(i)}$ to U_i. We now find open refinements $\{V_i\}$ and then $\{W_i\}$ such that

$$\overline{W}_i \subset V_i \subset \overline{V}_i \subset U_i,$$

the bar denoting closure in X. Each $\overline{V}_i$ being closed in X, it follows from our construction that $\varphi_i \overline{V}_i$ is closed in **E**, and so is $\varphi_i \overline{W}_i$. Using the theorem, and transporting functions on **E** to functions on X by means of the φ_i, we can find for each i a C^p-function $\psi_i: X \to \mathbf{R}$ with is 1 on $\overline{W}_i$ and 0 on $X - V_i$. We let $\psi = \sum \psi_i$ and $\theta_i = \psi_i/\psi$. Then the collection $\{\theta_i\}$ is the desired partition of unity.

II, §4. MANIFOLDS WITH BOUNDARY

Let **E** be a Banach space, and $\lambda: \mathbf{E} \to \mathbf{R}$ a continuous linear map into **R**. (This will also be called a **functional** on **E**.) We denote by $\mathbf{E}_\lambda^0$ the kernel of λ, and by $\mathbf{E}_\lambda^+$ (resp. $\mathbf{E}_\lambda^-$) the set of points $x \in \mathbf{E}$ such that $\lambda(x) \geqq 0$ (resp. $\lambda(x) \leqq 0$). We call $\mathbf{E}_\lambda^0$ a **hyperplane** and $\mathbf{E}_\lambda^+$ or $\mathbf{E}_\lambda^-$ a **half plane**.

If μ is another functional and $\mathbf{E}_\lambda^+ = \mathbf{E}_\mu^+$, then there exists a number $c > 0$ *such that* $\lambda = c\mu$. This is easily proved. Indeed, we see at once that the kernels of λ and μ must be equal. Suppose $\lambda \neq 0$. Let x_0 be such that $\lambda(x_0) > 0$. Then $\mu(x_0) > 0$ also. The functional

$$\lambda - (\lambda(x_0)/\mu(x_0))\mu$$

vanishes on the kernel of λ (or μ) and also on x_0. Therefore it is the 0 functional, and $c = \lambda(x_0)/\mu(x_0)$.

Let $\mathbf{E}$, $\mathbf{F}$ be Banach spaces, and let $\mathbf{E}_\lambda^+$ and $\mathbf{F}_\mu^+$ be two half planes in $\mathbf{E}$ and $\mathbf{F}$ respectively. Let U, V be two open subsets of these half planes respectively. We shall say that a mapping

$$f: U \to V$$

is a morphism of class C^p if the following condition is satisfied. Given a point $x \in U$, there exists an open neighborhood U_1 of x in $\mathbf{E}$, an open neighborhood V_1 of $f(x)$ in $\mathbf{F}$, and a morphism $f_1: U_1 \to V_1$ (in the sense of Chapter I) such that the restriction of f_1 to $U_1 \cap U$ is equal to f. (We assume that all morphisms are of class C^p with $p \geq 1$.)

If our half planes are full planes (i.e. equal to the vector spaces themselves), then our present definition is the same as the one used previously.

If we take as objects the open subsets of half planes in Banach spaces, and as morphisms the C^p-morphisms, then we obtain a category. The notion of isomorphism is therefore defined, and the definition of manifold by means of atlases and charts can be used as before. The manifolds of §1 should have been called **manifolds without boundary**, reserving the name of manifold for our new globalized objects. However, in most of this book, we shall deal exclusively with manifolds without boundary for simplicity. The following remarks will give readers the means of extending any result they wish (provided it is true) for the case of manifolds without boundaries to the case manifolds with.

First, concerning the notion of derivative, we have:

Proposition 4.1. *Let* $f: U \to \mathbf{F}$ *and* $g: U \to \mathbf{F}$ *be two morphisms of class* C^p ($p \geq 1$) *defined on an open subset* U *of* $\mathbf{E}$. *Assume that* f *and* g *have the same restriction to* $U \cap \mathbf{E}_\lambda^+$ *for some half plane* $\mathbf{E}_\lambda^+$, *and let*

$$x \in U \cap \mathbf{E}_\lambda^+.$$

Then $f'(x) = g'(x)$.

Proof. After considering the difference of f and g, we may assume without loss of generality that the restriction of f to $U \cap \mathbf{E}_\lambda^+$ is 0. It is then obvious that $f'(x) = 0$.

Proposition 4.2. *Let U be open in* $\mathbf{E}$. *Let μ be a non-zero functional on* $\mathbf{F}$ *and let* $f: U \to \mathbf{F}_\mu^+$ *be a morphism of class C^p with $p \geqq 1$. If x is a point of U such that $f(x)$ lies in $\mathbf{F}_\mu^0$ then $f'(x)$ maps* $\mathbf{E}$ *into* $\mathbf{F}_\mu^0$.

Proof. Without loss of generality, we may assume that $x = 0$ and $f(x) = 0$. Let W be a given neighborhood of 0 in $\mathbf{F}$. Suppose that we can find a small element $v \in \mathbf{E}$ such that $\mu f'(0)v \neq 0$. We can write (for small t):

$$f(tv) = tf'(0)v + o(t)w_t$$

with some element $w_t \in W$. By assumption, $f(tv)$ lies in $\mathbf{F}_\mu^+$. Applying μ we get

$$t\mu f'(0)v + o(t)\mu(w_t) \geqq 0.$$

Dividing by t, this yields

$$\mu f'(0)v \geqq \frac{o(t)}{t}\mu(w_t).$$

Replacing t by $-t$, we get a similar inequality on the other side. Letting t tend to 0 shows that $\mu f'(0)v = 0$, a contradiction.

Let U be open in some half plane $\mathbf{E}_\lambda^+$. We define the **boundary** of U (written ∂U) to be the intersection of U with $\mathbf{E}_\lambda^0$, and the **interior** of U (written $\mathrm{Int}(U)$) to be the complement of ∂U in U. Then $\mathrm{Int}(U)$ is open in $\mathbf{E}$.

It follows at once from our definition of differentiability that a half plane is C^∞-isomorphic with a product

$$\mathbf{E}_\lambda^+ \approx \mathbf{E}_\lambda^0 \times \mathbf{R}^+$$

where $\mathbf{R}^+$ is the set of real numbers $\geqq 0$, whenever $\lambda \neq 0$. The boundary of $\mathbf{E}_\lambda^+$ in that case is $\mathbf{E}_\lambda^0 \times 0$.

Proposition 4.3. *Let λ be a functional on* $\mathbf{E}$ *and μ a functional on* $\mathbf{F}$. *Let U be open in $\mathbf{E}_\lambda^+$ and V open in $\mathbf{F}_\mu^+$ and assume $U \cap \mathbf{E}_\lambda^0$, $V \cap \mathbf{F}_\mu^0$ are not empty. Let $f: U \to V$ be an isomorphism of class C^p $(p \geqq 1)$. Then $\lambda \neq 0$ if and only if $\mu \neq 0$. If $\lambda \neq 0$, then f induces a C^p-isomorphism of $\mathrm{Int}(U)$ on $\mathrm{Int}(V)$ and of ∂U on ∂V.*

Proof. By the functoriality of the derivative, we know that $f'(x)$ is a toplinear isomorphism for each $x \in U$. Our first assertion follows from the preceding proposition. We also see that no interior point of U maps on a boundary point of V and conversely. Thus f induces a bijection of ∂U on ∂V and a bijection of $\mathrm{Int}(U)$ on $\mathrm{Int}(V)$. Since these interiors are open in their respective spaces, our definition of derivative shows that f induces an isomorphism between them. As for the boundary, it is a submanifold of the full space, and locally, our definition of derivative, together with the product structure, shows that the restriction of f to ∂U must be an isomorphism on ∂V.

This last proposition shows that the boundary is a differentiable invariant, and thus that we can speak of the boundary of a manifold.

We give just two words of warning concerning manifolds with boundary. First, products do not exist in their category. Indeed, to get products, we are forced to define manifolds with **corners**, which would take us too far afield.

Second, in defining immersions or submanifolds, there is a difference in kind when we consider a manifold embedded in a manifold without boundary, or a manifold embedded in another manifold with boundary. Think of a closed interval embedded in an ordinary half plane. Two cases arise. The case where the interval lies inside the interior of the half plane is essentially distinct from the case where the interval has one end point touching the hyperplane forming the boundary of the half plane. (For instance, given two embeddings of the first type, there exists an automorphism of the half plane carrying one into the other, but there cannot exist an automorphism of the half plane carrying an embedding of the first type into one of the second type.)

We leave it to the reader to go systematically through the notions of tangent space, immersion, embedding (and later, tangent bundle, vector field, etc.) for arbitrary manifolds (with boundary). For instance, Proposition 2.2 shows at once how to get the tangent space functorially.

CHAPTER III

Vector Bundles

The collection of tangent spaces can be glued together to give a manifold with a natural projection, thus giving rise to the tangent bundle. The general glueing procedure can be used to construct more general objects known as vector bundles, which give powerful invariants of a given manifold. (For an interesting theorem see Mazur [Maz 61].) In this chapter, we develop purely formally certain functorial constructions having to do with vector bundles. In the chapters on differential forms and Riemannian metrics, we shall discuss in greater detail the constructions associated with multilinear alternating forms, and symmetric positive definite forms.

Partitions of unity are an essential tool when considering vector bundles. They can be used to combine together a random collection of morphisms into vector bundles, and we shall give a few examples showing how this can be done (concerning exact sequences of bundles).

III, §1. DEFINITION, PULL BACKS

Let X be a manifold (of class C^p with $p \geqq 0$) and let $\pi: E \to X$ be a morphism. Let $\mathbf{E}$ be a Banach space.

Let $\{U_i\}$ be an open covering of X, and for each i, suppose that we are given a mapping

$$\tau_i: \pi^{-1}(U_i) \to U_i \times \mathbf{E}$$

satisfying the following conditions:

VB 1. *The map τ_i is a C^p isomorphism commuting with the projection on U_i, that is, such that the following diagram is commutative:*

$$\pi^{-1}(U_i) \xrightarrow{\ \tau_i\ } U_i \times \mathbf{E}$$
$$U_i$$

In particular, we obtain an isomorphism on each fiber (written $\tau_i(x)$ or τ_{ix})

$$\pi_{ix}: \pi^{-1}(x) \to \mathbf{E}$$

VB 2. *For each pair of open sets U_i, U_j the map*

$$\tau_{jx} \circ \tau_{ix}^{-1}: \mathbf{E} \to \mathbf{E}$$

is a toplinear isomorphism.

VB 3. *If U_i and U_j are two members of the covering, then the map of $U_i \cap U_j$ into $L(\mathbf{E}, \mathbf{E})$ (actually $\mathrm{Laut}(\mathbf{E})$) given by*

$$x \mapsto (\tau_j \tau_i^{-1})_x$$

is a morphism.

Then we shall say that $\{(U_i, \tau_i)\}$ is a **trivializing covering** for π (or for E by abuse of language), and that $\{\tau_i\}$ are its **trivializing** maps. If $x \in U_i$, we say that τ_i (or U_i) trivializes at x. Two trivializing coverings for π are said to be **VB-equivalent** if taken together they also satisfy conditions **VB 2**, **VB 3**. An equivalence class of such trivializing coverings is said to determine a structure of **vector bundle** on π (or on E by abuse of language). We say that E is the **total space** of the bundle, and that X is its **base space**. If we wish to be very functorial, we shall write E_π and X_π for these spaces respectively. The fiber $\pi^{-1}(x)$ is also denoted by E_x or π_x. We also say that the vector bundle has **fiber E**, or is **modeled on E**. Note that from **VB 2**, the fiber $\pi^{-1}(x)$ above each point $x \in X$ can be given a structure of Banachable space, simply by transporting the Banach space structure of $\mathbf{E}$ to $\pi^{-1}(x)$ via τ_{ix}. Condition **VB 2** insures that using two different trivializing maps τ_{ix} or τ_{jx} will give the same structure of Banachable space (with equivalent norms, of course not the same norms).

Conversely, we could replace **VB 2** by a similar condition as follows.

VB 2′. *On each fiber $\pi^{-1}(x)$ we are given a structure of Banachable space, and for $x \in U_i$, the trivializing map*

$$\tau_{ix} \colon \pi^{-1}(x) = E_x \to \mathbf{E}$$

is a toplinear isomorphism.

Then it follows that $\tau_{jx} \circ \tau_{ix}^{-1} \colon \mathbf{E} \to \mathbf{E}$ is a toplinear isomorphism for each pair of open sets U_i, U_j and $x \in U_i \cap U_j$.

In the finite dimensional case, condition **VB 3** is implied by **VB 2**.

Proposition 1.1. *Let $\mathbf{E}$, $\mathbf{F}$ be finite dimensional vector spaces. Let U be open in some Banach space. Let*

$$f \colon U \times \mathbf{E} \to \mathbf{F}$$

be a morphism such that for each $x \in U$, the map

$$f_x \colon \mathbf{E} \to \mathbf{F}$$

given by $f_x(v) = f(x, v)$ is a linear map. Then the map of U into $L(\mathbf{E}, \mathbf{F})$ given by $x \mapsto f_x$ is a morphism.

Proof. We can write $\mathbf{F} = \mathbf{R}_1 \times \cdots \times \mathbf{R}_n$ (n copies of $\mathbf{R}$). Using the fact that $L(\mathbf{E}, \mathbf{F}) = L(\mathbf{E}, \mathbf{R}_1) \times \cdots \times L(\mathbf{E}, \mathbf{R}_n)$, it will suffice to prove our assertion when $\mathbf{F} = \mathbf{R}$. Similarly, we can assume that $\mathbf{E} = \mathbf{R}$ also. But in that case, the function $f(x, v)$ can be written $g(x)v$ for some map $g \colon U \to \mathbf{R}$. Since f is a morphism, it follows that as a function of each argument x, v it is also a morphism. Putting $v = 1$ shows that g is a morphism and concludes the proof.

Returning to the general definition of a vector bundle, we call the maps

$$\tau_{jix} = \tau_{jx} \circ \tau_{ix}^{-1}$$

the **transition** maps associated with the covering. They satisfy what we call the **cocycle condition**

$$\tau_{kjx} \circ \tau_{jix} = \tau_{kix}.$$

In particular, $\tau_{iix} = \mathrm{id}$ and $\tau_{jix} = \tau_{ijx}^{-1}$.

As with manifolds, we can recover a vector bundle from a trivializing covering.

Proposition 1.2. *Let X be a manifold, and $\pi: E \to X$ a mapping from some set E into X. Let $\{U_i\}$ be an open covering of X, and for each i suppose that we are given a Banach space $\mathbf{E}$ and a bijection (commuting with the projection on U_i),*

$$\tau_i: \pi^{-1}(U_i) \to U_i \times \mathbf{E},$$

such that for each pair i, j and $x \in U_i \cap U_j$, the map $(\tau_j \tau_i^{-1})_x$ is a toplinear isomorphism, and condition **VB 3** *is satisfied as well as the cocycle condition. Then there exists a unique structure of manifold on E such that π is a morphism, such that τ_i is an isomorphism making π into a vector bundle, and $\{(U_i, \tau_i)\}$ into a trivialising covering.*

Proof. By Proposition 3.10 of Chapter I and our condition **VB 3**, we conclude that the map

$$\tau_j \tau_i^{-1}: (U_i \cap U_j) \times \mathbf{E} \to (U_i \cap U_j) \times \mathbf{E}$$

is a morphism, and in fact an isomorphism since it has an inverse. From the definition of atlases, we conclude that E has a unique manifold structure such that the τ_i are isomorphisms. Since π is obtained locally as a composite of morphisms (namely τ_i and the projections of $U_i \times \mathbf{E}$ on the first factor), it becomes a morphism. On each fiber $\pi^{-1}(x)$, we can transport the topological vector space structure of any $\mathbf{E}$ such that x lies in U_i, by means of τ_{ix}. The result is independent of the choice of U_i since $(\tau_j \tau_i^{-1})_x$ is a toplinear isomorphism. Our proposition is proved.

Remark. It is relatively rare that a vector bundle is **trivial**, i.e. VB-isomorphic to a product $X \times \mathbf{E}$. By definition, it is always trivial locally. In the finite dimensional case, say when E has dimension n, a trivialization is equivalent to the existence of sections $\xi_1, \dots, \xi_n$ such that for each x, the vectors $\xi_1(x), \dots, \xi_n(x)$ form a basis of E_x. Such a choice of sections is called a **frame** of the bundle, and is used especially with the tangent bundle, to be defined below. In this book where we give proofs valid in the infinite dimensional case, frames will therefore not occur until we get to strictly finite dimensional phenomenon.

**The local representation of a vector bundle and
the vector component of a morphism**

For arbitrary vector bundles (and especially the tangent bundle to be defined below), we have a local representation of the bundle as a product in a chart. For many purposes, and especially the case of a morphism

$$f: Y \to E$$

of a manifold into the vector bundle, it is more convenient to use U to denote an open subset of a Banach space, and to let $\varphi: U \to X$ be an isomorphism of U with an open subset of X over which E has a trivialization $\tau: \pi^{-1}(\varphi U) \to U \times \mathbf{E}$. Suppose V is an open subset of Y such that $f(V) \subset \pi^{-1}(\varphi U)$. We then have the commutative diagram:

$$
\begin{array}{ccccc}
V & \xrightarrow{\ f\ } & \pi^{-1}(\varphi U) & \xrightarrow{\ \tau\ } & U \times \mathbf{E} \\
& & \downarrow & & \downarrow \\
& & \varphi U & \xrightarrow{\ \varphi^{-1}\ } & U
\end{array}
$$

The composite $\tau \circ f$ is a morphism of V into $U \times \mathbf{E}$, which has two components

$$\tau \circ f = (f_{U1}, f_{U2})$$

such that $f_{U1}: V \to U$ and $f_{U2}: V \to \mathbf{E}$. We call f_{U2} the **vector component of f in the vector bundle chart** $U \times \mathbf{E}$ over U. Sometimes to simplify the notation, we omit the subscript, and merely agree that $f_U = f_{U2}$ denotes this vector component; or to simplify the notation further, we may simply state that f itself denotes this vector component if a discussion takes place entirely in a chart. In this case, we say that $f = f_U$ **represents the morphism** in the vector bundle chart, or in the chart.

Vector bundle morphisms and pull backs

We now make the set of vector bundles into a category.

Let $\pi: E \to X$ and $\pi': E' \to X'$ be two vector bundles. A **VB-morphism** $\pi \to \pi'$ consists of a pair of morphisms

$$f_0: X \to X' \qquad \text{and} \qquad f: E \to E'$$

satisfying the following conditions.

VB Mor 1. *The diagram*

$$
\begin{array}{ccc}
E & \xrightarrow{\ f\ } & E' \\
\pi \downarrow & & \downarrow \pi' \\
X & \xrightarrow[\ f_0\ ]{} & X'
\end{array}
$$

is commutative, and the induced map for each $x \in X$

$$f_x : E_x \to E'_{f(x)}$$

is a continuous linear map.

VB Mor 2. *For each $x_0 \in X$ there exist trivializing maps*

$$\tau : \pi^{-1}(U) \to U \times \mathbf{E}$$

and

$$\tau' : \pi'^{-1}(U') \to U' \times \mathbf{E}'$$

at x_0 and $f(x_0)$ respectively, such that $f_0(U)$ is contained in U', and such that the map of U into $L(\mathbf{E}, \mathbf{E}')$ given by

$$x \mapsto \tau'_{f_0(x)} \circ f_x \circ \tau^{-1}$$

is a morphism.

As a matter of notation, we shall also use f to denote the VB-morphism, and thus write $f : \pi \to \pi'$. In most applications, f_0 is the identity. By Proposition 1.1, we observe that **VB Mor 2** is redundant in the finite dimensional case.

The next proposition is the analogue of Proposition 1.2 for VB-morphisms.

Proposition 1.3. *Let π, π' be two vector bundles over manifolds X, X' respectively. Let $f_0 : X \to X'$ be a morphism, and suppose that we are given for each $x \in X$ a continuous linear map*

$$f_x : \pi_x \to \pi'_{f_0(x)}$$

*such that, for each x_0, condition **VB Mor 2** is satisfied. Then the map f from π to π' defined by f_x on each fiber is a VB-morphism.*

Proof. One must first check that f is a morphism. This can be done under the assumption that π, π' are trivial, say equal to $U \times \mathbf{E}$ and $U' \times \mathbf{E}'$ (following the notation of **VB Mor 2**), with trivialising maps equal to the identity. Our map f is then given by

$$(x, v) \mapsto (f_0 x, f_x v).$$

Using Proposition 3.10 of Chapter I, we conclude that f is a morphism, and hence that (f_0, f) is a VB-morphism.

It is clear how to compose two VB-morphisms set theoretically. In fact, the composite of two VB-morphisms is a VB-morphism. There is no problem verifying condition **VB Mor 1**, and for **VB Mor 2**, we look at the situation locally. We encounter a commutative diagram of the following type:

$$
\begin{array}{ccccc}
\pi^{-1}(U) & \xrightarrow{\;f\;} & \pi'^{-1}(U') & \xrightarrow{\;g\;} & \pi''^{-1}(U'') \\
\downarrow{\tau} & & \downarrow{\tau'} & & \downarrow{\tau''} \\
U \times \mathbf{E} & \longrightarrow & U' \times \mathbf{E}' & \longrightarrow & U'' \times \mathbf{E}''
\end{array}
$$

and use Proposition 3.10 of Chapter I, to show that $g \circ f$ is a VB-morphism.

We therefore have a category, denoted by VB or VB^p, if we need to specify explicitly the order of differentiability.

The vector bundles over X from a subcategory $VB(X) = VB^p(X)$ (taking those VB-morphisms for which the map f_0 is the identity). If $\mathfrak{A}$ is a category of Banach spaces (for instance finite dimensional spaces), then we denote by $VB(X, \mathfrak{A})$ those vector bundles over X whose fibers lie in $\mathfrak{A}$.

A morphism from one vector bundle into another can be given locally. More precisely, suppose that U is an open subset of X and $\pi: E \to X$ a vector bundle over X. Let $E_U = \pi^{-1}(U)$ and

$$
\pi_U = \pi | E_U
$$

be the restriction of π to E_U. Then π_U is a vector bundle over U. Let $\{U_i\}$ be an open covering of the manifold X and let π, π' be two vector bundles over X. Suppose, given a VB-morphism

$$
f_i: \pi_{U_i} \to \pi'_{U_i}
$$

for each i, such that f_i and f_j agree over $U_i \cap U_j$ for each pair of indices i, j. Then there exists a unique VB-morphism $f: \pi \to \pi'$ which agrees with f_i on each U_i. The proof is trivial, but the remark will be used frequently in the sequel.

Using the discussion at the end of Chapter II, §2 and Proposition 2.7 of that chapter, we get immediately:

Proposition 1.4. *Let* $\pi: E \to Y$ *be a vector bundle, and* $f: X \to Y$ *a morphism. Then*

$$
f^*(\pi): f^*(E) \to X
$$

is a vector bundle, and the pair $(f, \pi^*(f))$ *is a* **VB**-*morphism*

$$
\begin{array}{ccc}
f^*(E) & \xrightarrow{\;\pi^*(f)\;} & E \\[4pt]
\Big\downarrow{\scriptstyle f^*(\pi)} & & \Big\downarrow{\scriptstyle \pi} \\[4pt]
X & \xrightarrow[\;f\;]{} & Y
\end{array}
$$

In Proposition 1.4, we could take f to be the inclusion of a submanifold. In that case, the pull-back is merely the restriction. As with open sets, we can then use the usual notation:

$$
E_X = \pi^{-1}(X) \qquad \text{and} \qquad \pi_X = \pi | E_X.
$$

Thus $\pi_X = f^*(\pi)$ in that case.

If X happens to be a point y of Y, then we have the constant map

$$
\pi_y \colon E_y \to y
$$

which will sometimes be identified with E_y.

If we identify each fiber $(f^*E)_x$ with $E_{f(x)}$ itself (a harmless identification since an element of the fiber at x is simply a pair (x, e) with e in $E_{f(x)}$), then we can describe the pull-back f^* of a vector bundle $\pi \colon E \to Y$ as follows. It is a vector bundle $f^*\pi \colon f^*E \to X$ satisfying the following properties:

PB 1. *For each* $x \in X$, *we have* $(f^*E)_x = E_{f(x)}$.

PB 2. *We have a commutative diagram*

$$
\begin{array}{ccc}
f^*(E) & \longrightarrow & E \\[4pt]
\Big\downarrow{\scriptstyle f^*(\pi)} & & \Big\downarrow{\scriptstyle \pi} \\[4pt]
X & \xrightarrow[\;f\;]{} & Y
\end{array}
$$

the top horizontal map being the identity on each fiber.

PB 3. *If* E *is trivial, equal to* $Y \times \mathbf{E}$, *then* $f^*E = X \times \mathbf{E}$ *and* $f^*\pi$ *is the projection.*

PB 4. *If* V *is an open subset of* Y *and* $U = f^{-1}(V)$, *then*

$$
f^*(E_V) = (f^*E)_U,
$$

and we have a commutative diagram:

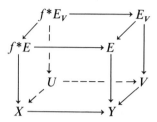

III, §2. THE TANGENT BUNDLE

Let X be a manifold of class C^p with $p \geq 1$. We shall define a functor T from the category of such manifolds into the category of vector bundles of class C^{p-1}.

For each manifold X we let $T(X)$ be the disjoint union of the tangent spaces $T_x(X)$. We have a natural projection

$$\pi: T(X) \to X$$

mapping $T_x(X)$ on x. We must make this into a vector bundle. If (U, φ) is a chart of X such that φU is open in the Banach space $\mathbf{E}$, then from the definition of the tangent vectors as equivalence classes of triples (U, φ, v) we get immediately a bijection

$$\tau_U: \pi^{-1}(U) = T(U) \to U \times \mathbf{E}$$

which commutes with the projection on U, that is such that

$$\pi^{-1}(U) \xrightarrow{\ \tau_U\ } U \times \mathbf{E}$$
$$\searrow \qquad \swarrow$$
$$U$$

is commutative. Furthermore, if (U_i, φ_i) and (U_j, φ_j) are two charts, and if we denote by φ_{ji} the map $\varphi_j \varphi_i^{-1}$ (defined on $\varphi_i(U_i \cap U_j)$), then we obtain a transition mapping

$$\tau_{ji} = (\tau_j \tau_i^{-1}): \varphi_i(U_i \cap U_j) \times \mathbf{E} \to \varphi_j(U_i \cap U_j) \times \mathbf{E}$$

by the formula

$$\tau_{ji}(x, v) = (\varphi_{ji} x, D\varphi_{ji}(x) \cdot v)$$

for $x \in U_i \cap U_j$ and $v \in \mathbf{E}$. Since the derivative $D\varphi_{ji} = \varphi'_{ji}$ is of class C^{p-1}

and is an isomorphism at x, we see immediately that all the conditions of Proposition 1.2 are verified (using Proposition 3.10 of Chapter I), thereby making $T(X)$ into a vector bundle of class C^{p-1}.

We see that the above construction can also be expressed as follows. If the manifold X is glued together from open sets $\{U_i\}$ in Banach spaces by means of transition mappings $\{\varphi_{ij}\}$, then we can glue together products $U_i \times \mathbf{E}$ by means of transition mappings $(\varphi_{ij}, D\varphi_{ij})$ where the derivative $D\varphi_{ij}$ can be viewed as a function of two variables (x, v). Thus locally, for open subsets U of Banach spaces, the tangent bundle can be identified with the product $U \times \mathbf{E}$. The reader will note that our definition coincides with the oldest definition employed by geometers, our tangent vectors being vectors which transform according to a certain rule (namely the derivative).

If $f: X \to X'$ is a C^p-morphism, we can define

$$Tf: T(X) \to T(X')$$

to be simply $T_x f$ on each fiber $T_x(X)$. In order to verify that Tf is a VB-morphism (of class C^{p-1}), it suffices to look at the situation locally, i.e. we may assume that X and X' are open in vector spaces $\mathbf{E}$, $\mathbf{E}'$, and that $T_x f = f'(x)$ is simply the derivative. Then the map Tf is given by

$$Tf(x, v) = \big(f(x), f'(x)v\big)$$

for $x \in X$ and $v \in \mathbf{E}$. Since f' is of class C^{p-1} by definition, we can apply Proposition 3.10 of Chapter I to conclude that Tf is also of class C^{p-1}. The functoriality property is trivially satisfied, and we have therefore defined the functor T as promised.

It will sometimes be notationally convenient to write f_* instead of Tf for the induced map, which is also called the **tangent** map. The bundle $T(X)$ is called the **tangent bundle** of X.

Remark. The above definition of the tangent bundle fits with Steenrod's point of view [Ste 51]. I don't understand why many differential geometers have systematically rejected this point of view, when they take the definition of a tangent vector as a differential operator.

III, §3. EXACT SEQUENCES OF BUNDLES

Let X be a manifold. Let $\pi': E' \to X$ and $\pi: E \to X$ be two vector bundles over X. Let $f: \pi' \to \pi$ be a VB-morphism. We shall say that the sequence

$$0 \to \pi' \xrightarrow{\ f\ } \pi$$

is **exact** if there exists a covering of X by open sets and for each open set U in this covering there exist trivializations

$$\tau': E'_U \to U \times E' \quad \text{and} \quad \tau: E_U \to U \times E$$

such that E can be written as a product $E = E' \times F$, making the following diagram commutative:

$$
\begin{array}{ccc}
E'_U & \xrightarrow{\ f\ } & E_U \\
{\scriptstyle \tau'}\downarrow & & \downarrow{\scriptstyle \tau} \\
U \times E' & \longrightarrow & U \times E' \times F
\end{array}
$$

(The bottom map is the natural one: Identity on U and the injection of E' on $E' \times 0$.)

Let $\pi_1: E_1 \to X$ be another vector bundle, and let $g: \pi_1 \to \pi$ be a VB-morphism such that $g(E_1)$ is contained in $f(E')$. Since f establishes a bijection between E' and its image $f(E')$ in E, it follows that there exists a unique map $g_1: E_1 \to E'$ such that $g = f \circ g_1$. We contend that g_1 is a VB-morphism. Indeed, to prove this we can work locally, and in view of the definition, over an open set U as above, we can write

$$g_1 = \tau'^{-1} \circ \mathrm{pr} \circ \tau \circ g$$

where pr is the projection of $U \times E' \times F$ on $U \times E'$. All the maps on the right-hand side of our equality are VB-morphisms; this proves our contention.

Let $\pi: E \to X$ be a vector bundle. A subset S of E will be called a **subbundle** if there exists an exact sequence $0 \to \pi' \to \pi$, also written

$$0 \to E' \xrightarrow{\ f\ } E,$$

such that $f(E') = S$. This gives S the structure of a vector bundle, and the previous remarks show that it is unique. In fact, given another exact sequence

$$0 \to E_1 \xrightarrow{\ g\ } E$$

such that $g(E_1) = S$, the natural map $f^{-1}g$ from E_1 to E' is a VB-isomorphism.

Let us denote by E/E' the union of all factor spaces E_x/E'_x. If we are dealing with an exact sequence as above, then we can give E/E' the structure of a vector bundle. We proceed as follows. Let $\{U_i\}$ be our covering, with trivialising maps τ'_i and τ_i. We can define for each i a bijection

$$\pi''_i: E_{U_i}/E'_{U_i} \to U_i \times F$$

obtained in a natural way from the above commutative diagram. (Without loss of generality, we can assume that the vector spaces $\mathbf{E}'$, $\mathbf{F}$ are constant for all i.) We have to prove that these bijections satisfy the conditions of Proposition 1.2.

Without loss of generality, we may assume that f is an inclusion (of the total space E' into E). For each pair i, j and $x \in U_i \cap U_j$, the toplinear automorphism $(\tau_j \tau_i^{-1})_x$ is represented by a matrix

$$\begin{pmatrix} h_{11}(x) & h_{12}(x) \\ h_{21}(x) & h_{22}(x) \end{pmatrix}$$

operating on the right on a vector $(v, w) \in \mathbf{E}' \times \mathbf{F}$. The map $(\tau_j'' \tau_i''^{-1})_x$ on $\mathbf{F}$ is induced by this matrix. Since $\mathbf{E}' = \mathbf{E}' \times 0$ has to be carried into itself by the matrix, we have $h_{12}(x) = 0$. Furthermore, since $(\tau_j \tau_i^{-1})_x$ has an inverse, equal to $(\tau_i \tau_j^{-1})_x$, it follows that $h_{22}(x)$ is a toplinear automorphism of $\mathbf{F}$, and represents $(\tau_j'' \tau_i''^{-1})_x$. Therefore condition **VB 3** is satisfied, and E/E' is a vector bundle.

The canonical map

$$E_U \to E_U/E_U'$$

is a morphism since it can be expressed in terms of τ, the projection, and τ''^{-1}. Consequently, we obtain a VB-morphism

$$g: \pi \to \pi''$$

in the canonical way (on the total spaces, it is the quotient mapping of E on E/E'). We shall call π'' the **factor bundle**.

Our map g satisfies the usual universal mapping property of a cokernel. Indeed, suppose that

$$\psi: E \to G$$

is a VB-morphism such that $\psi \circ f = 0$ (i.e. $\psi_x \circ f_x = 0$ on each fiber E_x'). We can then define set theoretically a canonical map

$$\psi_*: E/E' \to G,$$

and we must prove that it is a VB-morphism. This can be done locally. Using the above notation, we may assume that $E = U \times \mathbf{E}' \times \mathbf{F}$ and that g is the projection. In that case, ψ_* is simply the canonical injection of $U \times \mathbf{F}$ in $U \times \mathbf{E}' \times \mathbf{F}$ followed by ψ, and is therefore a VB-morphism.

We shall therefore call g the **cokernel** of f.

Dually, let $g: \pi \to \pi''$ be a given VB-morphism. We shall say that the sequence

$$\pi \xrightarrow{g} \pi'' \to 0$$

is **exact** if g is surjective, and if there exists a covering of X by open sets, and for each open set U in this covering there exist spaces $\mathbf{E}'$, $\mathbf{F}$ and trivializations

$$\tau: E_U \to U \times \mathbf{E}' \times \mathbf{F} \qquad \text{and} \qquad \tau'': E_U'' \to \mathbf{F}$$

making the following diagram commutative:

$$
\begin{array}{ccc}
E_U & \xrightarrow{\quad g \quad} & E_U'' \\
{\scriptstyle \tau} \downarrow & & \downarrow {\scriptstyle \tau''} \\
U \times \mathbf{E}' \times \mathbf{F} & \longrightarrow & U \times \mathbf{F}
\end{array}
$$

(The bottom map is the natural one: Identity on U and the projection of $\mathbf{E}' \times \mathbf{F}$ on $\mathbf{F}$.)

In the same way as before, one sees that the "kernel" of g, that is, the union of the kernels E_x' of each g_x, can be given a structure of vector bundle. This union E' will be called the **kernel** of g, and satisfies the usual universal mapping property.

Proposition 3.1. *Let X be a manifold and let*

$$f: \pi' \to \pi$$

be a VB-morphism of vector bundles over X. Assume that, for each $x \in X$, the continuous linear map

$$f_x: E_x' \to E_x$$

is injective and splits. Then the sequence

$$0 \to \pi' \xrightarrow{f} \pi$$

is exact.

Proof. We can assume that X is connected and that the fibers of E' and E are constant, say equal to the Banach spaces $\mathbf{E}'$ and $\mathbf{E}$. Let $a \in X$. Corresponding to the splitting of f_a we know that we have a product decomposition $\mathbf{E} = \mathbf{E}' \times \mathbf{F}$ and that there exists an open set U of X

containing a, together with trivialising maps

$$\tau: \pi^{-1}(U) \to U \times \mathbf{E} \qquad \text{and} \qquad \tau': \pi'^{-1}(U) \to U \times \mathbf{E}'$$

such that the composite map

$$\mathbf{E}' \xrightarrow{\tau_a'^{-1}} E_a' \xrightarrow{f_a} E_a \xrightarrow{\tau_a} \mathbf{E}' \times \mathbf{F}$$

maps $\mathbf{E}'$ on $\mathbf{E}' \times 0$.

For any point x in U, we have a map

$$(\tau f \tau'^{-1})_x: \mathbf{E}' \to \mathbf{E}' \times \mathbf{F},$$

which can be represented by a pair of continuous linear maps

$$(h_{11}(x), h_{21}(x)).$$

We define

$$h(x): \mathbf{E}' \times \mathbf{F} \to \mathbf{E}' \times \mathbf{F}$$

by the matrix

$$\begin{pmatrix} h_{11}(x) & 0 \\ h_{21}(x) & \text{id} \end{pmatrix},$$

operating on the right on a vector $(v, w) \in \mathbf{E}' \times \mathbf{F}$. Then $h(x)$ restricted to $\mathbf{E}' \times 0$ has the same action as $(\tau f \tau'^{-1})_x$.

The map $x \to h(x)$ is a morphism of U into $L(\mathbf{E}, \mathbf{E})$ and since it is continuous, it follows that for U small enough around our fixed point a, it maps U into the group of toplinear automorphisms of $\mathbf{E}$. This proves our proposition.

Dually to Proposition 3.1, we have:

Proposition 3.2. *Let X be a manifold and let*

$$g: \pi \to \pi''$$

be a VB-morphism of vector bundles over X. Assume that for each $x \in X$, the continuous linear map

$$g_x: E_x \to E_x''$$

is surjective and has a kernel that splits. Then the sequence

$$\pi \xrightarrow{g} \pi'' \to 0$$

is exact.

Proof. It is dual to the preceding one and we leave it to the reader.

In general, a sequence of VB-morphisms

$$0 \to \pi' \xrightarrow{\ f\ } \pi \xrightarrow{\ g\ } \pi'' \to 0$$

is said to be **exact** if both ends are exact, and if the image of f is equal to the kernel of g.

There is an important example of exact sequence. Let $f: X \to Y$ be an immersion. By the universal mapping property of pull backs, we have a canonical VB-morphism

$$T*f: T(X) \to f*T(Y)$$

of $T(X)$ into the pull back over X of the tangent bundle of Y. Furthermore, from the manner in which the pull back is obtained locally by taking products, and the definition of an immersion, one sees that the sequence

$$0 \to T(X) \xrightarrow{T*f} f*T(Y)$$

is exact. The factor bundle

$$f*T(Y)/\mathrm{Im}(T*f)$$

is called the **normal** bundle of f. It is denoted by $N(f)$, and its total space by $N_f(X)$ if we wish to distinguish between the two. We sometimes identify $T(X)$ with its image under $T*f$ and write

$$N(f) = f*T(Y)/T(X).$$

Dually, let $f: X \to Y$ be a submersion. Then we have an exact sequence

$$T(X) \xrightarrow{T*f} f*T(Y) \to 0$$

whose kernel could be called the **subbundle** of f, or the **bundle along the fiber**.

There is an interesting case where we can describe the kernel more precisely. Let

$$\pi: E \to X$$

be a vector bundle. Then we can form the pull back of E over itself, that is, $\pi*E$, and we contend that we have an exact sequence

$$0 \to \pi*E \to T(E) \to \pi*T(X) \to 0.$$

To define the map on the left, we look at the subbundle of π more closely. For each $x \in X$ we have an inclusion

$$E_x \to E,$$

whence a natural injection

$$T(E_x) \to T(E).$$

The local product structure of a bundle shows that the union of the $T(E_x)$ as x ranges over X gives the subbundle set theoretically. On the other hand, the total space of π^*E consists of pairs of vectors (v, w) lying over the same base point x, that is, the fiber at x of π^*E is simply $E_x \times E_x$. Since $T(E_x)$ has a natural identification with $E_x \times E_x$, we get for each x a bijection

$$(\pi^*E)_x \to T(E_x)$$

which defines our map from π^*E to $T(E)$. Considering the map locally in terms of the local product structure shows at once that it gives a VB-isomorphism between π^*E and the subbundle of π, as desired.

III, §4. OPERATIONS ON VECTOR BUNDLES

We consider subcategories of Banach spaces $\mathfrak{A}$, $\mathfrak{B}$, $\mathfrak{C}$ and let

$$\lambda \colon \mathfrak{A} \times \mathfrak{B} \to \mathfrak{C}$$

be a functor in, say, two variables, which is, say, contravariant in the first and covariant in the second. (Everything we shall do extends in the obvious manner to functors of several variables, letting $\mathfrak{A}$, $\mathfrak{B}$ stand for n-tuples.)

Example. We took a functor in two variables for definiteness, and to illustrate both variances. However, we could consider a functor in one or more than two variables. For instance, let us consider the functor

$$\mathbf{E} \mapsto L(\mathbf{E}, \mathbf{R}) = L(\mathbf{E}) = \mathbf{E}^\vee,$$

which we call the **dual**. It is a contravariant functor in one variable. On the other hand, the functor

$$\mathbf{E} \mapsto L_a^r(\mathbf{E}, \mathbf{F})$$

of continuous multilinear maps of $\mathbf{E} \times \cdots \times \mathbf{E}$ into a Banach space $\mathbf{F}$ is contravariant in $\mathbf{E}$ and covariant in $\mathbf{F}$. The functor $\mathbf{E} \mapsto L_a^r(\mathbf{E}, \mathbf{R})$ gives rise later to what we call differential forms. We shall treat such forms systematically in Chapter V, §3.

If $f: \mathbf{E}' \to \mathbf{E}$ and $g: \mathbf{F} \to \mathbf{F}'$ are two continuous linear maps, with f a morphism of $\mathfrak{A}$ and g a morphism of $\mathfrak{B}$, then by definition, we have a map

$$L(\mathbf{E}', \mathbf{E}) \times L(\mathbf{F}, \mathbf{F}') \to L(\lambda(\mathbf{E}, \mathbf{F}), \lambda(\mathbf{E}', \mathbf{F}')),$$

assigning $\lambda(f, g)$ to (f, g).

We shall say that λ is of **class** C^p if the following condition is satisfied. Give a manifold U, and two morphisms

$$\varphi: U \to L(\mathbf{E}', \mathbf{E}) \qquad \text{and} \qquad \psi: U \to L(\mathbf{F}, \mathbf{F}'),$$

then the composite

$$U \to L(\mathbf{E}', \mathbf{E}) \times L(\mathbf{F}, \mathbf{F}') \to L(\lambda(\mathbf{E}, \mathbf{F}), \lambda(\mathbf{E}', \mathbf{F}'))$$

is also a morphism. (One could also say that λ is **differentiable**.)

Theorem 4.1. *Let λ be a functor as above, of class C^p, $p \geq 0$. Then for each manifold X, there exists a functor λ_X, on vector bundles (of class C^p)*

$$\lambda_X: \mathrm{VB}(X, \mathfrak{A}) \times \mathrm{VB}(X, \mathfrak{B}) \to \mathrm{VB}(X, \mathfrak{C})$$

satisfying the following properties. For any bundles α, β in $\mathrm{VB}(X, \mathfrak{A})$ and $\mathrm{VB}(X, \mathfrak{B})$ respectively, and VB-morphisms

$$f: \alpha' \to \alpha \qquad \text{and} \qquad g: \beta \to \beta'$$

in the respective categories, and for each $x \in X$, we have:

OP 1. $\lambda_X(\alpha, \beta)_x = \lambda(\alpha_x, \beta_x)$.

OP 2. $\lambda_X(f, g)_x = \lambda(f_x, g_x)$.

OP 3. *If α is the trivial bundle $X \times \mathbf{E}$ and β the trivial bundle $X \times \mathbf{F}$, then $\lambda_X(\alpha, \beta)$ is the trivial bundle $X \times \lambda(\mathbf{E}, \mathbf{F})$.*

OP 4. *If $h: Y \to X$ is a C^p-morphism, then*

$$\lambda_Y^*(h^*\alpha, h^*\beta) = h^*\lambda_X(\alpha, \beta).$$

Proof. We may assume that X is connected, so that all the fibers are toplinearly isomorphic to a fixed space. For each open subset U of X we let the total space $\lambda_U(E_\alpha, E_\beta)$ of $\lambda_U(\alpha, \beta)$ be the union of the sets

$$\{x\} \times \lambda(\alpha_x, \beta_x)$$

(identified harmlessly throughout with $\lambda(\alpha_x, \beta_x)$), as x ranges over U. We can find a covering $\{U_i\}$ of X with trivializing maps $\{\tau_i\}$ for α, and $\{\sigma_i\}$ for β,

$$\tau_i \colon \alpha^{-1}(U_i) \to U_i \times \mathbf{E},$$

$$\sigma_i \colon \beta^{-1}(U_i) \to U_i \times \mathbf{F}.$$

We have a bijection

$$\lambda(\tau_i^{-1}, \sigma_i) \colon \lambda_{U_i}(E_\alpha, E_\beta) \to U_i \times \lambda(\mathbf{E}, \mathbf{F})$$

obtained by taking on each fiber the map

$$\lambda(\tau_{ix}^{-1}, \sigma_{ix}) \colon \lambda(\alpha_x, \beta_x) \to \lambda(\mathbf{E}, \mathbf{F}).$$

We must verify that **VB 3** is satisfied. This means looking at the map

$$x \to \lambda(\tau_{jx}^{-1}, \sigma_{jx}) \circ \lambda(\tau_{ix}^{-1}, \sigma_{ix})^{-1}.$$

The expression on the right is equal to

$$\lambda(\tau_{ix}\tau_{jx}^{-1}, \sigma_{jx}\sigma_{ix}^{-1}).$$

Since λ is a functor of class $\mathbf{C}^p$, we see that we get a map

$$U_i \cap U_j \to L\big(\lambda(\mathbf{E}, \mathbf{F}), \lambda(\mathbf{E}, \mathbf{F})\big)$$

which is a C^p-morphism. Furthermore, since λ is a functor, the transition mappings are in fact toplinear isomorphisms, and **VB2**, **VB 3** are proved.

The proof of the analogous statement for $\lambda_X(f, g)$, to the effect that it is a VB-morphism, proceeds in an analogous way, again using the hypothesis that λ is of class $\mathbf{C}^p$. Condition **OP 3** is obviously satisfied, and **OP 4** follows by localizing. This proves our theorem.

The next theorem gives us the uniqueness of the operation λ_X.

Theorem 4.2. *If μ is another functor of class C^p with the same variance as λ, and if we have a natural transformation of functors $t\colon \lambda \to \mu$, then for each X, the mapping*

$$t_X \colon \lambda_X \to \mu_X,$$

defined on each fiber by the map

$$t(\alpha_x, \beta_x)\colon \lambda(\alpha_x, \beta_x) \to \mu(\alpha_x, \beta_x),$$

is a natural transformation of functors (in the VB-*category).*

Proof. For simplicity of notation, assume that λ and μ are both functors of one variable, and both covariant. For each open set $U = U_i$ of a trivializing covering for β, we have a commutative diagram:

$$
\begin{array}{ccc}
U \times \lambda(\mathbf{E}) & \xrightarrow{\ \text{id} \times t(\mathbf{E})\ } & U \times \mu(\mathbf{E}) \\[4pt]
\Big\uparrow{\lambda_U(\sigma)} & & \Big\uparrow{\mu_U(\sigma)} \\[4pt]
\lambda_U(\beta) & \xrightarrow[\ t_U\ ]{} & \mu_U(\beta)
\end{array}
$$

The vertical maps are trivializing VB-isomorphisms, and the top horizontal map is a VB-morphism. Hence t_U is a VB-morphism, and our assertion is proved.

In particular, for $\lambda = \mu$ and $t = \text{id}$ we get the uniqueness of our functor λ_X.

(In the proof of Theorem 4.2, we do not use again explicitly the hypotheses that λ, μ are differentiable.)

In practice, we omit the subscript X on λ, and write λ for the functor on vector bundles.

Examples. Let $\pi\colon E \to X$ be a vector bundle. We take λ to be the dual, that is $\mathbf{E} \mapsto \mathbf{E}^{\vee} = L(\mathbf{E}, \mathbf{R})$. Then $\lambda(E)$ is denoted by $E^{\vee}$, and is called the **dual bundle**. The fiber at each point $x \in X$ is the dual space $E_x^{\vee}$. The dual bundle of the tangent bundle is called the **cotangent bundle** $T^{\vee}X$.

Similarly, instead of taking $L(E)$, we could take $L_a^r(E)$ to be the bundle of alternating multilinear forms on E. The fiber at each point is the space $L_a^r(E_x)$ consisting of all r-multilinear alternating continuous functions on E_x. When $E = TX$ is the tangent bundle, the sections of $L_a^r(TX)$ are called **differential forms** of degree r. Thus a 1-form is a section of $E^{\vee}$. Differential forms will be treated later in detail.

Recall that $\text{End}(\mathbf{E}) = L(\mathbf{E}, \mathbf{E})$. In the theory of curvature, we shall deal with both functors

$$E \mapsto L^4(\mathbf{E}) = L^4(\mathbf{E}, \mathbf{R}) \qquad \text{and} \qquad \mathbf{E}^3 \mapsto L^2(\mathbf{E}, \text{End}(\mathbf{E})) = L^2(\mathbf{E}, L(\mathbf{E}, \mathbf{E})).$$

In fact, if $R \in L^2(\mathbf{E}, L(\mathbf{E}, \mathbf{E}))$, then for each pair of elements $v, w \in \mathbf{E}$ and $z \in \mathbf{E}$, we see that $R(v, w) \in L(\mathbf{E}, \mathbf{E})$ and $R(v, w)z \in \mathbf{E}$, so we get a 3-linear map

$$(v, w, z) \mapsto R(v, w)z.$$

We shall apply both functors to the tangent bundle in Chapter IX.

For another type of operation, we have the **direct sum** (also called the **Whitney sum**) of two bundles α, β over X. It is denoted by $\alpha \oplus \beta$, and the fiber at a point x is

$$(\alpha \oplus \beta)_x = \alpha_x \oplus \beta_x.$$

Of course, the finite direct sum of vector spaces can be identified with their finite direct products, but we write the above operation as a direct sum in order not to confuse it with the following direct product.

Let $\alpha \colon E_\alpha \to X$ and $\beta \colon E_\beta \to Y$ be two vector bundles in $VB(X)$ and $VB(Y)$ respectively. Then the map

$$\alpha \times \beta \colon E_\alpha \times E_\beta \to X \times Y$$

is a vector bundle, and it is this operation which we call the **direct product** of α and β.

Let X be a manifold, and λ a functor of class C^p with $p \geq 1$. The **tensor bundle** of type λ over X is defined to be $\lambda_x(T(X))$, also denoted by $\lambda T(X)$ or $T_\lambda(X)$. The sections of this bundle are called **tensor fields** of type λ, and the set of such sections is denoted by $\Gamma_\lambda(X)$. Suppose that we have a trivialization of $T(X)$, say

$$T(X) = X \times \mathbf{E}.$$

Then $T_\lambda(X) = X \times \lambda(\mathbf{E})$. A section of $T_\lambda(X)$ in this representation is completely described by the projection on the second factor, which is a morphism

$$f \colon X \to \lambda(\mathbf{E}).$$

We shall call it the **local representation** of the tensor field (in the given trivialization). If ξ is the tensor field having f as its local representation, then

$$\xi(x) = (x, f(x)).$$

Let $f \colon X \to Y$ be a morphism of class C^p ($p \geq 1$). Let ω be a tensor field of type L^r over Y, which could also be called a **multilinear tensor field**. For each $y \in Y$, $\omega(y)$ (also written ω_y) is a continuous multilinear function on $T_y(Y)$:

$$\omega_y \colon T_y \times \cdots \times T_y \to \mathbf{R}.$$

For each $x \in X$, we can define a continuous multilinear map

$$f_x^*(\omega): T_x \times \cdots \times T_x \to \mathbf{R}$$

by the composition of maps $(T_x f)^r$ and $\omega_{f(x)}$:

$$T_x \times \cdots \times T_x \to T_{f(x)} \times \cdots \times T_{f(x)} \to \mathbf{R}.$$

We contend that the map $x \mapsto f_x^*(\omega)$ is a tensor field over X, of the same type as ω. To prove this, we may work with local representation. Thus we can assume that we work with a morphism

$$f: U \to V$$

of one open set in a Banach space into another, and that

$$\omega: V \to L^r(\mathbf{F})$$

is a morphism, V being open in $\mathbf{F}$. If U is open in $\mathbf{E}$, then $f^*(\omega)$ (now denoting a local representation) becomes a mapping of U into $L^r(\mathbf{E})$, given by the formula

$$f_x^*(\omega) = L^r(f'(x)) \cdot \omega(f(x)).$$

Since $L^r: L(\mathbf{E}, \mathbf{F}) \to L(L^r(\mathbf{F}), L^r(\mathbf{E}))$ is of class C^∞, it follows that $f^*(\omega)$ is a morphism of the same class as ω. This proves what we want.

Of course, the same argument is valid for the other functors L_s^r and L_a^r (symmetric and alternating continuous multilinear maps). Special cases will be considered in later chapters. If λ denotes any one of our three functors, then we see that we have obtained a mapping (which is in fact linear)

$$f^*: \Gamma_\lambda(Y) \to \Gamma_\lambda(X)$$

which is clearly functorial in f. We use the notation f^* instead of the more correct (but clumsy) notation f_λ or $\Gamma_\lambda(f)$. No confusion will arise from this.

III, §5. SPLITTING OF VECTOR BUNDLES

The next proposition expresses the fact that the VB-morphisms of one bundle into another (over a fixed morphism) form a module over the ring of functions.

Proposition 5.1. *Let X, Y be manifolds and $f_0: X \to Y$ a morphism. Let α, β be vector bundles over X, Y respectively, and let f, $g: \alpha \to \beta$ be two VB-morphisms over f_0. Then the map $f + g$ defined by the formula*

$$(f + g)_x = f_x + g_x$$

is also a VB-morphism. Furthermore, if $\psi: Y \to \mathbf{R}$ is a function on Y, then the map ψf defined by

$$(\psi f)_x = \psi(f_0(x)) f_x$$

is also a VB-morphism.

Proof. Both assertions are immediate consequences of Proposition 3.10 of Chapter I.

We shall consider mostly the situation where $X = Y$ and f_0 is the identity, and will use it, together with partitions of unity, to glue VB-morphisms together.

Let α, β be vector bundles over X and let $\{(U_i, \psi_i)\}$ be a partition of unity on X. Suppose given for each U_i a VB-morphism

$$f_i: \alpha | U_i \to \beta | U_i.$$

Each one of the maps $\psi_i f_i$ (defined as in Proposition 5.1) is a VB-morphism. Furthermore, we can extend $\psi_i f_i$ to a VB-morphism of α into β simply by putting

$$(\psi_i f_i)_x = 0$$

for all $x \notin U_i$. If we now define

$$f: \alpha \to \beta$$

by the formula

$$f_x(v) = \sum \psi_i(x) f_{ix}(v)$$

for all pairs (x, v) with $v \in \alpha_x$, then the sum is actually finite, at each point x, and again by Proposition 5.1, we see that f is a VB-morphism. We observe that if each f_i is the identity, then $f = \sum \psi_i f_i$ is also the identity.

Proposition 5.2. *Let X be a manifold admitting partitions of unity. Let $0 \to \alpha \xrightarrow{f} \beta$ be an exact sequence of vector bundles over X. Then there exists a surjective VB-morphism $g: \beta \to \alpha$ whose kernel splits at each point, such that $g \circ f = \mathrm{id}$.*

Proof. By the definition of exact sequence, there exists a partition of unity $\{(U_i, \psi_i)\}$ on X such that for each i, we can split the sequence over U_i. In other words, there exists for each i a VB-morphism

$$g_i: \beta|U_i \to \alpha|U_i$$

which is surjective, whose kernel splits, and such that $g_i \circ f_i = \mathrm{id}_i$. We let $g = \sum \psi_i g_i$. Then g is a VB-morphism of β into α by what we have just seen, and

$$g \circ f = \sum \psi_i g_i f_i = \mathrm{id}.$$

It is trivial that g is surjective because $g \circ f = \mathrm{id}$. The kernel of g_x splits at each point x because it has a closed complement, namely $f_x \alpha_x$. This concludes the proof.

If γ is the kernel of β, then we have $\beta \approx \alpha \oplus \gamma$.

A vector bundle π over X will be said to be of **finite type** if there exists a finite trivialization for π (i.e. a trivialization $\{(U_i, \tau_i)\}$ such that i ranges over a finite set).

If k is an integer ≥ 1 and $\mathbf{E}$ a topological vector space, then we denote by $\mathbf{E}^k$ the direct product of $\mathbf{E}$ with itself k times.

Proposition 5.3. *Let X be a manifold admitting partitions of unity. Let π be a vector bundle of finite type in $\mathrm{VB}(X, \mathbf{E})$, where $\mathbf{E}$ is a Banach space. Then there exists an integer $k > 0$ and a vector bundle α in $\mathrm{VB}(X, \mathbf{E}^k)$ such that $\pi \oplus \alpha$ is trivializable.*

Proof. We shall prove that there exists an exact sequence

$$0 \to \pi \xrightarrow{f} \beta$$

with $E_\beta = X \times \mathbf{E}^k$. Our theorem will follow from the preceding proposition.

Let $\{(U_i, \tau_i)\}$ be a finite trivialization of π with $i = 1, \ldots, k$. Let $\{(U_i, \psi_i)\}$ be a partition of unity. We define

$$f: E_\pi \to X \times \mathbf{E}^k$$

as follows. If $x \in X$ and v is in the fiber of E_π at x, then

$$f_x(v) = \big(x, \psi_1(x)\tau_1(v), \ldots, \psi_k(x)\tau_k(v)\big).$$

The expression on the right makes sense, because in case x does not lie in U_i then $\psi_i(x) = 0$ and we do not have to worry about the expression $\tau_i(v)$. If x lies in U_i, then $\tau_i(v)$ means $\tau_{ix}(v)$.

Given any point x, there exists some index i such that $\psi_i(x) > 0$ and hence f is injective. Furthermore, for this x and this index i, f_x maps E_x onto a closed subspace of $\mathbf{E}^k$, which admits a closed complement, namely

$$\mathbf{E} \times \cdots \times 0 \times \cdots \times \mathbf{E}$$

with 0 in the i-th place. This proves our proposition.

CHAPTER IV

Vector Fields and Differential Equations

In this chapter, we collect a number of results all of which make use of the notion of differential equation and solutions of differential equations.

Let X be a manifold. A vector field on X assigns to each point x of X a tangent vector, differentiably. (For the precise definition, see §2.) Given x_0 in X, it is then possible to construct a unique curve $\alpha(t)$ starting at x_0 (i.e. such that $\alpha(0) = x_0$) whose derivative at each point is the given vector. It is not always possible to make the curve depend on time t from $-\infty$ to $+\infty$, although it is possible if X is compact.

The structure of these curves presents a fruitful domain of investigation, from a number of points of view. For instance, one may ask for topological properties of the curves, that is those which are invariant under topological automorphisms of the manifold. (Is the curve a closed curve, is it a spiral, is it dense, etc.?) More generally, following standard procedures, one may ask for properties which are invariant under any given interesting group of automorphisms of X (discrete groups, Lie groups, algebraic groups, Riemannian automorphisms, ad lib.).

We do not go into these theories, each of which proceeds according to its own flavor. We give merely the elementary facts and definitions associated with vector fields, and some simple applications of the existence theorem for their curves.

Throughout this chapter, we assume all manifolds to be Hausdorff, of class C^p with $p \geq 2$ from §2 on, and $p \geq 3$ from §3 on. This latter condition insures that the tangent bundle is of class C^{p-1} with $p - 1 \geq 1$ (or 2).

We shall deal with mappings of several variables, say $f(t, x, y)$, the first of which will be a real variable. We identify $D_1 f(t, x, y)$ with

$$\lim_{h \to 0} \frac{f(t + h, x, y) - f(t, x, y)}{h}.$$

IV, §1. EXISTENCE THEOREM FOR DIFFERENTIAL EQUATIONS

Let E be a Banach space and U an open subset of E. In this section we consider vector fields locally. The notion will be globalized later, and thus for the moment, we define (the local representation of) a **time-dependent vector field** on U to be a C^p-morphism ($p \geq 0$)

$$f : J \times U \to E,$$

where J is an open interval containing 0 in $\mathbf{R}$. We think of f as assigning to each point x in U a vector $f(t, x)$ in E, depending on time t.

Let x_0 be a point of U. An **integral curve** for f with **initial condition** x_0 is a mapping of class C^r ($r \geq 1$)

$$\alpha : J_0 \to U$$

of an open subinterval of J containing 0, into U, such that $\alpha(0) = x_0$ and such that

$$\alpha'(t) = f(t, \alpha(t)).$$

Remark. Let $\alpha : J_0 \to U$ be a continuous map satisfying the condition

$$\alpha(t) = x_0 + \int_0^t f(u, \alpha(u))\, du.$$

Then α is differentiable, and its derivative is $f(t, \alpha(t))$. Hence α is of class C^1. Furthermore, we can argue recursively, and conclude that if f is of class C^p, then so is α. Conversely, if α is an integral curve for f with initial condition x_0, then it obviously satisfies our integral relation.

Let

$$f : J \times U \to E$$

be as above, and let x_0 be a point of U. By a **local flow** for f at x_0 we mean a mapping

$$\alpha : J_0 \times U_0 \to U$$

where J_0 is an open subinterval of J containing 0, and U_0 is an open subset of U containing x_0, such that for each x in U_0 the map

$$\alpha_x(t) = \alpha(t, x)$$

is an integral curve for f with initial condition x (i.e. such that $\alpha(0, x) = x$).

As a matter of notation, when we have a mapping with two arguments, say $\varphi(t, x)$, then we denote the separate mappings in each argument when the other is kept fixed by $\varphi_x(t)$ and $\varphi_t(x)$. The choice of letters will always prevent ambiguity.

We shall say that f satisfies a **Lipschitz condition** on U **uniformly with respect to** J if there exists a number $K > 0$ such that

$$|f(t, x) - f(t, y)| \leq K|x - y|$$

for all x, y in U and t in J. We call K a **Lipschitz constant**. If f is of class C^1, it follows at once from the mean value theorem that f is Lipschitz on some open neighborhood $J_0 \times U_0$ of a given point $(0, x_0)$ of U, and that it is bounded on some such neighborhood.

We shall now prove that under a Lipschitz condition, local flows exist and are unique locally. In fact, we prove more, giving a uniformity property for such flows. If b is real > 0, then we denote by J_b the open interval $-b < t < b$.

Proposition 1.1. *Let J be an open interval of $\mathbf{R}$ containing 0, and U open in the Banach space $\mathbf{E}$. Let x_0 be a point of U, and $a > 0$, $a < 1$ a real number such that the closed ball $\bar{B}_{3a}(x_0)$ lies in U. Assume that we have a continuous map*

$$f \colon J \times U \to \mathbf{E}$$

which is bounded by a constant $L \geq 1$ on $J \times U$, and satisfies a Lipschitz condition on U uniformly with respect to J, with constant $K \geq 1$. If $b < a/LK$, then for each x in $\bar{B}_a(x_0)$ there exists a unique flow

$$\alpha \colon J_b \times B_a(x_0) \to U.$$

If f is of class C^p ($p \geq 1$), then so is each integral curve α_x.

Proof. Let I_b be the closed interval $-b \leq t \leq b$, and let x be a fixed point in $\bar{B}_a(x_0)$. Let M be the set of continuous maps

$$\alpha \colon I_b \to \bar{B}_{2a}(x_0)$$

of the closed interval into the closed ball of center x_0 and radius $2a$, such that $\alpha(0) = x$. Then M is a complete metric space if we define as usual the distance between maps α, β to be

$$\sup_{t \in I_b} |\alpha(t) - \beta(t)|.$$

We shall now define a mapping

$$S: M \to M$$

of M into itself. For each α in M, we let $S\alpha$ be defined by

$$(S\alpha)(t) = x + \int_0^t f(u, \alpha(u)) \, du.$$

Then $S\alpha$ is certainly continuous, we have $S\alpha(0) = x$, and the distance of any point on $S\alpha$ from x is bounded by the norm of the integral, which is bounded by

$$b \sup |f(u, y)| \leq bL < a.$$

Thus $S\alpha$ lies in M.

We contend that our map S is a shrinking map. Indeed,

$$|S\alpha - S\beta| \leq b \sup |f(u, \alpha(u)) - f(u, \beta(u))|$$
$$\leq bK |\alpha - \beta|,$$

thereby proving our contention.

By the shrinking lemma (Chapter I, Lemma 5.1) our map has a unique fixed point α, and by definition, $\alpha(t)$ satisfies the desired integral relation. Our remark above concludes the proof.

Corollary 1.2. *The local flow α in Proposition 1.1 is continuous. Furthermore, the map $x \mapsto \alpha_x$ of $\bar{B}_a(x_0)$ into the space of curves is continuous, and in fact satisfies a Lipschitz condition.*

Proof. The second statement obviously implies the first. So fix x in $\bar{B}_a(x_0)$ and take y close to x in $\bar{B}_a(x_0)$. We let S_x be the shrinking map of the theorem, corresponding to the initial condition x. Then

$$\|\alpha_x - S_y \alpha_x\| = \|S_x \alpha_x - S_y \alpha_x\| \leq |x - y|.$$

Let $C = bK$ so $0 < C < 1$. Then

$$\|\alpha_x - S_y^n \alpha_x\| \leq \|\alpha_x - S_y \alpha_x\| + \|S_y \alpha_x - S_y^2 \alpha_x\| + \cdots + \|S_y^{n-1} \alpha_x - S_y^n \alpha_x\|$$
$$\leq (1 + C + \cdots + C^{n-1})|x - y|.$$

Since the limit of $S_y^n \alpha_x$ is equal to α_y as n goes to infinity, the continuity of the map $x \mapsto \alpha_x$ follows at once. In fact, the map satisfies a Lipschitz condition as stated.

It is easy to formulate a uniqueness theorem for integral curves over their whole domain of definition.

Theorem 1.3 (Uniqueness Theorem). *Let U be open in* **E** *and let* $f: U \to E$ *be a vector field of class* $C^p, p \geqq 1$. *Let*

$$\alpha_1: J_1 \to U \qquad and \qquad \alpha_2: J_2 \to U$$

be two integral curves for f with the same initial condition x_0. *Then* α_1 *and* α_2 *are equal on* $J_1 \cap J_2$.

Proof. Let Q be the set of numbers b such that $\alpha_1(t) = \alpha_2(t)$ for

$$0 \leqq t < b.$$

Then Q contains some number $b > 0$ by the local uniqueness theorem. If Q is not bounded from above, the equality of $\alpha_1(t)$ and $\alpha_2(t)$ for all $t > 0$ follows at once. If Q is bounded from above, let b be its least upper bound. We must show that b is the right end point of $J_1 \cap J_2$. Suppose that this is not the case. Define curves β_1 and β_2 near 0 by

$$\beta_1(t) = \alpha_1(b + t) \qquad and \qquad \beta_2(t) = \alpha_2(b + t).$$

Then β_1 and β_2 are integral curves of f with the initial conditions $\alpha_1(b)$ and $\alpha_2(b)$ respectively. The values $\beta_1(t)$ and $\beta_2(t)$ are equal for small negative t because b is the least upper bound of Q. By continuity it follows that $\alpha_1(b) = \alpha_2(b)$, and finally we see from the local uniqueness theorem that

$$\beta_1(t) = \beta_2(t)$$

for all t in some neighborhood of 0, whence α_1 and α_2 are equal in a neighborhood of b, contradicting the fact that b is a least upper bound of Q. We can argue the same way towards the left end points, and thus prove our statement.

For each $x \in U$, let $J(x)$ be the union of all open intervals containing 0 on which integral curves for f are defined, with initial condition equal to x. The uniqueness statement allows us to define the integral curve uniquely on all of $J(x)$.

Remark. The choice of 0 as the initial time value is made for convenience. From the uniqueness statement one obtains at once (making a time translation) the analogous statement for an integral curve defined on any open interval; in other words, if J_1, J_2 do not necessarily contain 0, and t_0 is a point in $J_1 \cap J_2$ such that $\alpha_1(t_0) = \alpha_2(t_0)$, and also we have the differential equations

$$\alpha_1'(t) = f(\alpha_1(t)) \quad \text{and} \quad \alpha_2'(t) = f(\alpha_2(t)),$$

then α_1 and α_2 are equal on $J_1 \cap J_2$.

In practice, one meets vector fields which may be time dependent, and also depend on parameters. We discuss these to show that their study reduces to the study of the standard case.

Time-dependent vector fields

Let J be an open interval, U open in a Banach space $\mathbf{E}$, and

$$f: J \times U \to \mathbf{E}$$

a C^p map, which we view as depending on time $t \in J$. Thus for each t, the map $x \mapsto f(t, x)$ is a vector field on U. Define

$$\bar{f}: J \times U \to \mathbf{R} \times \mathbf{E}$$

by

$$\bar{f}(t, x) = (1, f(t, x)),$$

and view $\bar{f}$ as a time-independent vector field on $J \times U$. Let $\bar{\alpha}$ be its flow, so that

$$\bar{\alpha}'(t, s, x) = \bar{f}(\bar{\alpha}(t, s, x)), \qquad \bar{\alpha}(0, s, x) = (s, x).$$

We note that $\bar{\alpha}$ has its values in $J \times U$ and thus can be expressed in terms of two components. In fact, it follows at once that we can write $\bar{\alpha}$ in the form

$$\bar{\alpha}(t, s, x) = (t + s, \bar{\alpha}_2(t, s, x)).$$

Then $\bar{\alpha}_2$ satisfies the differential equation

$$D_1\bar{\alpha}_2(t, s, x) = f(t + s, \bar{\alpha}_2(t, s, x))$$

as we see from the definition of $\bar{f}$. Let

$$\beta(t, x) = \bar{\alpha}_2(t, 0, x).$$

Then β is a flow for f, that is β satisfies the differential equation

$$D_1\beta(t, x) = f(t, \beta(t, x)), \qquad \beta(0, x) = x.$$

Given $x \in U$, any value of t such that α is defined at (t, x) is also such that $\bar{\alpha}$ is defined at $(t, 0, x)$ because α_x and β_x are integral curves of the same vector field, with the same initial condition, hence are equal. Thus the study of time-dependent vector fields is reduced to the study of time-independent ones.

Dependence on parameters

Let V be open in some space $\mathbf{F}$ and let

$$g: J \times V \times U \to \mathbf{E}$$

be a map which we view as a time-dependent vector field on U, also depending on parameters in V. We define

$$G: J \times V \times U \to \mathbf{F} \times \mathbf{E}$$

by

$$G(t, z, y) = \bigl(0, g(t, z, y)\bigr)$$

for $t \in J$, $z \in V$, and $y \in U$. This is now a time-dependent vector field on $V \times U$. A local flow for G depends on three variables, say $\beta(t, z, y)$, with initial condition $\beta(0, z, y) = (z, y)$. The map β has two components, and it is immediately clear that we can write

$$\beta(t, z, y) = \bigl(z, \alpha(t, z, y)\bigr)$$

for some map α depending on three variables. Consequently α satisfies the differential equation

$$D_1 \alpha(t, z, y) = g\bigl(t, z, \alpha(t, z, y)\bigr), \qquad \alpha(0, z, y) = y,$$

which gives the flow of our original vector field g depending on the parameters $z \in V$. This procedure reduces the study of differential equations depending on parameters to those which are independent of parameters.

We shall now investigate the behavior of the flow with respect to its second argument, i.e. with respect to the points of U. We shall give two methods for this. The first depends on approximation estimates, and the second on the implicit mapping theorem in function spaces.

Let J_0 be an open subinterval of J containing 0, and let

$$\varphi: J_0 \to U$$

be of class C^1. We shall say that φ is an ε-**approximate solution** of f on J_0 if

$$|\varphi'(t) - f(t, \varphi(t))| \leq \varepsilon$$

for all t in J_0.

Proposition 1.4. *Let φ_1 and φ_2 be two ε_1- and ε_2-approximate solutions of f on J_0 respectively, and let $\varepsilon = \varepsilon_1 + \varepsilon_2$. Assume that f is Lipschitz with constant K on U uniformly in J_0, or that $D_2 f$ exists and is bounded by K on $J \times U$. Let t_0 be a point of J_0. Then for any t in J_0, we have*

$$|\varphi_1(t) - \varphi_2(t)| \leq |\varphi_1(t_0) - \varphi_2(t_0)| e^{K|t - t_0|} + \frac{\varepsilon}{K} e^{K|t - t_0|}.$$

Proof. By assumption, we have

$$|\varphi_1'(t) - f(t, \varphi_1(t))| \leq \varepsilon_1,$$
$$|\varphi_2'(t) - f(t, \varphi_2(t))| \leq \varepsilon_2.$$

From this we get

$$|\varphi_1'(t) - \varphi_2'(t) + f(t, \varphi_2(t)) - f(t, \varphi_1(t))| \leq \varepsilon.$$

Say $t \geq t_0$ to avoid putting bars around $t - t_0$. Let

$$\psi(t) = |\varphi_1(t) - \varphi_2(t)|,$$
$$\omega(t) = |f(t, \varphi_1(t)) - f(t, \varphi_2(t))|.$$

Then, after integrating from t_0 to t, and using triangle inequalities we obtain

$$|\psi(t) - \psi(t_0)| \leq \varepsilon(t - t_0) + \int_{t_0}^{t} \omega(u)\, du$$

$$\leq \varepsilon(t - t_0) + K \int_{t_0}^{t} \psi(u)\, du$$

$$\leq K \int_{t_0}^{t} [\psi(u) + \varepsilon/K]\, du,$$

and finally the recurrence relation

$$\psi(t) \leq \psi(t_0) + K \int_{t_0}^{t} [\psi(u) + \varepsilon/K]\, du.$$

On any closed subinterval of J_0, our map ψ is bounded. If we add ε/K to both sides of this last relation, then we see that our proposition will follow from the next lemma.

Lemma 1.5. *Let g be a positive real valued function on an interval, bounded by a number L. Let t_0 be in the interval, say $t_0 \leq t$, and assume that there are numbers $A, K \geq 0$ such that*

$$g(t) \leq A + K \int_{t_0}^{t} g(u)\, du.$$

Then for all integers $n \geq 1$ we have

$$g(t) \leq A\left[1 + \frac{K(t - t_0)}{1!} + \cdots + \frac{K^{n-1}(t - t_0)^{n-1}}{(n - 1)!}\right] + \frac{LK^n(t - t_0)^n}{n!}.$$

Proof. The statement is an assumption for $n = 1$. We proceed by induction. We integrate from t_0 to t, multiply by K, and use the recurrence relation. The statement with $n + 1$ then drops out of the statement with n.

Corollary 1.6. *Let $f: J \times U \to E$ be continuous, and satisfy a Lipschitz condition on U uniformly with respect to J. Let x_0 be a point of U. Then there exists an open subinterval J_0 of J containing 0, and an open subset of U containing x_0 such that f has a unique flow*

$$\alpha: J_0 \times U_0 \to U.$$

We can select J_0 and U_0 such that α is continuous and satisfies a Lipschitz condition on $J_0 \times U_0$.

Proof. Given x, y in U_0 we let $\varphi_1(t) = \alpha(t, x)$ and $\varphi_2(t) = \alpha(t, y)$, using Proposition 1.6 to get J_0 and U_0. Then $\varepsilon_1 = \varepsilon_2 = 0$. For s, t in J_0 we obtain

$$|\alpha(t, x) - \alpha(s, y)| \leq |\alpha(t, x) - \alpha(t, y)| + |\alpha(t, y) - \alpha(s, y)|$$

$$\leq |x - y|e^K + |t - s|L,$$

if we take J_0 of small length, and L is a bound for f. Indeed, the term containing $|x - y|$ comes from Proposition 1.4, and the term containing $|t - s|$ comes from the definition of the integral curve by means of an integral and the bound L for f. This proves our corollary.

Corollary 1.7. *Let J be an open interval of $\mathbf{R}$ containing 0 and let U be open in $\mathbf{E}$. Let $f: J \times U \to \mathbf{E}$ be a continuous map, which is Lipschitz*

*on U uniformly for every compact subinterval of J. Let $t_0 \in J$ and let
φ_1, φ_2 be two morphisms of class C^1 such that $\varphi_1(t_0) = \varphi_2(t_0)$ and
satisfying the relation*

$$\varphi'(t) = f(t, \varphi(t))$$

for all t in J. Then $\varphi_1(t) = \varphi_2(t)$.

Proof. We can take $\varepsilon = 0$ in the proposition.

The above corollary gives us another proof for the uniqueness of
integral curves. Given $f: J \times U \to E$ as in this corollary, we can define
an integral curve α for f on a maximal open subinterval of J having a
given value $\alpha(t_0)$ for a fixed t_0 in J. Let J be the open interval (a, b) and
let (a_0, b_0) be the interval on which α is defined. We want to know when
$b_0 = b$ (or $a_0 = a$), that is when the integral curve of f can be continued
to the entire interval over which f itself is defined.

There are essentially two reasons why it is possible that the integral
curve cannot be extended to the whole domain of definition J, or cannot
be extended to infinity in case f is independent of time. One possibility
is that the integral curve tends to get out of the open set U, as on the
following picture:

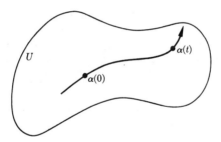

This means that as t approaches b_0, say, the curve $\alpha(t)$ approaches a
point which does not lie in U. Such an example can actually be constructed artificially. If we are in a situation when a curve can be extended to infinity, just remove a point from the open set lying on the
curve. Then the integral curve on the resulting open set cannot be
continued to infinity. The second possibility is that the vector field is
unbouned. The next corollary shows that these possibilities are the only
ones. In other words, if an integral curve does not tend to get out of the
open set, and if the vector field is bounded, then the curve can be
continued as far as the original data will allow a priori.

Corollary 1.8. *Let J be the open interval (a, b) and let U be open in **E**.
Let $f: J \times U \to \mathbf{E}$ be a continuous map, which is Lipschitz on U, uni-*

formly for every compact subset of J. Let α be an integral curve of f, defined on a maximal open subinterval (a_0, b_0) of J. Assume:

(i) *There exists ε > 0 such that $\overline{\alpha((b_0 - \varepsilon, b_0))}$ is contained in U.*
(ii) *There exists a number B > 0 such that $|f(t, \alpha(t))| \leq B$ for all t in $(b_0 - \varepsilon, b_0)$.*

Then $b_0 = b$.

Proof. From the integral expression for α, namely

$$\alpha(t) = \alpha(t_0) + \int_{t_0}^{t} f(u, \alpha(u))\, du,$$

we see that for t_1, t_2 in $(b_0 - \varepsilon, b_0)$ we have

$$|\alpha(t_1) - \alpha(t_2)| \leq B|t_1 - t_2|.$$

From this it follows that the limit

$$\lim_{t \to b_0} \alpha(t)$$

exists, and is equal to an element x_0 of U (by hypothesis (i)). Assume that $b_0 \neq b$. By the local existence theorem, there exists an integral curve β of f defined on an open interval containing b_0 such that $\beta(b_0) = x_0$ and $\beta'(t) = f(t, \beta(t))$. Then $\beta' = \alpha'$ on an open interval to the left of b_0, and hence α, β differ by a constant on this interval. Since their limit as $t \to b_0$ are equal, this constant is 0. Thus we have extended the domain of definition of α to a larger interval, as was to be shown.

The next proposition describes the solutions of **linear differential equations** depending on parameters.

Proposition 1.9. *Let J be an open interval of* **R** *containing 0, and let V be an open set in a Banach space. Let* **E** *be a Banach space. Let*

$$g: J \times V \to L(\mathbf{E}, \mathbf{E})$$

be a continuous map. Then there exists a unique map

$$\lambda: J \times V \to L(\mathbf{E}, \mathbf{E})$$

which, for each $x \in V$, is a solution of the differential equation

$$D_1 \lambda(t, x) = g(t, x)\lambda(t, x), \qquad \lambda(0, x) = \text{id}.$$

This map λ is continuous.

Remark. In the present case of a linear differential equation, it is not necessary to shrink the domain of definition of its flow. Note that the differential equation is on the space of continuous linear maps. The corresponding linear equation on **E** itself will come out as a corollary.

Proof of Proposition 1.9. Let us first fix $x \in V$. Consider the differential equation

$$D_1 \lambda(t, x) = g(t, x)\lambda(t, x),$$

with initial condition $\lambda(0, x) = \text{id}$. This is a differential equation on $L(\mathbf{E}, \mathbf{E})$, where $f(t, z) = g_x(t)z$ for $z \in L(\mathbf{E}, \mathbf{E})$, and we write $g_x(t)$ instead of $g(t, x)$. Let the notation be as in Corollary 1.8. Then hypothesis (i) is automatically satisfied since the open set U is all of $L(\mathbf{E}, \mathbf{E})$. On every compact subinterval of J, g_x is bounded, being continuous. Omitting the index x for simplicity, we have

$$\lambda(t) = \text{id} + \int_0^t g(u)\lambda(u)\, du,$$

whence for $t \geq 0$, say,

$$|\lambda(t)| \leq 1 + B \int_0^t |\lambda(u)|\, du.$$

Using Lemma 1.5, we see that hypothesis (ii) of Corollary 1.8 is also satisfied. Hence the integral curve is defined on all of J.

We shall now prove the continuity of λ. Let $(t_0, x_0) \in J \times V$. Let I be a compact interval contained in J, and containing t_0 and 0. As a function of t, $\lambda(t, x_0)$ is continuous (even differentiable). Let $C > 0$ be such that $|\lambda(t, x_0)| \leq C$ for all $t \in I$. Let V_1 be an open neighborhood of x_0 in V such that g is bounded by a constant $K > 0$ on $I \times V_1$.

For $(t, x) \in I \times V_1$ we have

$$|\lambda(t, x) - \lambda(t_0, x_0)| \leq |\lambda(t, x) - \lambda(t, x_0)| + |\lambda(t, x_0) - \lambda(t_0, x_0)|.$$

The second term on the right is small when t is close to t_0. We investigate the first term on the right, and shall estimate it by viewing $\lambda(t, x)$ and $\lambda(t, x_0)$ as approximate solutions of the differential equation satisfied by $\lambda(t, x)$. We find

$$|D_1 \lambda(t, x_0) - g(t, x)\lambda(t, x_0)|$$

$$= |D_1 \lambda(t, x_0) - g(t, x)\lambda(t, x_0) + g(t, x_0)\lambda(t, x_0) - g(t, x_0)\lambda(t, x_0)|$$

$$\leq |g(t, x_0) - g(t, x)||\lambda(t, x_0)| \leq |g(t, x_0) - g(t, x)|\, C.$$

By the usual proof of uniform continuity applied to the compact set $I \times \{x_0\}$, given $\varepsilon > 0$, there exists an open neighborhood V_0 of x_0 contained in V_1, such that for all $(t, x) \in I \times V_0$ we have

$$|g(t, x) - g(t, x_0)| < \varepsilon/C.$$

This implies that $\lambda(t, x_0)$ is an ε-approximate solution of the differential equation satisfied by $\lambda(t, x)$. We apply Proposition 1.4 to the two curves

$$\varphi_0(t) = \lambda(t, x_0) \qquad \text{and} \qquad \varphi_x(t) = \lambda(t, x)$$

for each $x \in V_0$. We use the fact that $\lambda(0, x) = \lambda(0, x_0) = \text{id}$. We then find

$$|\lambda(t, x) - \lambda(t, x_0)| < \varepsilon K_1$$

for some constant $K_1 > 0$, thereby proving the continuity of λ at (t_0, x_0).

Corollary 1.10. *Let the notation be as in Proposition 1.9. For each $x \in V$ and $z \in E$ the curve*

$$\beta(t, x, z) = \lambda(t, x)z$$

with initial condition $\beta(0, x, z) = z$ is a solution of the differential equation

$$D_1\beta(t, x, z) = g(t, x)\beta(t, x, z).$$

Furthermore, β is continuous in its three variables.

Proof. Obvious.

Theorem 1.11 (Local Smoothness Theorem). *Let J be an open interval in $\mathbf{R}$ containing 0 and U open in the Banach space $\mathbf{E}$. Let*

$$f: J \times U \to \mathbf{E}$$

be a C^p-morphism with $p \geqq 1$, and let $x_0 \in U$. There exists a unique local flow for f at x_0. We can select an open subinterval J_0 of J containing 0 and an open subset U_0 of U containing x_0 such that the unique local flow

$$\alpha: J_0 \times U_0 \to U$$

is of class C^p, and such that $D_2\alpha$ satisfies the differential equation

$$\boxed{D_1 D_2 \alpha(t, x) = D_2 f(t, \alpha(t, x)) D_2 \alpha(t, x)}$$

on $J_0 \times U_0$ with initial condition $D_2\alpha(0, x) = \text{id}$.

Proof. Let

$$g: J \times U \to L(\mathbf{E}, \mathbf{E})$$

be given by $g(t, x) = D_2 f(t, \alpha(t, x))$. Select J_1 and U_0 such that α is bounded and Lipschitz on $J_1 \times U_0$ (by Corollary 1.6), and such that g is continuous and bounded on $J_1 \times U_0$. Let J_0 be an open subinterval of J_1 containing 0 such that its closure $\bar{J}_0$ is contained in J_1.

Let $\lambda(t, x)$ be the solution of the differential equation on $L(\mathbf{E}, \mathbf{E})$ given by

$$D_1 \lambda(t, x) = g(t, x)\lambda(t, x), \qquad \lambda(0, x) = \mathrm{id},$$

as in Proposition 1.9. We contend that $D_2 \alpha$ exists and is equal to λ on $J_0 \times U_0$. This will prove that $D_2 \alpha$ is continuous, on $J_0 \times U_0$.

Fix $x \in U_0$. Let

$$\theta(t, h) = \alpha(t, x + h) - \alpha(t, x).$$

Then

$$
\begin{aligned}
D_1 \theta(t, h) &= D_1 \alpha(t, x + h) - D_1 \alpha(t, x) \\
&= f(t, \alpha(t, x + h)) - f(t, \alpha(t, x)).
\end{aligned}
$$

By the mean value theorem, we obtain

$$
\begin{aligned}
&|D_1 \theta(t, h) - g(t, x)\theta(t, h)| \\
&\quad = |f(t, \alpha(t, x + h)) - f(t, \alpha(t, x)) - D_2 f(t, \alpha(t, x))\theta(t, h)| \\
&\quad \leq |h| \sup |D_2 f(t, y) - D_2 f(t, \alpha(t, x))|,
\end{aligned}
$$

where y ranges over the segment between $\alpha(t, x)$ and $\alpha(t, x + h)$. By the compactness of $\bar{J}_0$ it follows that our last expression is bounded by $|h| \psi(h)$ where $\psi(h)$ tends to 0 with h, uniformly for t in $\bar{J}_0$. Hence we obtain

$$|\theta'(t, h) - g(t, x)\theta(t, h)| \leq |h| \psi(h),$$

for all t in $\bar{J}_0$. This shows that $\theta(t, h)$ is an $|h| \psi(h)$ approximate solution for the differential equation satisfied by $\lambda(t, x)h$, namely

$$D_1 \lambda(t, x)h - g(t, x)\lambda(t, x)h = 0,$$

with the initial condition $\lambda(0, x)h = h$. We note that $\theta(t, h)$ has the same initial condition, $\theta(0, h) = h$. Taking $t_0 = 0$ in Proposition 1.4, we obtain the estimate

$$|\theta(t, h) - \lambda(t, x)h| \leq C_1 |h| \psi(h)$$

for all t in $\bar{J}_0$. This proves that $D_2\alpha$ is equal to λ on $J_0 \times U_0$, and is therefore continuous on $J_0 \times U_0$.

We have now proved that $D_1\alpha$ and $D_2\alpha$ exist and are continuous on $J_0 \times U_0$, and hence that α is of class C^1 on $J_0 \times U_0$.

Furthermore, $D_2\alpha$ satisfies the differential equation given in the statement of our theorem on $J_0 \times U_0$. Thus our theorem is proved when $p = 1$.

A flow which satisfies the properties stated in the theorem will be called **locally of class** C^p.

Consider now again the linear equation of Proposition 1.9. We reformulate it to eliminate formally the parameters, namely we define a vector field

$$G: J \times V \times L(\mathbf{E}, \mathbf{E}) \to F \times L(\mathbf{E}, \mathbf{E})$$

to be the map such that

$$G(t, x, \omega) = \bigl(0, g(t, x)\omega\bigr)$$

for $\omega \in L(\mathbf{E}, \mathbf{E})$. The flow for this vector field is then given by the map Λ such that

$$\Lambda(t, x, \omega) = \bigl(x, \lambda(t, x)\omega\bigr).$$

If g is of class C^1 we can now conclude that the flow Λ is locally of class C^1, and hence putting $\omega = \mathrm{id}$, that λ is locally of class C^1.

We apply this to the case when $g(t, x) = D_2 f\bigl(t, \alpha(t, x)\bigr)$, and to the solution $D_2\alpha$ of the differential equation

$$D_1(D_2\alpha)(t, x) = g(t, x)D_2\alpha(t, x)$$

locally at each point $(0, x)$, $x \in U$. Let $p \geqq 2$ be an integer and assume our theorem proved up to $p - 1$, so that we can assume α locally of class C^{p-1}, and f of class C^p. Then g is locally of class C^{p-1}, whence $D_2\alpha$ is locally C^{p-1}. From the expression

$$D_1\alpha(t, x) = f\bigl(t, \alpha(t, x)\bigr)$$

we conclude that $D_1\alpha$ is C^{p-1}, whence α is locally C^p.

If f is C^∞, and if we knew that α is of class C^p for every integer p **on its domain of definition**, then we could conclude that α is C^∞; in other words, there is no shrinkage in the inductive application of the local theorem. We shall do this at the end of the section.

We shall now give another proof for the local smoothness of the flow, which depends on a simple application of the implicit mapping theorem in Banach spaces, and was found independently by Pugh and Robbin [Ro 68]. One advantage of this proof is that it extends to H^p vector fields, as noted by Ebin and Marsden [EbM 70].

Let U be open in $\mathbf{E}$ and let $f: U \to \mathbf{E}$ be a C^p map. Let $b > 0$ and let I_b be the closed interval of radius b centered at 0. Let

$$F = C^0(I_b, \mathbf{E})$$

be the Banach space of continuous maps of I_b into $\mathbf{E}$. We let V be the subset of $\mathbf{F}$ consisting of all continuous curves

$$\sigma: I_b \to U$$

mapping I_b into our open set U. Then it is clear that V is open in $\mathbf{F}$ because for each curve σ the image $\sigma(I_b)$ is compact, hence at a finite distance from the complement of U, so that any curve close to it is also contained in U.

We define a map

$$T: U \times V \to \mathbf{F}$$

by

$$T(x, \sigma) = x + \int_0 f \circ \sigma - \sigma.$$

Here we omit the dummy variable of integration, and x stands for the constant curve with value x. If we evaluate the curve $T(x, \sigma)$ at t, then by definition we have

$$T(x, \sigma)(t) = x + \int_0^t f(\sigma(u))\, du - \sigma(t).$$

Lemma 1.12. *The map T is of class C^p, and its second partial derivative is given by the formula*

$$D_2 T(x, \sigma) = \int_0 Df \circ \sigma - I$$

where I is the identity. In terms of t, this reads

$$D_2 T(x, \sigma)h(t) = \int_0^t Df(\sigma(u))h(u)\, du - h(t).$$

Proof. It is clear that the first partial derivative $D_1 T$ exists and is continuous, in fact C^∞, being linear in x up to a translation. To determine the second partial, we apply the definition of the derivative. The derivative of the map $\sigma \mapsto \sigma$ is of course the identity. We have to get the derivative with respect to σ of the integral expression. We have for small h

$$\left\| \int_0 f \circ (\sigma + h) - \int_0 f \circ \sigma - \int_0 (Df \circ \sigma)h \right\|$$

$$\leqq \int_0 |f \circ (\sigma + h) - f \circ \sigma - (Df \circ \sigma)h|.$$

We estimate the expression inside the integral at each point u, with u between 0 and the upper variable of integration. From the mean value theorem, we get

$$|f(\sigma(u) + h(u)) - f(\sigma(u)) - Df(\sigma(u))h(u)| \leqq \|h\| \sup |Df(z_u) - Df(\sigma(u))|$$

where the sup is taken over all points z_u on the segment between $\sigma(u)$ and $\sigma(u) + h(u)$. Since Df is continuous, and using the fact that the image of the curve $\sigma(I_b)$ is compact, we conclude (as in the case of uniform continuity) that as $\|h\| \to 0$, the expression

$$\sup |Df(z_u) - Df(\sigma(u))|$$

also goes to 0. (Put the ε and δ in yourself.) By definition, this gives us the derivative of the integral expression in σ. The derivative of the final term is obviously the identity, so this proves that $D_2 T$ is given by the formula which we wrote down.

This derivative does not depend on x. It is continuous in σ. Namely, we have

$$D_2 T(x, \tau) - D_2 T(x, \sigma) = \int_0 [Df \circ \tau - Df \circ \sigma].$$

If σ is fixed and τ is close to σ, then $Df \circ \tau - Df \circ \sigma$ is small, as one proves easily from the compactness of $\sigma(I_b)$, as in the proof of uniform continuity. Thus $D_2 T$ is continuous. By Proposition 3.5 of Chapter I, we now conclude that T is of class C^1.

The derivative of $D_2 T$ with respect to σ can again be computed as before if Df is itself of class C^1, and thus by induction, if f is of class C^p we conclude that $D_2 T$ is of class C^{p-1} so that by the same reference, we conclude that T itself is of class C^p. This proves our lemma.

We observe that a solution of the equation

$$T(x, \sigma) = 0$$

is precisely an integral curve for the vector field, with initial condition equal to x. Thus we are in a situation where we want to apply the implicit mapping theorem.

Lemma 1.13. *Let $x_0 \in U$. Let $a > 0$ be such that Df is bounded, say by a number $C_1 > 0$, on the ball $B_a(x_0)$ (we can always find such a since Df is continuous at x_0). Let $b < 1/C_1$. Then $D_2 T(x, \sigma)$ is invertible for all (x, σ) in $B_a(x_0) \times V$.*

Proof. We have an estimate

$$\left| \int_0^t Df(\sigma(u)) h(u) \, du \right| \leq b C_1 \|h\|.$$

This means that

$$|D_2 T(x, \sigma) + I| < 1,$$

and hence that $D_2 T(x, \sigma)$ is invertible, as a continuous linear map, thus proving Lemma 1.13.

We are ready to reprove the local smoothness theorem by the present means, when p is an integer, namely:

Theorem 1.14. *Let p be a positive integer, and let $f: U \to \mathbf{E}$ be a C^p vector field. Let $x_0 \in U$. Then there exist numbers $a, b > 0$ such that the local flow*

$$\alpha: J_b \times B_a(x_0) \to U$$

is of class C^p.

Proof. We take a so small and then b so small that the local flow exists and is uniquely determined by Proposition 1.1. We then take b smaller and a smaller so as to satisfy the hypotheses of Lemma 1.13. We can then apply the implicit mapping theorem to conclude that the map $x \mapsto \alpha_x$ is of class C^p. Of course, we have to consider the flow α and still must show that α itself is of class C^p. It will suffice to prove that $D_1 \alpha$ and $D_2 \alpha$ are of class C^{p-1}, by Proposition 3.5 of Chapter I. We first consider the case $p = 1$.

We could derive the continuity of α from Corollary 1.2 but we can also get it as an immediate consequence of the continuity of the map $x \mapsto \alpha_x$. Indeed, fixing (s, y) we have

$$|\alpha(t, x) - \alpha(s, y)| \leq |\alpha(t, x) - \alpha(t, y)| + |\alpha(t, y) - \alpha(s, y)|$$

$$\leq \|\alpha_x - \alpha_y\| + |\alpha_y(t) - \alpha_y(s)|.$$

Since α_y is continuous (being differentiable), we get the continuity of α.

Since

$$D_1 \alpha(t, x) = f(\alpha(t, x)),$$

we conclude that $D_1 \alpha$ is a composite of continuous maps, whence continuous.

Let φ be the derivative of the map $x \mapsto \alpha_x$, so that

$$\varphi: B_a(x_0) \to L(\mathbf{E}, C^0(I_b, \mathbf{E})) = L(\mathbf{E}, \mathbf{F})$$

is of class C^{p-1}. Then

$$\alpha_{x+w} - \alpha_x = \varphi(x)w + |w|\psi(w),$$

where $\psi(w) \to 0$ as $w \to 0$. Evaluating at t, we find

$$\alpha(t, x + w) - \alpha(t, x) = (\varphi(x)w)(t) + |w|\psi(w)(t),$$

and from this we see that

$$D_2 \alpha(t, x)w = (\varphi(x)w)(t).$$

Then

$$|D_2 \alpha(t, x)w - D_2 \alpha(s, y)w|$$
$$\leq |(\varphi(x)w)(t) - (\varphi(y)w)(t)| + |(\varphi(y)w)(t) - (\varphi(y)w)(s)|.$$

The first term on the right is bounded by

$$|\varphi(x) - \varphi(y)||w|$$

so that

$$|D_2 \alpha(t, x) - D_2 \alpha(t, y)| \leq |\varphi(x) - \varphi(y)|.$$

We shall prove below that

$$|(\varphi(y)w)(t) - (\varphi(y)w)(s)|$$

is uniformly small with respect to w when s is close to t. This proves the continuity of $D_2 \alpha$, and concludes the proof that α is of class C^1.

The following proof that $|(\varphi(y)w)(t) - (\varphi(y)w)(s)|$ is uniformly small was shown to be by Professor Yamanaka. We have

(1)
$$\alpha(t, x) = x + \int_0^t f(\alpha(u, x)) \, du.$$

Replacing x with $x + \lambda w$ ($w \in E$, $\lambda \neq 0$), we obtain

$$(2) \qquad \alpha(t, x + \lambda w) = x + \lambda w + \int_0^t f(\alpha(u, x + \lambda w))\, du.$$

Therefore

$$(3) \qquad \frac{\alpha(t, x + \lambda w) - \alpha(t, x)}{\lambda} = w + \int_0^t \frac{1}{\lambda}\left[f(\alpha(u, x + \lambda w)) - f(\alpha(u, x))\right] du.$$

On the other hand, we have already seen in the proof of Theorem 1.14 that

$$(4) \qquad \alpha(t, x + \lambda w) - \alpha(t, x) = \lambda(\varphi(x)w)(t) + |\lambda||w|\psi(\lambda w)(t).$$

Substituting (4) in (3), we obtain:

$$(\varphi(x)w)(t) + \frac{|\lambda|}{\lambda}|w|\psi(\lambda w)(t) = w + \int_0^t \frac{1}{\lambda}\left[f(\alpha(u, x + \lambda w)) - f(\alpha(u, x))\right] du$$

$$= w + \int_0^t \int_0^1 G(u, \lambda, v)\, dv\, du,$$

where

$$G(u, \lambda, v) = Df(\alpha(u, x) + v\varepsilon_1(\lambda))((\varphi(x)w)(u) + \varepsilon_2(\lambda))$$

with

$$\varepsilon_1(\lambda) = \lambda(\varphi(x)w)(u) + |\lambda||w|\psi(\lambda w)(u), \qquad \varepsilon_2(\lambda) = \frac{|\lambda|}{\lambda}\psi(\lambda w)(u).$$

Letting $\lambda \to 0$, we have

$$(5) \qquad (\varphi(x)w)(t) = w + \int_0^t Df(\alpha(u, x))(\varphi(x)w)(u)\, du.$$

By (5) we have

$$\left|(\varphi(x)w)(t) - (\varphi(x)w)(s)\right| \leq \left|\int_s^t Df(\alpha(u, x))(\varphi(x)w)(u)\, du\right|$$

$$\leq bC_1|\varphi(x)|\cdot|w|\cdot|t - s|,$$

from which we immediately obtain the desired uniformity.

Returning to our main concern, the flow, we have

$$\alpha(t, x) = x + \int_0^t f(\alpha(u, x))\, du.$$

We can differentiate under the integral sign with respect to the parameter x and thus obtain

$$D_2\alpha(t, x) = I + \int_0^t Df(\alpha(u, x))D_2\alpha(u, x)\, du,$$

where I is a constant linear map (the identity). Differentiating with respect to t yields the linear differential equation satisfied by $D_2\alpha$, namely

$$\boxed{D_1 D_2\alpha(t, x) = Df(\alpha(t, x))D_2\alpha(t, x)}$$

and this differential equation depends on time and parameters. We have seen earlier how such equations can be reduced to the ordinary case. We now conclude that locally, by induction, $D_2\alpha$ is of class C^{p-1} since Df is of class C^{p-1}. Since

$$D_1\alpha(t, x) = f(\alpha(t, x)),$$

we conclude by induction that $D_1\alpha$ is C^{p-1}. Hence α is of class C^p by Proposition 3.5 of Chapter I. Note that each time we use induction, the domain of the flow may shrink. We have proved Theorem 1.14, when p is an integer.

We now give the arguments needed to globalize the smoothness. We may limit ourselves to the time-independent case. We have seen that the time-dependent case reduces to the other.

Let U be open in a Banach space E, and let $f: U \to E$ be a C^p vector field. We let $J(x)$ be the domain of the integral curve with initial condition equal to w.

Let $\mathfrak{D}(f)$ be the set of all points (t, x) in $\mathbf{R} \times U$ such that t lies in $J(x)$. Then we have a map

$$\alpha: \mathfrak{D}(f) \to U$$

defined on all of $\mathfrak{D}(f)$, letting $\alpha(t, x) = \alpha_x(t)$ be the integral curve on $J(x)$ having x as initial condition. We call this the **flow** determined by f, and we call $\mathfrak{D}(f)$ its **domain of definition**.

Lemma 1.15. *Let $f: U \to E$ be a C^p vector field on the open set U of E, and let α be its flow. Abbreviate $\alpha(t, x)$ by tx, if (t, x) is in the domain of definition of the flow. Let $x \in U$. If t_0 lies in $J(x)$, then*

$$J(t_0 x) = J(x) - t_0$$

(translation of $J(x)$ by $-t_0$), and we have for all t in $J(x) - t_0$:

$$t(t_0 x) = (t + t_0)x.$$

Proof. The two curves defined by

$$t \mapsto \alpha\bigl(t, \alpha(t_0, x)\bigr) \qquad \text{and} \qquad t \mapsto \alpha(t + t_0, x)$$

are integral curves of the same vector field, with the same initial condition $t_0 x$ at $t = 0$. Hence they have the same domain of definition $J(t_0 x)$. Hence t_1 lies in $J(t_0 x)$ if and only if $t_1 + t_0$ lies in $J(x)$. This proves the first assertion. The second assertion comes from the uniqueness of the integral curve having given initial condition, whence the theorem follows.

Theorem 1.16 (Global Smoothness of the Flow). *If f is of class C^p (with $p \leq \infty$), then its flow is of class C^p on its domain of definition.*

Proof. First let p be an integer ≥ 1. We know that the flow is locally of class C^p at each point $(0, x)$, by the local theorem. Let $x_0 \in U$ and let $J(x_0)$ be the maximal interval of definition of the integral curve having x_0 as initial condition. Let $\mathfrak{D}(f)$ be the domain of definition of the flow, and let α be the flow. Let Q be the set of numbers $b > 0$ such that for each t with $0 \leq t < b$ there exists an open interval J containing t and an open set V containing x_0 such that $J \times V$ is contained in $\mathfrak{D}(f)$ and such that α is of class C^p on $J \times V$. Then Q is not empty by the local theorem. If Q is not bounded from above, then we are done looking toward the right end point of $J(x_0)$. If Q is bounded from above, we let b be its least upper bound. We must prove that b is the right end point of $J(x_0)$. Suppose that this is not the case. Then $\alpha(b, x_0)$ is defined. Let $x_1 = \alpha(b, x_0)$. By the local theorem, we have a unique local flow at x_1, which we denote by β:

$$\beta: J_a \times B_a(x_1) \to U, \qquad \beta(0, x) = x,$$

defined for some open interval $J_a = (-a, a)$ and open ball $B_a(x_1)$ of radius a centered at x_1. Let δ be so small that whenever $b - \delta < t < b$ we have

$$\alpha(t, x_0) \in B_{a/4}(x_1).$$

We can find such δ because

$$\lim_{t \to b} \alpha(t, x_0) = x_1$$

by continuity. Select a point t_1 such that $b - \delta < t_1 < b$. By the hypothesis on b, we can select an open interval J_1 containing t_1 and an open set

U_1 containing x_0 so that

$$\alpha: J_1 \times U_1 \to B_{a/2}(x_1)$$

maps $J_1 \times U_1$ into $B_{a/2}(x_1)$. We can do this because α is continuous at (t_1, x_0), being in fact C^p at this point. If $|t - t_1| < a$ and $x \in U_1$, we define

$$\varphi(t, x) = \beta(t - t_1, \alpha(t_1, x)).$$

Then

$$\varphi(t_1, x) = \beta(0, \alpha(t_1, x)) = \alpha(t_1, x)$$

and

$$\begin{aligned} D_1 \varphi(t, x) &= D_1 \beta(t - t_1, \alpha(t_1, x)) \\ &= f(\beta(t - t_1, \alpha(t_1, x))) \\ &= f(\varphi(t, x)). \end{aligned}$$

Hence both φ_x and α_x are integral curves for f with the same value at t_1. They coincide on any interval on which they are defined by the uniqueness theorem. If we take δ very small compared to a, say $\delta < a/4$, we see that φ is an extension of α to an open set containing (t_1, α_0), and also containing (b, x_0). Furthermore, φ is of class C^p, thus contradicting the fact that b is strictly smaller than the end point of $J(x_0)$. Similarly, one proves the analogous statement on the other side, and we therefore see that $\mathfrak{D}(f)$ is open in $\mathbf{R} \times U$ and that α is of class C^p on $\mathfrak{D}(f)$, as was to be shown.

The idea of the above proof is very simple geometrically. We go as far to the right as possible in such a way that the given flow α is of class C^p locally at (t, x_0). At the point $\alpha(b, x_0)$ we then use the flow β to extend differentiably the flow α in case b is not the right-hand point of $J(x_0)$. The flow β at $\alpha(b, x_0)$ has a fixed local domain of definition, and we simply take t close enough to b so that β gives an extension of α, as described in the above proof.

Of course, if f is of class C^∞, then we have shown that α is of class C^p for each positive integer p, and therefore the flow is also of class C^∞.

In the next section, we shall see how these arguments globalize even more to manifolds.

IV, §2. VECTOR FIELDS, CURVES, AND FLOWS

Let X be a manifold of class C^p with $p \geq 2$. We recall that X is assumed to be Hausdorff. Let $\pi: T(X) \to X$ be its tangent bundle. Then $T(X)$ is of class C^{p-1}, $p \geq 1$.

By a (time-independent) **vector field** on X we mean a cross section of the tangent bundle, i.e. a morphism (of class C^{p-1})

$$\xi: X \to T(X)$$

such that $\xi(x)$ lies in the tangent space $T_x(X)$ for each $x \in X$, or in other words, such that $\pi\xi = \mathrm{id}$.

If $T(X)$ is trivial, and say X is an **E**-manifold, so that we have a VB-isomorphism of $T(X)$ with $X \times \mathbf{E}$, then the morphism ξ is completely determined by its projection on the second factor, and we are essentially in the situaiton of the preceding paragraph, except for the fact that our vector field is independent of time. In such a product representation, the projection of ξ on the second factor will be called the **local representation** of ξ. It is a C^{p-1}-morphism

$$f: X \to \mathbf{E}$$

and $\xi(x) = (x, f(x))$. We shall also say that ξ is **represented by** f **locally** if we work over an open subset U of X over which the tangent bundle admits a trivialisation. We then frequently use ξ itself to denote this local representation.

Let J be an open interval of $\mathbf{R}$. The tangent bundle of J is then $J \times \mathbf{R}$ and we have a canonical section ι such that $\iota(t) = 1$ for all $t \in J$. We sometimes write ι_t instead of $\iota(t)$.

By a **curve** in X we mean a morphism (always of class ≥ 1 unless otherwise specified)

$$\alpha: J \to X$$

from an open interval in $\mathbf{R}$ into X. If $g: X \to Y$ is a morphism, then $g \circ \alpha$ is a curve in Y. From a given curve α, we get an induced map on the tangent bundles:

$$
\begin{array}{ccc}
J \times \mathbf{R} & \xrightarrow{\;\alpha_*\;} & T(X) \\
\downarrow & & \downarrow{\scriptstyle \pi} \\
J & \xrightarrow[\;\alpha\;]{} & X
\end{array}
$$

and $\alpha_* \circ \iota$ will be denoted by α' or by $d\alpha/dt$ if we take its value at a point t in J. Thus α' is a curve in $T(X)$, of class C^{p-1} if α is of class C^p. Unless otherwise specified, it is always understood in the sequel that we start with enough differentiability to begin with so that we never end up with maps of class < 1. Thus to be able to take derivatives freely we have to take X and α of class C^p with $p \geq 2$.

If $g: X \to Y$ is a morphism, then

$$(g \circ \alpha)'(t) = g_* \alpha'(t).$$

This follows at once from the functoriality of the tangent bundle and the definitions.

Suppose that J contains 0, and let us consider curves defined on J and such that $\alpha(0)$ is equal to a fixed point x_0. We could say that two such curves α_1, α_2 are **tangent** at 0 if $\alpha_1'(0) = \alpha_2'(0)$. The reader will verify immediately that there is a natural bijection between tangency classes of curves with $\alpha(0) = x_0$ and the tangent space $T_{x_0}(X)$ of X at x_0. The tangent space could therefore have been defined alternatively by taking equivalence classes of curves through the point.

Let ξ be a vector field on X and x_0 a point of X. An **integral curve** for the vector field ξ with **initial condition** x_0, or starting at x_0, is a curve (of class C^{p-1})

$$\alpha: J \to X$$

mapping an open interval J of $\mathbf{R}$ containing 0 into X, such that $\alpha(0) = x_0$ and such that

$$\alpha'(t) = \xi(\alpha(t))$$

for all $t \in J$. Using a local representation of the vector field, we know from the preceding section that integral curves exist locally. The next theorem gives us their global existence and uniqueness.

Theorem 2.1. *Let* $\alpha_1: J_1 \to X$ *and* $\alpha_2: J_2 \to X$ *be two integral curves of the vector field* ξ *on* X, *with the same initial condition* x_0. *Then* α_1 *and* α_2 *are equal on* $J_1 \cap J_2$.

Proof. Let J^* be the set of points t such that $\alpha_1(t) = \alpha_2(t)$. Then J^* certainly contains a neighborhood of 0 by the local uniqueness theorem. Furthermore, since X is Hausdorff, we see that J^* is closed. We must show that it is open. Let t^* be in J^* and define β_1, β_2 near 0 by

$$\beta_1(t) = \alpha_1(t^* + t),$$

$$\beta_2(t) = \alpha_2(t^* + t).$$

Then β_1 and β_2 are integral curves of ξ with initial condition $\alpha_1(t^*)$ and $\alpha_2(t^*)$ respectively, so by the local uniqueness theorem, β_1 and β_2 agree in a neighborhood of 0 and thus α_1, α_2 agree in a neighborhood of t^*, thereby proving our theorem.

It follows from Theorem 2.1 that the union of the domains of all integral curves of ξ with a given initial condition x_0 is an open interval which we denote by $J(x_0)$. Its end points are denoted by $t^+(x_0)$ and $t^-(x_0)$ respectively. (We do not exclude $+\infty$ and $-\infty$.)

Let $\mathfrak{D}(\xi)$ be the subset of $\mathbf{R} \times X$ consisting of all points (t, x) such that

$$t^-(x) < t < t^+(x).$$

A (global) **flow** for ξ is a mapping

$$\alpha: \mathfrak{D}(\xi) \to X,$$

such that for each $x \in X$, the map $\alpha_x: J(x) \to X$ given by

$$\alpha_x(t) = \alpha(t, x)$$

defined on the open interval $J(x)$ is a morphism and is an integral curve for ξ with initial condition x. When we select a chart at a point x_0 of X, then one sees at once that this definition of flow coincides with the definition we gave locally in the previous section, for the local representation of our vector field.

Given a point $x \in X$ and a number t, we say that tx is **defined** if (t, x) is in the domain of α, and we denote $\alpha(t, x)$ by tx in that case.

Theorem 2.2. *Let ξ be a vector field on X, and α its flows. Let x be a point of X. If t_0 lies in $J(x)$, then*

$$J(t_0 x) = J(x) - t_0$$

(translation of $J(x)$ by $-t_0$), and we have for all t in $J(x) - t_0$:

$$t(t_0 x) = (t + t_0)x.$$

Proof. Our first assertion follows immediately from the maximality assumption concerning the domains of the integral curves. The second is equivalent to saying that the two curves given by the left-hand side and right-hand side of the last equality are equal. They are both integral curves for the vector field, with initial condition $t_0 x$ and must therefore be equal.

In particular, if t_1, t_2 are two numbers such that $t_1 x$ is defined and $t_2(t_1 x)$ is also defined, then so is $(t_1 + t_2)x$ and they are equal.

Theorem 2.3. *Let ξ be a vector field on X, and x a point of X. Assume that $t^+(x) < \infty$. Given a compact set $A \subset X$, there exists $\varepsilon > 0$ such that for all $t > t^+(x) - \varepsilon$, the point tx does not lie in A, and similarly for t^-.*

Proof. Suppose such ε does not exist. Then we can find a sequence t_n of real numbers approaching $t^+(x)$ from below, such that $t_n x$ lies in A. Since A is compact, taking a subsequence if necessary, we may assume that $t_n x$ converges to a point in A. By the local existence theorem, there exists a neighborhood U of this point y and a number $\delta > 0$ such that $t^+(z) > \delta$ for all $z \in U$. Taking n large, we have

$$t^+(x) < \delta + t_n$$

and $t_n x$ is in U. Then by Theorem 2.2,

$$t^+(x) = t^+(t_n x) + t_n > \delta + t_n > t^+(x)$$

contradiction.

Corollary 2.4. *If X is compact, and ξ is a vector field on X, then*

$$\mathfrak{D}(\xi) = \mathbf{R} \times X.$$

It is also useful to give one other criterion when $\mathfrak{D}(\xi) = \mathbf{R} \times X$, even when X is not compact. Such a criterion must involve some structure stronger than the differentiable structure (essentially a metric of some sort), because we can always dig holes in a compact manifold by taking away a point.

Proposition 2.5. *Let $\mathbf{E}$ be a Banach space, and X an $\mathbf{E}$-manifold. Let ξ be a vector field on X. Assume that there exist numbers $a > 0$ and $K > 0$ such that every point x of X admits a chart (U, φ) at x such that the local representation f of the vector field on this chart is bounded by K, and so is its derivative f'. Assume also that φU contains a ball of radius a around φx. Then $\mathfrak{D}(\xi) = \mathbf{R} \times X$.*

Proof. This follows at once from the global continuation theorem, and the uniformity of Proposition 1.1.

We shall prove finally that $\mathfrak{D}(\xi)$ is open and that α is a morphism.

Theorem 2.6. *Let ξ be a vector field of class C^{p-1} on the C^p-manifold X $(2 \leqq p \leqq \infty)$. Then $\mathfrak{D}(\xi)$ is open in $\mathbf{R} \times X$, and the flow α for ξ is a C^{p-1}-morphism.*

Proof. Let first p be an integer $\geqq 2$. Let $x_0 \in X$. Let J^* be the set of points in $J(x_0)$ for which there exists a number $b > 0$ and an open neighborhood U of x_0 such that $(t - b, t + b)$ U is contained in $\mathfrak{D}(\xi)$, and such that the restriction of the flow α to this product is a C^{p-1}-morphism. Then J^* is open in $J(x_0)$, and certainly contains 0 by the local theorem. We must therefore show that J^* is closed in $J(x_0)$.

Let s be in its closure. By the local theorem, we can select a neighborhood V of $sx_0 = \alpha(s, x_0)$ so that we have a unique local flow

$$\beta: J_a \times V \to X$$

for some number $a > 0$, with initial condition $\beta(0, x) = x$ for all $x \in V$, and such that this local flow β is C^{p-1}.

The integral curve with initial condition x_0 is certainly continuous on $J(x_0)$. Thus tx_0 approaches sx_0 as t approaches s. Let V_1 be a given small neighborhood of sx_0 contained in V. By the definition of J^*, we can find an element t_1 in J^* very close to s, and a small number b (compared to a) and a small neighborhood U of x_0 such that α maps the product

$$(t_1 - b, t_1 + b) \times U$$

into V_1, and is C^{p-1} on this product. For $t \in J_a + t_1$ and $x \in U$, we define

$$\varphi(t, x) = \beta(t - t_1, \alpha(t_1, x)).$$

Then $\varphi(t_1, x) = \beta(0, \alpha(t_1, x)) = \alpha(t_1, x)$, and

$$
\begin{aligned}
D_1 \varphi(t, x) D_1 \beta(t - t_1, \alpha(t_1, x)) \\
= \xi(\beta(t - t_1, \alpha(t_1, x))) \\
= \xi(\varphi(t, x)).
\end{aligned}
$$

Hence both φ_x, α_x are integral curves for ξ, with the same value at t_1. They coincide on any interval on which they are defined, so that φ_x is a continuation of α_x to a bigger interval containing s. Since α is C^{p-1} on the product $(t_1 - b, t_1 + b) \times U$, we conclude that φ is also C^{p-1} on $(J_a + t_1) \times U$. From this we see that $\mathfrak{D}(\xi)$ is open in $\mathbf{R} \times X$, and that α is of class C^{p-1} on its full domain $\mathfrak{D}(\xi)$. If $p = \infty$, then we can now conclude that α is of class C^r for each positive integer r on $\mathfrak{D}(\xi)$, and hence is C^∞, as desired.

Corollary 2.7. *For each $t \in \mathbf{R}$, the set of $x \in X$ such that (t, x) is contained in the domain $\mathfrak{D}(\xi)$ is open in X.*

Corollary 2.8. *The functions $t^+(x)$ and $t^-(x)$ are upper and lower semicontinuous respectively.*

Theorem 2.9. *Let ξ be a vector field on X and α its flow. Let $\mathfrak{D}_t(\xi)$ be the set of points x of X such that (t, x) lies in $\mathfrak{D}(\xi)$. Then $\mathfrak{D}_t(\xi)$ is open for each $t \in \mathbf{R}$, and α_t is an isomorphism of $\mathfrak{D}_t(\xi)$ onto an open subset of X. In fact, $\alpha_t(\mathfrak{D}_t) = \mathfrak{D}_{-t}$ and $\alpha_t^{-1} = \alpha_{-t}$.*

Proof. Immediate from the preceding theorem.

Corollary 2.10. *If x_0 is a point of X and t is in $J(x_0)$, then there exists an open neighborhood U of x_0 such that t lies in $J(x)$ for all $x \in U$, and the map*

$$x \mapsto tx$$

is an isomorphism of U onto an open neighborhood of tx_0.

Critical points

Let ξ be a vector field. A **critical point** of ξ is a point x_0 such that $\xi(x_0) = 0$. Critical points play a significant role in the study of vector fields, notably in the Morse theory. We don't go into this here, but just make a few remarks to show at the basic level how they affect the behavior of integral curves.

Proposition 2.11. *If α is an integal curve of a C^1 vector field, ξ, and α passes through a critical point, then α is constant, that is $\alpha(t) = x_0$ for all t.*

Proof. The constant curve through x_0 is an integral curve for the vector field, and the uniqueness theorem shows that it is the only one.

Some smoothness of the vector field in addition to continuity must be assumed for the uniqueness. For instance, the following picture illustrates a situation where the integral curves are not unique. They consist in translations of the curve $y = x^3$ in the plane. The vector field is continuous but not locally Lipschitz.

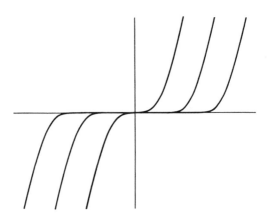

Proposition 2.12. *Let ξ be a vector field and α an integral curve for ξ. Assume that all $t \geq 0$ are in the domain of α, and that*

$$\lim_{t \to 0} \alpha(t) = x_1$$

exists. Then x_1 is a critical point for ξ, that is $\xi(x_1) = 0$.

Proof. Selecting t large, we may assume that we are dealing with the local representation f of the vector field near x_1. Then for $t' > t$ large, we have

$$\alpha(t') - \alpha(t) = \int_t^{t'} f(\alpha(u))\, du.$$

Write $f(\alpha(u)) = f(x_1) + g(u)$, where $\lim g(u) = 0$. Then

$$|f(x_1)||t' - t| \leq |\alpha(t') - \alpha(t)| + |t' - t| \sup|g(u)|,$$

where the sup is taken for u large, and hence for small values of $g(u)$. Dividing by $|t' - t|$ shows that $f(x_1)$ is arbitrarily small, hence equal to 0, as was to be shown.

Proposition 2.13. *Suppose on the other hand that x_0 is not a critical point of the vector field ξ. Then there exists a chart at x_0 such that the local representation of the vector field on this chart is constant.*

Proof. In an arbitrary chart the vector field has a representation as a morphism

$$\xi \colon U \to E$$

near x_0. Let α be its flow. We wish to "straighten out" the integral curves of the vector field according to the next figure.

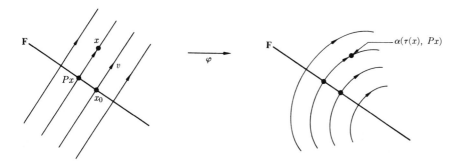

In other words, let $v = \xi(x_0)$. We want to find a local isomorphism φ at x_0 such that

$$\varphi'(x)v = \xi(\varphi(x)).$$

We inspire ourselves from the picture. Without loss of generality, we may assume that $x_0 = 0$. Let λ be a functional such that $\lambda(v) \neq 0$. We decompose $\mathbf{E}$ as a direct sum

$$\mathbf{E} = \mathbf{F} \oplus \mathbf{R}v,$$

where $\mathbf{F}$ is the kernel of λ. Let P be the projection on $\mathbf{F}$. We can write any x near 0 in the form

$$x = Px + \tau(x)v,$$

where

$$\tau(x) = \frac{\lambda(x)}{\lambda(v)}.$$

We then bend the picture on the left to give the picture on the right using the flow α of ξ, namely we define

$$\varphi(x) = \alpha(\tau(x), Px).$$

This means that starting at Px, instead of going linearly in the direction of v for a time $\tau(x)$, we follow the flow (integral curve) for this amount of time. We find that

$$\varphi'(x) = D_1\alpha(\tau(x), Px)\frac{\lambda}{\lambda(v)} + D_2\alpha(\tau(x), Px)P.$$

Hence $\varphi'(0) = \mathrm{id}$, so by the inverse mapping theorem, φ is a local isomorphism at 0. Furthermore, since $Pv = 0$ by definition, we have

$$\varphi'(x)v = D_1\alpha(\tau(x), Px) = \xi(\varphi(x)),$$

thus proving Proposition 2.13.

IV, §3. SPRAYS

Second-order vector fields and differential equations

Let X be a manifold of class C^p with $p \geq 3$. Then its tangent bundle $T(X)$ is of class C^{p-1}, and the tangent bundle of the tangent bundle $T(T(X))$ is of class C^{p-2}, with $p - 2 \geq 1$.

Let $\alpha: J \to X$ be a curve of class C^q $(q \leq p)$. A **lifting** of α into $T(X)$ is a curve $\beta: J \to T(X)$ such that $\pi\beta = \alpha$. We shall always deal with $q \geq 2$ so that a lift will be assumed of class $q - 1 \geq 1$. Such lifts always exist, for instance the curve α' discussed in the previous section, called the **canonical lifting** of α.

A **second-order** vector field over X is a vector field F on the tangent bundle $T(X)$ (of class C^{p-1}) such that, if $\pi: TX \to X$ denotes the canonical projection of $T(X)$ on X, then

$$\pi_* \circ F = \text{id.,} \qquad \text{that is} \quad \pi_* F(v) = v \qquad \text{for all } v \text{ in } T(X).$$

Observe that the succession of symbols makes sense, because

$$\pi_*: TT(X) \to T(X)$$

maps the double tangent bundle into $T(X)$ itself.

A vector field F on TX is a second-order vector field on X if and only if it satisfies the following condition: Each integral curve β of F is equal to the canonical lifting of $\pi\beta$, in other words

$$(\pi\beta)' = \beta.$$

Here, $\pi\beta$ is the canonical projection of β on X, and if we put the argument t, then our formula reads

$$(\pi\beta)'(t) = \beta(t)$$

for all t in the domain of β. The proof is immediate from the definitions, because

$$(\pi\beta)' = \pi_*\beta' = \pi_* \circ F \circ \beta$$

We then use the fact that given a vector $v \in TX$, there is an integral curve $\beta = \beta_v$ with $\beta_v(0) = v$ (initial condition v).

Let $\alpha: J \to X$ be a curve in X, defined on an interval J. We define α to be a **geodesic with respect to** F if the curve

$$\alpha': J \to TX$$

is an integral curve of F. Since $\pi\alpha' = \alpha$, that is α' lies above α in TX, we can express the geodesic condition equivalently by stating that α satisfies the relation

$$\alpha'' = F(\alpha').$$

This relation for curves α in X is called the **second-order differential**

equation for the curve α, determined by F. Observe that by definition, if β is an integral curve of F in TX, then $\pi\beta$ is a geodesic for the second order vector field F.

Next we shall give the representation of the second order vector field and of the integral curves in a chart.

Representation in charts

Let U be open in the Banach space $\mathbf{E}$, so that $T(U) = U \times \mathbf{E}$, and $T(T(U)) = (U \times \mathbf{E}) \times (\mathbf{E} \times \mathbf{E})$. Then $\pi: U \times \mathbf{E} \to U$ is simply the projection, and we have a commutative diagram:

$$
\begin{array}{ccc}
(U \times \mathbf{E}) \times (\mathbf{E} \times \mathbf{E}) & \xrightarrow{\ \pi_* \ } & U \times \mathbf{E} \\
\downarrow & & \downarrow \\
U \times \mathbf{E} & \xrightarrow[\ \pi \ ]{} & U
\end{array}
$$

The map π_* on each fiber $\mathbf{E} \times \mathbf{E}$ is constant, and is simply the projection of $\mathbf{E} \times \mathbf{E}$ on the first factor $\mathbf{E}$, that is

$$\pi_*(x, v, u, w) = (x, u).$$

Any vector field on $U \times \mathbf{E}$ has a local representation

$$f: U \times \mathbf{E} \to \mathbf{E} \times \mathbf{E}$$

which has therefore two components, $f = (f_1, f_2)$, each f_i mapping $U \times \mathbf{E}$ into $\mathbf{E}$. The next statement describes second order vector fields locally in the chart.

Let U be open in the Banach space $\mathbf{E}$, and let $T(U) = U \times \mathbf{E}$ be the tangent bundle. A C^{p-2}-morphism

$$f: U \times \mathbf{E} \to \mathbf{E} \times \mathbf{E}$$

is the local representation of a second order vector field on U if and only if

$$f(x, v) = (v, f_2(x, v)).$$

The above statement is merely making explicit the relation $\pi_* F = \mathrm{id}$, in the chart. If we write $f = (f_1, f_2)$, then we see that

$$f_1(x, v) = v.$$

We express the above relations in terms of integral curves as follows. Let $\beta = \beta(t)$ be an integral curve for the vector field F on TX. In the chart, the curve has two components

$$\beta(t) = \big(x(t), v(t)\big) \in U \times \mathbf{E}.$$

By definition, if f is the local representation of F, we must have

$$\frac{d\beta}{dt} = \left(\frac{dx}{dt}, \frac{dv}{dt}\right) = f(x, v) = (v, f_2(x, v)).$$

Consequently, our differential equation can be rewritten in the following manner:

$$\frac{dx}{dt} = v(t),$$

$$\frac{d^2x}{dt^2} = \frac{dv}{dt} = f_2\left(x, \frac{dx}{dt}\right),$$

which is of course familiar.

Sprays

We shall be interested in special kinds of second-order differential equations. Before we discuss these, we make a few technical remarks.

Let s be a real number, and $\pi\colon E \to X$ be a vector bundle. If v is in E, so in E_x for some x in X, then sv is again in E_x since E_x is a vector space. We write s_E for the mapping of E into itself given by this scalar multiplication. This mapping is in fact a VB-morphism, and even a VB-isomorphism if $s \neq 0$. Then

$$T(s_E) = (s_E)_*\colon T(E) \to T(E)$$

is the usual induced map on the tangent bundle of E.

Now let $E = TX$ be the tangent bundle itself. Then our map s_{TX} satisfies the property

$$(s_{TX})_* \circ s_{TTX} = s_{TTX} \circ (s_{TX})_*,$$

which follows from the linearity of s_{TX} on each fiber, and can also be seen directly from the representation on charts given below.

We define a **spray** to be a second-order vector field which satisfies the homogeneous quadratic condition:

SPR 1. *For all $s \in \mathbf{R}$ and $v \in T(X)$, we have*

$$F(sv) = (s_{TX})_* sF(v).$$

It is immediate from the conditions defining sprays (second-order vector field satisfying **SPR 1**) that **sprays form a convex set**! Hence if we can exhibit sprays over open subsets of Banach spaces, then we can glue them together by means of partitions of unity, and we obtain at once the following global existence theorem.

Theorem 3.1. *Let X be a manifold of class C^p ($p \geq 3$). If X admits partitions of unity, then there exists a spray over X.*

Representations in a chart

Let U be open in $\mathbf{E}$, so that $TU = U \times \mathbf{E}$. Then

$$TTU = (U \times \mathbf{E}) \times (\mathbf{E} \times \mathbf{E}),$$

and the representations of s_{TU} and $(s_{TU})_*$ in the chart are given by the maps

$$s_{TU} \colon (x, v) \mapsto (x, sv) \qquad \text{and} \qquad (s_{TU})_* \colon (x, v, u, w) \mapsto (x, sv, u, sw).$$

Thus

$$s_{TTU} \circ (s_{TU})_* \colon (x, v, u, w) \mapsto (x, sv, su, s^2 w).$$

We may now give the local condition for a second-order vector field F to be a spray.

Proposition 3.2. *In a chart $U \times \mathbf{E}$ for TX, let $f \colon U \times \mathbf{E} \to \mathbf{E} \times \mathbf{E}$ represent F, with $f = (f_1, f_2)$. Then f represents a spray if and only if, for all $s \in \mathbf{R}$ we have*

$$f_2(x, sv) = s^2 f_2(x, v).$$

Proof. The proof follows at once from the definitions and the formula giving the chart representation of $s(s_{TX})_*$.

Thus we see that the condition **SPR 1** (in addition to being a second-order vector field), simply means that f_2 is homogeneous of degree 2 in the variable v. By the remark in Chapter I, §3, it follows that f_2 is a quadratic map in its second variable, and specifically, this quadratic map is given by

$$f_2(x, v) = \tfrac{1}{2} D_2^2 f_2(x, 0)(v, v).$$

Thus the spray is induced by a symmetric bilinear map given at each point x in a chart by

$$B(x) = \tfrac{1}{2} D_2^2 f_2(x, 0).$$

Conversely, suppose given a morphism

$$U \to L^2_{\text{sym}}(\mathbf{E}, \mathbf{E}) \qquad \text{given by} \qquad x \mapsto B(x)$$

from U into the space of symmetric bilinear maps $\mathbf{E} \times \mathbf{E} \to \mathbf{E}$. Thus for each v, $w \in \mathbf{E}$ the value of $B(x)$ at (v, w) is denoted by $B(x; v, w)$ or $B(x)(v, w)$. Define

$$f_2(x, v) = B(x; v, v).$$

Then f_2 is quadratic in its second variable, and the map f defined by

$$f(x, v) = (v, B(x; v, v)) = (v, f_2(x, v))$$

represents a spray over U. We call B the **symmetric bilinear map associated with the spray**.

We recall the trivial fact from linear algebra that the bilinear map B is determined purely algebraically from the quadratic map, by the formula

$$B(v, w) = \tfrac{1}{2}[f_2(v + w) - f_2(v) - f_2(w)].$$

We have suppressed the x from the notation to focus on the relevant second variable v. Thus the quadratic map and the symmetric bilinear map determine each other uniquely.

The above discussion has been local, over an open set U in a Banach space. In Proposition 3.4 and the subsequent discussion of connections, we show how to globalize the bilinear map B intrinsically on the manifold.

Examples. As a trivial special case, we can always take $f_2(x, v) = (v, 0)$ to represent the second component of a spray in the chart.

In the chapter on Riemannian metrics, we shall see how to construct a spray in a natural fashion, depending on the metric.

In the chapter on covariant derivatives we show how a spray gives rise to such derivatives.

Next, let us give the transformation rule for a spray under a change of charts, i.e. an isomorphism

$$h: U \to V.$$

On TU, the map Th is represented by a morphism (its vector component)

$$H: U \times \mathbf{E} \to \mathbf{E} \times \mathbf{E} \qquad \text{given by} \qquad H(x, v) = (h(x), h'(x)v).$$

We then have one further lift to the double tangent bundle TTU, and we may represent the diagram of maps symbolically as follows:

$$
\begin{array}{ccc}
(U \times \mathbf{E}) \times (\mathbf{E} \times \mathbf{E}) & \xrightarrow{(H, H')} & (V \times \mathbf{E}) \times (\mathbf{E} \times \mathbf{E}) \\
\Big\downarrow \nearrow{\scriptstyle f_{U,2}} & & \Big\downarrow \nearrow{\scriptstyle f_{V,2}} \\
U \times \mathbf{E} & \xrightarrow{H = (h, h')} & V \times \mathbf{E} \\
\Big\downarrow & & \Big\downarrow \\
U & \xrightarrow{\quad h \quad} & V
\end{array}
$$

Then the derivative $H'(x, v)$ is given by the Jacobian matrix operating on column vectors ${}^t(u, w)$ with $u, w \in \mathbf{E}$, namely

$$
H'(x, v) = \begin{pmatrix} h'(x) & 0 \\ h''(x)v & h'(x) \end{pmatrix} \qquad \text{so} \qquad H'(x, v)\begin{pmatrix} u \\ w \end{pmatrix} = \begin{pmatrix} h'(x) & 0 \\ h''(x)v & h'(x) \end{pmatrix}\begin{pmatrix} u \\ w \end{pmatrix}.
$$

Thus the top map on elements in the diagram is given by

$$
(H, H'): (x, v, u, w) \mapsto \big(h(x), h'(x)v, h'(x)u, h''(x)(u, v) + h'(x)w\big).
$$

For the application, we put $u = v$ because $f_1(x, v) = v$, and $w = f_{U,2}(x, v)$, where f_U and f_V denote the representations of the spray over U and V respectively. It follows that f_U and f_V are related by the formula

$$
f_V\big(h(x), h'(x)v\big) = \big(h'(x)v, h''(x)(v, v) + h'(x)f_{U,2}(x, v)\big).
$$

Therefore we obtain:

Proposition 3.3. Change of variable formula for the quadratic part of a spray:

$$
f_{V,2}\big(h(x), h'(x)v\big) = h''(x)(v, v) + h'(x)f_{U,2}(x, v),
$$

$$
B_V\big(h(x); h'(x)v, h'(x)w\big) = h''(x)(v, w) + h'(x)B_U(x; v, w).
$$

Proposition 3.3 admits a converse:

Proposition 3.4. *Suppose we are given a covering of the manifold X by open sets corresponding to charts U, V, ..., and for each U we are given a morphism*

$$
B_U: U \to L^2_{\mathrm{sym}}(\mathbf{E}, \mathbf{E})
$$

which transforms according to the formula of Proposition 3.3 under an isomorphism $h: U \to V$. Then there exists a unique spray whose associated bilinear map in the chart U is given by B_U.

Proof. We leave the verification to the reader.

Remarks. Note that $B_U(x; v, w)$ does not transform like a tensor of type $L^2_{sym}(\mathbf{E}, \mathbf{E})$, i.e. a section of the bundle $L^2_{sym}(TX, TX)$. There are several ways of defining the bilinear map B intrinsically. One of them is via second order bundles, or bundles of second order jets, and to extend the terminology we have established previously to such bundles, and even higher order jet bundles involving higher derivatives, as in [Po 62]. Another way will be done below, via connections. For our immediate purposes, it suffices to have the above discussion on second-order differential equations together with Proposition 3.3 and 3.4. I used sprays (as recommended by Palais) in the earliest version of this book [La 62]. In [Lo 69] the bilinear map B_U is expressed in terms of second order jets. For applications to symmetric spaces, see [He 78] (expanded version of the similar book from 1962), and [Lo 69]. See also [Pa 57] for early Lie group applications. The basics of differential topology and geometry were being established in the early sixties. Cf. the bibliographical notes from [Lo 69] at the end of his first chapter.

Connections

We now show how to define the bilinear map B intrinsically and directly.
Matters will be clearer if we start with an arbitrary vector bundle

$$p: E \to X$$

over a manifold X. As it happens we also need the notion of a fiber bundle when the fibers are not necessarily vector spaces, so don't have a linear structure. Let $f: Y \to X$ be a morphism. We say that f (or Y over X) is a **fiber bundle** if f is surjective, and if each point x of X has an open neighborhood U, and there is some manifold Z and an isomorphism $h: f^{-1}(U) \to U \times Z$ such that the following diagram is commutative:

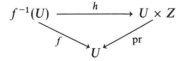

Thus locally, $f: Y \to X$ looks like the projection from a product space. The reason why we need a fiber bundle is that the tangent bundle

$$\pi_E: TE \to E$$

is a vector bundle over E, but the composite $f = p \circ \pi_E: TE \to X$ is only a fiber bundle over X, a fact which is obvious by picking trivializations in charts. Indeed, if U is a chart in X, and if $U \times \mathbf{F} \to U$ is a vector bundle chart for E, with fiber $\mathbf{F}$, and $Y = TE$, then we have a natural isomorphism of fiber bundles over U:

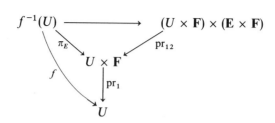

Note that U being a chart in X implies that $U \times \mathbf{E} \to U$ is a vector bundle chart for the tangent bundle TU over U.

The tangent bundle TE has two natural maps making it a vector bundle:

$\pi_E \colon TE \to E$ is a vector bundle over E;

$T(p) \colon TE \to TX$ is a vector bundle over TX.

Therefore we have a natural morphism of fiber bundle (not vector bundle) over X:

$$\bigl(\pi_E, T(p)\bigr) \colon TE \to E \oplus TX \qquad \text{given by} \qquad W \mapsto \bigl(\pi_E W, T(p)W\bigr)$$

for $W \in TE$. If $W \in T_e E$ with $e \in E_x$, then $\pi_E W \in E_x$ and $T(p)W \in T_x X$.

After these preliminaries, we define a **connection** to be a morphism of fiber bundles over X, from the direct sum $E \oplus TX$ into TE:

$$H \colon E \oplus TX \to TE$$

such that

$$\bigl(\pi_E, T(p)\bigr) \circ H = \mathrm{id}_{E \oplus TX},$$

and such that H is bilinear, in other words $H_x \colon E_x \oplus T_x X \to TE$ is bilinear.

Consider a chart U as in the above diagram, so

$$TU = U \times \mathbf{E} \qquad \text{and} \qquad T(U \times \mathbf{F}) = (U \times \mathbf{F}) \times (\mathbf{E} \times \mathbf{F}).$$

Then our map H has a coordinate representation

$$H(x, e, v) = \bigl(x, e, H_1(x, e, v), H_2(x, e, v)\bigr) \qquad \text{for } e \in \mathbf{F} \text{ and } v \in \mathbf{E}.$$

The fact that $\bigl(\pi_E, T(p)\bigr) \circ H = \mathrm{id}_{E \oplus TX}$ implies at once that $H_1(x, e, v) = v$. The bilinearity condition implies that for fixed x, the map

$$(e, v) \mapsto H_2(x, e, v)$$

is bilinear as a map $\mathbf{F} \times \mathbf{E} \to \mathbf{E}$. We shall therefore denote this map by $B(x)$, and we write in the chart

$$H(x, e, v) = \big(x, e, v, B(x)(e, v)\big) \quad \text{or also} \quad \big(x, e, v, B(x, e, v)\big).$$

Now take the special case when $E = TX$. We say that the connection is **symmetric** if the bilinear map B is symmetric. Suppose this is the case, then we may define the corresponding quadratic map $TX \to TTX$ by

$$f_2(x, v) = B(x, v, v).$$

Globally, this amounts to defining a morphism

$$F: TX \to TTX \qquad \text{such that} \qquad F = H \circ \text{diagonal}$$

where the diagonal is taken in $TX \oplus TX$, in each fiber. Thus

$$F(v) = H(v, v) \text{ for } v \in T_x X.$$

Then F is a vector field on TX, and the condition $(\pi_*, \pi_*) \circ H = \text{id}$ on $TX \oplus TX$ implies that F is a second-order vector field on X, in other words, F defines a spray. It is obvious that all sprays can be obtained in this fashion. Thus we have shown how to describe geometrically the bilinear map associated with a spray.

Going back to the general case of a vector bundle E unrelated to TX, we note that the image of a connection H is a vector subbundle over E. Let V denote the kernel of the map $T(p): TE \to TX$. We leave it to the reader to verify in charts that V is a vector subbundle of TE over E, and that the image of H is a complementary subbundle. One calls V the **vertical subbundle**, canonically defined, and one calls H the **horizontal subbundle** determined by the connection.

IV, §4. THE FLOW OF A SPRAY AND THE EXPONENTIAL MAP

The condition we have taken to define a spray is equivalent to other conditions concerning the integral curves of the second-order vector field F. We shall list these conditions systematically. We shall use the following relation. If $\alpha: J \to X$ is a curve, and α_1 is the curve defined by $\alpha_1(t) = \alpha(st)$, then

$$\alpha_1'(t) = s\alpha'(st),$$

this being the chain rule for differentiation.

If v is a vector in TX, let β_v be the unique integral curve of F with initial condition v (i.e. such that $\beta_v(0) = v$). In the next three conditions, the sentence should begin with "for each v in TX".

SPR 2. *A number t is in the domain of β_{sv} if and only if st is in the domain of β_v and then*

$$\beta_{sv}(t) = s\beta_v(st).$$

SPR 3. *If s, t are numbers, st is in the domain of β_v if and only if s is in the domain of β_{tv}, and then*

$$\pi\beta_{tv}(s) = \pi\beta_v(st).$$

SPR 4. *A number t is in the domain of β_v if and only if 1 is in the domain of β_{tv}, and then*

$$\pi\beta_v(t) = \pi\beta_{tv}(1).$$

We shall now prove the equivalence between all four conditions.

Assume **SPR 1**, and let s be fixed. For all t such that st is in the domain of β_v, the curve $\beta_v(st)$ is defined and we have

$$\frac{d}{dt}(s\beta_v(st)) = s_* s\beta_v'(st) = s_* sF(\beta_v(st)) = F(s\beta_v(st)).$$

Hence the curve $s\beta_v(st)$ is an integral curve for F, with initial condition $s\beta_v(0) = sv$. By uniqueness we must have

$$s\beta_v(st) = \beta_{sv}(t).$$

This proves **SPR 2**.

Assume **SPR 2**. Since β_v is an integral curve of F for each v, with initial condition v, we have by definition

$$\beta_{sv}'(0) = F(sv).$$

Using our assumption, we also have

$$\beta_{sv}'(t) = \frac{d}{dt}(s\beta_v(st)) = s_* s\beta_v'(st).$$

Put $t = 0$. Then **SPR 1** follows because β_{sv} and β_v are integral curves of F with initial conditions sv and v respectively.

It is obvious that **SPR 2** implies **SPR 3**. Conversely, assume **SPR 3**. To prove **SPR 2**, we have

$$\beta_{sv}(t) = (\pi\beta_{sv})'(t) = \frac{d}{dt}\pi\beta_v(st) = s(\pi\beta_v)'(st) = s\beta_v(st),$$

which proves **SPR 2**.

Assume **SPR 4**. Then st is in the domain of β_r if and only if 1 is in the domain of β_{stv}, and s is in the domain of β_{tv} if and only if 1 is in the domain of β_{stv}. This proves the first assertion of **SPR 3**, and again by **SPR 4**, assuming these relations, we get **SPR 3**.

It is similarly clear that **SPR 3** implies **SPR 4**.

Next we consider further properties of the integral curves of a spray. Let F be a spray on X. As above, we let β_v be the integral curve with initial condition v. Let $\mathfrak{D}$ be the set of vectors v in $T(X)$ such that β_v is defined at least on the interval $[0, 1]$. We know from Corollary 2.7 that $\mathfrak{D}$ is an open set in $T(X)$, and by Theorem 2.6 the map

$$v \mapsto \beta_v(1)$$

is a morphism of $\mathfrak{D}$ into $T(X)$. We now define the **exponential** map

$$\exp: \mathfrak{D} \to X$$

to be

$$\exp(v) = \pi\beta_v(1).$$

Then exp is a C^{p-2}-morphism. We also call $\mathfrak{D}$ the **domain of the expo-nential map (associated with** F**)**.

If $x \in X$ and 0_x denotes the zero vector in T_x, then from **SPR 1**, taking $s = 0$, we see that $F(0_x) = 0$. Hence

$$\exp(0_x) = x.$$

Thus our exponential map coincides with π on the zero cross section, and so induces an isomorphism of the cross section onto X. It will be convenient to denote the zero cross section of a vector bundle E over X by $\zeta_E(X)$ or simply ζX if the reference to E is clear. Here, E is the tangent bundle.

We denote by $\exp_x$ the restriction of exp to the tangent space T_x. Thus

$$\exp_x: T_x \to X.$$

Theorem 4.1. *Let X be a manifold and F a spray on X. Then*

$$\exp_x : T_x \to X$$

induces a local isomorphism at 0_x, and in fact $(\exp_x)_*$ is the identity at 0_x.

Proof. We prove the second assertion first because the main assertion follows from it by the inverse mapping theorem. Furthermore, since T_x is a vector space, it suffices to determine the derivative of $\exp_x$ on rays, in other words, to determine the derivative with respect to t of a curve $\exp_x(tv)$. This is done by using **SPR 3**, and we find

$$\frac{d}{dt}\pi\beta_{tv} = \beta_{tv}.$$

Evaluating this at $t = 0$ and taking into account that β_w has w as initial condition for any w gives us

$$(\exp_x)_*(0_x) = \mathrm{id}.$$

This concludes the proof of Theorem 4.1.

Helgason gave a general formula for the differential of the exponential map on analytic manifolds [He 61], reproduced in [He 78], Chapter I, Theorem 6.5. We shall study the differential of the exponential map in connection with Jacobi fields, in Chapter IX, §2.

Next we describe all geodesics.

Proposition 4.2. *The images of straight segments through the origin in T_x, under the exponential map $\exp_x$, are geodesics. In other words, if $v \in T_x$ and we let*

$$\alpha(v, t) = \alpha_v(t) = \exp_x(tv),$$

then α_v is a geodesic. Conversely, let $\alpha: J \to X$ be a C^2 geodesic defined on an interval J containing 0, and such that $\alpha(0) = x$. Let $\alpha'(0) = v$. Then $\alpha(t) = \exp_x(tv)$.

Proof. The first statement by definition means that α'_v is an integral curve of the spray F. Indeed, by the **SPR** conditions, we know that

$$\alpha(v, t) = \alpha_v(t) = \pi\beta_{tv}(1) = \pi\beta_v(t),$$

and $(\pi\beta_v)' = \beta_v$ is indeed an integral curve of the spray. Thus our assertion that the curves $t \mapsto \exp(tv)$ are geodesics is obvious from the definition of the exponential map and the **SPR** conditions.

Conversely, given a geodesic $\alpha: J \to X$, by definition α' satisfies the differential equation

$$\alpha''(t) = F(\alpha'(t)).$$

The two curves $t \mapsto \alpha(t)$ and $t \mapsto \exp_x(tv)$ satisfy the same differential equation and have the same initial conditions, so the two curves are equal. This proves the second statement and concludes the proof of the proposition.

Remark. From the theorem, we note that a C^1 curve in X is a geodesic if and only if, after a linear reparametrization of its interval of definition, it is simply $t \mapsto \exp_x(tv)$ for some x and some v.

We call the map $(v, t) \mapsto \alpha(v, t)$ the **geodesic flow** on X. It is defined on an open subset of $TX \times \mathbf{R}$, with $\alpha(v, 0) = x$ if $v \in T_x X$. Note that since $\pi(s\beta_v(t)) = \pi\beta_v(t)$ for $s \in \mathbf{R}$, we obtain from **SPR 2** the property

$$\alpha(sv, t) = \alpha(v, st)$$

for the geodesic flow. Precisely, t is in the domain of α_{sv} if and only if st is in the domain of α_v, and in that case the formula holds. As a slightly more precise version of Theorem 4.1 in this light, we obtain:

Corollary 4.3. *Let F be a spray on X, and let $x_0 \in X$. There exists an open neighborhood U of x_0, and an open neighborhood V of 0_{x_0} in TX satisfying the following condition. For every $x \in U$ and $v \in V \cap T_x X$, there exists a unique geodesic*

$$\alpha_v: (-2, 2) \to X$$

such that

$$\alpha_v(0) = x \qquad and \qquad \alpha_v'(0) = v.$$

Observe that in a chart, we may pick V as a product

$$V = U \times V_2(0) \subset U \times \mathbf{E}$$

where $V_2(0)$ is a neighborhood of 0 in $\mathbf{E}$. Then the geodesic flow is defined on $U \times V_2(0) \times J$, where $J = (-2, 2)$. We picked $(-2, 2)$ for concreteness. What we really want is that 0 and 1 lie in the interval. Any bounded interval J containing 0 and 1 could have been selected in the statement of the corollary. Then of course, U and V (or $V_2(0)$) depend on J.

IV, §5. EXISTENCE OF TUBULAR NEIGHBORHOODS

Let X be a submanifold of a manifold Y. A **tubular neighborhood** of X in Y consists of a vector bundle $\pi: E \to X$ over X, an open neighborhood Z of the zero section $\zeta_E X$ in E, and an isomorphism

$$f: Z \to U$$

of Z onto an open set in Y containing X, which commutes with ζ:

We shall call f the **tubular map** and Z or its image $f(Z)$ the corresponding **tube** (in E or Y respectively). The bottom map j is simply the inclusion. We could obviously assume that it is an embedding and define tubular neighborhoods for embeddings in the same way. We shall say that our tubular neighborhood is **total** if $Z = E$. In this section, we investigate conditions under which such neighborhoods exist. We shall consider the uniqueness problem in the next section.

Theorem 5.1. *Let Y be of class C^p ($p \geq 3$) and admit partitions of unity. Let X be a closed submanifold. Then there exists a tubular neighborhood of X in Y, of class C^{p-2}.*

Proof. Consider the exact sequence of tangent bundles:

$$0 \to T(X) \to T(Y)|X \to N(X) \to 0.$$

We know that this sequence splits, and thus there exists some splitting

$$T(Y)|X = T(X) \oplus N(X)$$

where $N(X)$ may be identified with a subbundle of $T(Y)|X$. Following Palais, we construct a spray ξ on $T(Y)$ using Theorem 3.1 and obtain the corresponding exponential map. We shall use its restriction to $N(X)$, denoted by $\exp|N$. Thus

$$\exp|N: \mathfrak{D} \cap N(X) \to Y.$$

We contend that this map is a local isomorphism. To prove this, we may work locally. Corresponding to the submanifold, we have a product decomposition $U = U_1 \times U_2$, with $X = U_1 \times 0$. If U is open in $\mathbf{E}$, then we may take U_1, U_2 open in $\mathbf{F}_1$, $\mathbf{F}_2$ respectively. Then the injection of $N(X)$ in $T(Y)|X$ may be represented locally by an exact sequence

$$0 \to U_1 \times \mathbf{F}_2 \xrightarrow{\varphi} U_1 \times \mathbf{F}_1 \times \mathbf{F}_2,$$

and the inclusion of $T(Y)|X$ in $T(Y)$ is simply the inclusion

$$U_1 \times \mathbf{F}_1 \times \mathbf{F}_2 \to U_1 \times U_2 \times \mathbf{F}_1 \times \mathbf{F}_2.$$

We work at the point $(x_1, 0)$ in $U_1 \times \mathbf{F}_2$. We must compute the derivative of the composite map

$$U_1 \times \mathbf{F}_2 \xrightarrow{\varphi} U_1 \times U_2 \times \mathbf{F}_1 \times \mathbf{F}_2 \xrightarrow{\exp} Y$$

at $(x_1, 0)$. We can do this by the formula for the partial derivatives. Since the exponential map coincides with the projection on the zero cross section, its "horizontal" partial derivative is the identity. By Theorem 4.1 we know that its "vertical" derivative is also the identity. Let

$$\psi = (\exp) \circ \overline{\varphi}$$

(where $\overline{\varphi}$ is simply φ followed by the inclusion). Then for any vector (w_1, w_2) in $\mathbf{F}_1 \times \mathbf{F}_2$ we get

$$D\psi(x_1, 0) \cdot (w_1, w_2) = (w_1, 0) + \varphi_{x_1}(w_2),$$

where φ_{x_1} is the linear map given by φ on the fiber over x_1. By hypothesis, we know that $\mathbf{F}_1 \times \mathbf{F}_2$ is the direct sum of $\mathbf{F}_1 \times 0$ and of the image of φ_{x_1}. This proves that $D\psi(x_1, 0)$ is a toplinear isomorphism, and in fact proves that **the exponential map restricted to a normal bundle is a local isomorphism** on the zero cross section.

We have thus shown that there exists a vector bundle $E \to X$, an open neighborhood Z of the zero section in E, and a mapping $f: Z \to Y$ which, for each x in ζ_E, is a local isomorphism at x. We must show that Z can be shrunk so that f restricts to an isomorphism. To do this we follow Godement ([God 58], p. 150). We can find a locally finite open covering of X by open sets U_i in Y such that, for each i we have inverse isomorphisms

$$f_i: Z_i \to U_i \qquad \text{and} \qquad g_i: U_i \to Z_i$$

between U_i and open sets Z_i in Z, such that each Z_i contains a point x of X, such that f_i, g_i are the identity on X (viewed as a subset of both Z and Y) and such that f_i is the restriction of f to Z_i. We now find a locally finite covering $\{V_i\}$ of X by open sets of Y such that $\overline{V}_i \subset U_i$, and let $V = \bigcup V_i$. We let W be the subset of elements $y \in V$ such that, if y lies in an intersection $\overline{V}_i \cap \overline{V}_j$, then $g_i(y) = g_j(y)$. Then W certainly contains X. We contend that W contains an open subset containing X.

Let $x \in X$. There exists an open neighborhood G_x of x in Y which meets only a finite number of $\overline{V}_i$, say $\overline{V}_{i_1}, \ldots, \overline{V}_{i_r}$. Taking G_x small enough, we can assume that x lies in each one of these, and that G_x is contained in each one of the sets $\overline{U}_{i_1}, \ldots, \overline{U}_{i_r}$. Since x lies in each $\overline{V}_{i_1}, \ldots, \overline{V}_{i_r}$, it is contained in $U_{i_1}, \ldots, U_{i_r}$ and our maps $g_{i_1}, \ldots, g_{i_r}$ take the same value at x, namely x itself. Using the fact that $f_{i_1}, \ldots, f_{i_r}$ are restrictions of f, we see at once that our finite number of maps $g_{i_1}, \ldots, g_{i_r}$ must agree on G_x if we take G_x small enough.

Let G be the union of the G_x. Then G is open, and we can define a map

$$g: G \to g(G) \subset Z$$

by taking g equal to g_i on $G \cap V_i$. Then $g(G)$ is open in Z, and the restriction of f to $g(G)$ is an inverse for g. This proves that f, g are inverse isomorphisms on G and $g(G)$, and concludes the proof of the theorem.

A vector bundle $E \to X$ will be said to be **compressible** if, given an open neighborhood Z of the zero section, there exists an isomorphism

$$\varphi: E \to Z_1$$

of E with an open subset Z_1 of Z containing the zero section, which commutes with the projection on X:

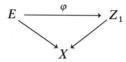

It is clear that if a bundle is compressible, and if we have a tubular neighborhood defined on Z, then we can get a total tubular neighborhood defined on E. We shall see in the chapter on Riemannian metrics that certain types of vector bundles are compressible (Hilbert bundles, assuming that the base manifold admits partitions of unity).

IV, §6. UNIQUENESS OF TUBULAR NEIGHBORHOODS

Let X, Y be two manifolds, and $F: \mathbf{R} \times X \to Y$ a morphism. We shall say that F is an **isotopy** (of embeddings) if it satisfies the following conditions. First, for each $t \in \mathbf{R}$, the map F_t given by $F_t(x) = F(t, x)$ is an embedding. Second, there exist numbers $t_0 < t_1$ such that $F_t = F_{t_0}$ for all $t \leq t_0$ and $F_{t_1} = F_t$ for all $t \geq t_1$. We then say that the interval $[t_0, t_1]$ is a **proper domain** for the isotopy, and the constant embeddings on the left and right will also be denoted by $F_{-\infty}$ and $F_{+\infty}$ respectively. We say that two embeddings $f: X \to Y$ and $g: X \to Y$ are **isotopic** if there exists an isotopy F_t as above such that $f = F_{t_0}$ and $g = F_{t_1}$ (notation as above). We write $f \approx g$ for f isotopic to g.

Using translations of intervals, and multiplication by scalars, we can always transform an isotopy to a new one whose proper domain is contained in the interval $(0, 1)$. Furthermore, the relation of isotopy between embeddings is an equivalence relation. It is obviously symmetric and reflexive, and for transitivity, suppose $f \approx g$ and $g \approx h$. We can choose the ranges of these isotopies so that the first one ends and stays constant at g before the second starts moving. Thus it is clear how to compose isotopies in this case.

If $s_0 < s_1$ are two numbers, and $\sigma: \mathbf{R} \to \mathbf{R}$ is a function (morphism) such that $\sigma(s) = t_0$ for $s \leq s_0$ and $\sigma(s) = t_1$ for $s \geq s_1$, and σ is monotone increasing, then from a given isotopy F_t we obtain another one, $G_t = F_{\sigma(t)}$. Such a function σ can be used to smooth out a piece of isotopy given only on a closed interval.

Remark. We shall frequently use the following trivial fact: If $f_t: X \to Y$ is an isotopy, and if $g: X_1 \to X$ and $h: Y \to Y_1$ are two embeddings, then the composite map

$$hf_t g: X_1 \to Y_1$$

is also an isotopy.

Let Y be a manifold and X a submanifold. Let $\pi: E \to X$ be a vector bundle, and Z an open neighborhood of the zero section. An isotopy $f_t: Z \to Y$ of open embeddings such that each f_t is a tubular neighborhood of X will be called an **isotopy of tubular neighborhoods**. In what follows, the domain will usually be all of E.

Proposition 6.1. *Let X be a manifold. Let $\pi: E \to X$ and $\pi_1: E_1 \to X$ be two vector bundles over X. Let*

$$f: E \to E_1$$

be a tubular neighborhood of X in E_1 (identifying X with its zero section in E_1). Then there exists an isotopy

$$f_t: E \to E_1$$

with proper domain $[0, 1]$ such that $f_1 = f$ and f_0 is a VB-isomorphism. (If f, π, π_1 are of class C^p then f_t can be chosen of class C^{p-1}.)

Proof. We define F by the formula

$$F_t(e) = t^{-1} f(te)$$

for $t \neq 0$ and $e \in E$. Then F_t is an embedding since it is composed of embeddings (the scalar multiplications by t, t^{-1} are in fact VB-isomorphism).

We must investigate what happens at $t = 0$.

Given $e \in E$, we find an open neighborhood U_1 of πe over which E_1 admits a trivialization $U_1 \times E_1$. We then find a still smaller open neighborhood U of πe and an open ball B around 0 in the typical fiber E of E such that E admits a trivialization $U \times E$ over U, and such that the representation $\bar{f}$ of f on $U \times B$ (contained in $U \times E$) maps $U \times B$ into $U_1 \times E_1$. This is possible by continuity. On $U \times B$ we can represent $\bar{f}$ by two morphisms,

$$\bar{f}(x, v) = \big(\varphi(x, v), \psi(x, v)\big)$$

and $\varphi(x, 0) = x$ while $\psi(x, 0) = 0$. Observe that for all t sufficiently small, te is contained in $U \times B$ (in the local representation).

We can represent F_t locally on $U \times B$ as the mapping

$$\bar{F}_t(x, v) = \big(\varphi(x, tv), t^{-1}\psi(x, tv)\big).$$

The map φ is then a morphism in the three variables x, v, and t even at $t = 0$. The second component of $\bar{F}_t$ can be written

$$t^{-1}\psi(x, tv) = t^{-1} \int_0^1 D_2\psi(x, stv)\cdot(tv)\, ds$$

and thus t^{-1} cancels t to yield simply

$$\int_0^1 D_2\psi(x, stv)\cdot v\, ds.$$

This is a morphism in t, even at $t = 0$. Furthermore, for $t = 0$, we obtain

$$\bar{F}_0(x, v) = \big(x, D_2\psi(x, 0)v\big).$$

Since f was originally assumed to be an embedding, it follows that $D_2\psi(x, 0)$ is a toplinear isomorphism, and therefore F_0 is a VB-isomorphism. To get our isotopy in standard form, we can use a function $\sigma: \mathbf{R} \to \mathbf{R}$ such that $\sigma(t) = 0$ for $t \leq 0$ and $\sigma(t) = 1$ for $t \geq 1$, and σ is monotone increasing. This proves our proposition.

Theorem 6.2. *Let X be a submanifold of Y. Let*

$$\pi: E \to X \qquad and \qquad \pi_1: E_1 \to X$$

be two vector bundles, and assume that E is compressible. Let $f: E \to Y$ and $g: E_1 \to Y$ be two tubular neighborhoods of X in Y. Then there exists a C^{p-1}-isotopy

$$f_t: E \to Y$$

of tubular neighborhoods with proper domain $[0, 1]$ and a VB-isomorphism $\lambda: E \to E_1$ such that $f_1 = f$ and $f_0 = g\lambda$.

Proof. We observe that $f(E)$ and $g(E_1)$ are open neighborhoods of X in Y. Let $U = f^{-1}(f(E) \cap g(E_1))$ and let $\varphi: E \to U$ be a compression. Let ψ be the composite map

$$E \xrightarrow{\varphi} U \xrightarrow{f|U} Y$$

$\psi = (f|U) \circ \varphi$. Then ψ is a tubular neighborhood, and $\psi(E)$ is contained in $g(E_1)$. Therefore $g^{-1}\psi: E \to E_1$ is a tubular neighborhood of the same type considered in the previous proposition. There exists an isotopy of tubular neighborhoods of X:

$$G_t: E \to E_1$$

such that $G_1 = g^{-1}\psi$ and G_0 is a VB-isomorphism. Considering the isotopy gG_t, we find an isotopy of tubular neighborhoods

$$\psi_t: E \to Y$$

such that $\psi_1 = \psi$ and $\psi_0 = g\omega$ where $\omega: E \to E_1$ is a VB-isomorphism. We have thus shown that ψ and $g\omega$ are isotopic (by an isotopy of tubular neighborhoods). Similarly, we see that ψ and $f\mu$ are isotopic for some VB-isomorphism

$$\mu: E \to E.$$

Consequently, adjusting the proper domains of our isotopies suitably, we get an isotopy of tubular neighborhoods going from $g\omega$ to $f\mu$, say F_t. Then $F_t\mu^{-1}$ will give us the desired isotopy from $g\omega\mu^{-1}$ to f, and we can put $\lambda = \omega\mu^{-1}$ to conclude the proof.

(By the way, the uniqueness proof did not use the existence theorem for differential equations.)

CHAPTER V

Operations on Vector Fields and Differential Forms

If $E \to X$ is a vector bundle, then it is of considerable interest to investigate the special operation derived from the functor "multilinear alternating forms." Applying it to the tangent bundle, we call the sections of our new bundle differential forms. One can define formally certain relations between functions, vector fields, and differential forms which lie at the foundations of differential and Riemannian geometry. We shall give the basic system surrounding such forms. In order to have at least one application, we discuss the fundamental 2-form, and in the next chapter connect it with Riemannian metrics in order to construct canonically the spray associated with such a metric.

We assume throughout that our manifolds are Hausdorff, and sufficiently differentiable so that all of our statements make sense.

V, §1. VECTOR FIELDS, DIFFERENTIAL OPERATORS, BRACKETS

Let X be a manifold of class C^p and φ a function defined on an open set U, that is a morphism

$$\varphi \colon U \to \mathbf{R}.$$

Let ξ be a vector field of class C^{p-1}. Recall that

$$T_x\varphi \colon T_x(U) \to T_x(\mathbf{R}) = \mathbf{R}$$

is a continuous linear map. With it, we shall define a new function to be denoted by $\xi\varphi$ or $\xi(\varphi)$. (There will be no confusion with this notation and composition of mappings.)

Proposition 1.1. *There exists a unique function $\xi\varphi$ on U of class C^{p-1} such that*

$$(\xi\varphi)(x) = (T_x\varphi)\xi(x).$$

If U is open in the Banach space $\mathbf{E}$ and ξ denotes the local representation of the vector field on U, then

$$(\xi\varphi)(x) = \varphi'(x)\xi(x).$$

Proof. The first formula certainly defines a mapping of U into $\mathbf{R}$. The local formula defines a C^{p-1}-morphism on U. It follows at once from the definitions that the first formula expresses invariantly in terms of the tangent bundle the same mapping as the second. Thus it allows us to define $\xi\varphi$ as a morphism globally, as desired.

Let $\mathfrak{F}^p$ denote the ring of functions (of class C^p). Then our operation $\varphi \mapsto \xi\varphi$ gives rise to a linear map

$$\delta_\xi: \mathfrak{F}^p(U) \to \mathfrak{F}^{p-1}(U), \qquad \text{defined by} \quad \delta_\xi\varphi = \xi\varphi.$$

A mapping

$$\delta: R \to S$$

from a ring R into an R-algebra S is called a **derivation** if it satisfies the usual formalism: Linearity, and $\delta(ab) = a\delta(b) + \delta(a)b$.

Proposition 1.2. *Let X be a manifold and U open in X. Let ξ be a vector field over X. If $\delta_\xi = 0$, then $\xi(x) = 0$ for all $x \in U$. Each δ_ξ is a derivation of $\mathfrak{F}^p(U)$ into $\mathfrak{F}^{p-1}(U)$.*

Proof. Suppose $\xi(x) \neq 0$ for some x. We work with the local representations, and take φ to be a continuous linear map of $\mathbf{E}$ into $\mathbf{R}$ such that $\varphi(\xi(x)) \neq 0$, by Hahn–Banach. Then $\varphi'(y) = \varphi$ for all $y \in U$, and we see that $\varphi'(x)\xi(x) \neq 0$, thus proving the first assertion. The second is obvious from the local formula.

From Proposition 1.2 we deduce that if two vector fields induce the same differential operator on the functions, then they are equal.

Given two vector fields ξ, η on X, we shall now define a new vector field $[\xi, \eta]$, called their **bracket product**.

Proposition 1.3. *Let ξ, η be two vector fields of class C^{p-1} on X. Then there exists a unique vector field $[\xi, \eta]$ of class C^{p-2} such that for each open set U and function φ on U we have*

$$[\xi, \eta]\varphi = \xi(\eta(\varphi)) - \eta(\xi(\varphi)).$$

If U is open in **E** *and* ξ, η *are the local representations of the vector fields, then* $[\xi, \eta]$ *is given by the local formula*

$$[\xi, \eta]\varphi(x) = \varphi'(x)\big(\eta'(x)\xi(x) - \xi'(x)\eta(x)\big).$$

Thus the local representation of $[\xi, \eta]$ *is given by*

$$[\xi, \eta](x) = \eta'(x)\xi(x) - \xi'(x)\eta(x).$$

Proof. By Proposition 1.2, any vector field having the desired effect on functions is uniquely determined. We check that the local formula gives us this effect locally. Differentiating formally, we have (using the law for the derivative of a product):

$$(\eta\varphi)'\xi - (\xi\varphi)'\eta = (\varphi'\eta)'\xi - (\varphi'\xi)\eta$$
$$= \varphi'\eta'\xi + \varphi''\eta\xi - \varphi'\xi'\eta - \varphi''\xi\eta.$$

The terms involving φ'' must be understood correctly. For instance, the first such term at a point x is simply

$$\varphi''(x)\big(\eta(x), \xi(x)\big)$$

remembering that $\varphi''(x)$ is a bilinear map, and can thus be evaluated at the two vectors $\eta(x)$ and $\xi(x)$. However, we know that $\varphi''(x)$ is symmetric. Hence the two terms involving the second derivative of φ cancel, and give us our formula.

Corollary 1.4. *The bracket* $[\xi, \eta]$ *is bilinear in both arguments, we have* $[\xi, \eta] = -[\eta, \xi]$, *and Jacobi's identity*

$$[\xi, [\eta, \zeta]] + [\eta, [\zeta, \xi]] + [\zeta, [\xi, \eta]] = 0.$$

If φ *is a function, then*

$$[\xi, \varphi\eta] = (\xi\varphi)\eta + \varphi[\xi, \eta],$$
$$[\varphi\xi, \eta] = \varphi[\xi, \eta] - (\eta\varphi)\xi.$$

Proof. The first two assertions are obvious. The third comes from the definition of the bracket. We apply the vector field on the left of the equality to a function φ. All the terms cancel out (the reader will write it out as well or better than the author). The last two formulas are immediate.

We make some comments concerning the functoriality of vector fields. Let

$$f: X \to Y$$

be an isomorphism. Let ξ be a vector field over X. Then we obtain an induced vector field $f_*\xi$ over Y, defined by the formula

$$(f_*\xi)(f(x)) = Tf(\xi(x)).$$

It is the vector field making the following diagram commutative:

$$
\begin{array}{ccc}
TX & \xrightarrow{\ Tf\ } & TY \\
\xi \Big\uparrow\Big\downarrow & & \Big\uparrow\Big\downarrow f_*\xi \\
X & \xrightarrow[\ f\ ]{} & Y
\end{array}
$$

We shall also write f^* for $(f^{-1})_*$ when applied to a vector field. Thus we have the formulas

$$\boxed{f_*\xi = Tf \circ \xi \circ f^{-1}} \qquad \text{and} \qquad \boxed{f^*\xi = Tf^{-1} \circ \xi \circ f.}$$

If f is not an isomorphism, then one cannot in general define the direct or inverse image of a vector field as done above. However, let ξ be a vector field over X, and let η be a vector field over Y. If for each $x \in X$ we have

$$Tf(\xi(x)) = \eta(f(x)),$$

then we shall say that f maps ξ into η, or that ξ and η are f-**related**. If this is the case, then we may denote by $f_*\xi$ the map from $f(X)$ into TY defined by the above formula.

Let ξ_1, ξ_2 be vector fields over X, and let η_1, η_2 be vector fields over Y. If ξ_i is f-related to η_i for $i = 1, 2$ then as maps on $f(X)$ we have

$$\boxed{f_*[\xi_1, \xi_2] = [\eta_1, \eta_2].}$$

We may write suggestively the formula in the form

$$\boxed{f_*[\xi_1, \xi_2] = [f_*\xi_1, f_*\xi_2].}$$

Of course, this is meaningless in general, since $f_* \xi_1$ may not be a vector field on Y. When f is an isomorphism, then it is a correct formulation of the other formula. In any case, it suggests the correct formula.

To prove the formula, we work with the local representations, when $X = U$ is open in $\mathbf{E}$, and $Y = V$ is open in $\mathbf{F}$. Then ξ_i, η_i are maps of U, V into the spaces $\mathbf{E}$, $\mathbf{F}$ respectively. For $x \in X$ we have

$$(f_*[\xi_1, \xi_2])(x) = f'(x)\big(\xi_2'(x)\xi_1(x) - \xi_1'(x)\xi_2(x)\big).$$

On the other hand, by assumption, we have

$$\eta_i(f(x)) = f'(x)\xi_i(x),$$

so that

$$
\begin{aligned}
[\eta_1, \eta_2](f(x)) &= \eta_2'(f(x))\eta_1(f(x)) - \eta_1'(f(x))\eta_2(f(x)) \\
&= \eta_2'(f(x))f'(x)\xi_1(x) - \eta_1'(f(x))f'(x)\xi_2(x) \\
&= (\eta_2 \circ f)'(x)\xi_1(x) - \big(\eta_1 \circ f('(x)\xi_2(x))\big) \\
&= f''(x) \cdot \xi_2(x) \cdot \xi_1(x) + f'(x)\xi_2'(x)\xi_1(x) \\
&\quad - f''(x) \cdot \xi_1(x) \cdot \xi_2(x) - f'(x)\xi_1'(x)\xi_2(x).
\end{aligned}
$$

Since $f''(x)$ is symmetric, two terms cancel, and the remaining two terms give the same value as $(f_*[\xi_1, \xi_2])(x)$, as was to be shown.

The bracket between vector fields gives an infinitesimal criterion for commutativity in various contexts. We give here one theorem of a general nature as an example of this phenomenon.

Theorem 1.5. *Let ξ, η be vector fields on X, and assume that $[\xi, \eta] = 0$. Let α and β be the flows for ξ and η respectively. Then for real values t, s we have*

$$\alpha_t \circ \beta_s = \beta_s \circ \alpha_t.$$

Or in other words, for any $x \in X$ we have

$$\alpha(t, \beta(s, x)) = \beta(s, \alpha(t, x)),$$

in the sense that if for some value of t a value of s is in the domain of one of these expressions, then it is in the domain of the other and the two expressions are equal.

Proof. For a fixed value of t, the two curves in s given by the right- and left-hand side of the last formula have the same initial condition, namely $\alpha_t(x)$. The curve on the right

$$s \longmapsto \beta(s, \alpha(t, x))$$

is by definition the integral curve of η. The curve on the left

$$s \longmapsto \alpha(t, \beta(s, x))$$

is the image under α_t of the integral curve for η having initial condition x. Since x is fixed, let us denote $\beta(s, x)$ simply by $\beta(s)$. What we must show is that the two curves on the right and on the left satisfy the same differential equation.

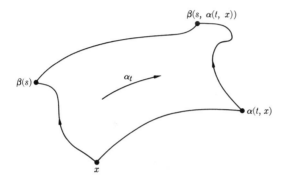

In the above figure, we see that the flow α_t shoves the curve on the left to the curve on the right. We must compute the tangent vectors to the curve on the right. We have

$$\frac{d}{ds}(\alpha_t(\beta(s))) = D_2\alpha(t, \beta(s))\beta'(s)$$

$$= D_2\alpha(t, \beta(s))\eta(\beta(s)).$$

Now fix s, and denote this last expression by $F(t)$. We must show that if

$$G(t) = \eta(\alpha(t, \beta(s))),$$

then

$$F(t) = G(t).$$

We have trivially $F(0) = G(0)$, in other words the curves F and G have the same initial condition. On the other hand,

$$F'(t) = \xi'(\alpha(t, \beta(s)))D_2\alpha(t, \beta(s))\eta(\beta(s))$$

and

$$G'(t) = \eta'(\alpha(t, \beta(s)))\xi(\alpha(t, \beta(s)))$$

$$= \xi'(\alpha(t, \beta(s)))\eta(\alpha(t, \beta(s))) \quad \text{(because } [\xi, \eta] = 0\text{)}.$$

Hence we see that our two curves F and G satisfy the same differential equation, whence they are equal. This proves our theorem.

Vector fields ξ, η such that $[\xi, \eta] = 0$ are said to **commute**. One can generalize the process of straightening out vector fields to a finite number of commuting vector fields, using the same method of proof, using Theorem 1.5. As another application, one can prove that if the Lie algebra of a connected Lie group is commutative, then the group is commutative. Cf. the section on Lie groups.

V, §2. LIE DERIVATIVE

Let λ be a differentiable functor on Banach spaces. For convenience, take λ to be covariant and in one variable. What we shall say in the rest of this section would hold in the same way (with slightly more involved notation) if λ had several variables and were covariant in some and contravariant in others.

Given a manifold X, we can take $\lambda(T(X))$. It is a vector bundle over X, which we denote by $T_\lambda(X)$ as in Chapter III. Its sections $\Gamma_\lambda(X)$ are the tensor fields of type λ.

Let ξ be a vector field on X, and U open in X. It is then possible to associate with ξ a map

$$\mathscr{L}_\xi \colon \Gamma_\lambda(U) \to \Gamma_\lambda(U)$$

(with a loss of two derivatives). This is done as follows.

Given a point x of U and a local flow α for ξ at x, we have for each t sufficiently small a local isomorphism α_t in a neighborhood of our point x. Recall that locally, $\alpha_t^{-1} = \alpha_{-t}$. If η is a tensor field of type λ, then the composite mapping $\eta \circ \alpha_t$ has its range in $T_\lambda(X)$. Finally, we can take the tangent map $T(\alpha_{-t}) = (\alpha_{-t})_*$ to return to $T_\lambda(X)$ in the fiber above x. We thus obtain a composite map

$$F(t, x) = (\alpha_{-t})_* \circ \eta \circ \alpha_t(x) = (\alpha_t^* \eta)(x),$$

which is a morphism, locally at x. We take its derivative with respect to t and evaluate it at 0. After looking at the situation locally in a trivialization of $T(X)$ and $T_\lambda(X)$ at x, one sees that the map one obtains gives a section of $T_\lambda(U)$, that is a tensor field of type λ over U. This is our map $\mathscr{L}_\xi$. To summarize,

$$\mathscr{L}_\xi \eta = \frac{d}{dt}\bigg|_{t=0} (\alpha_{-t})_* \circ \eta \circ \alpha_t.$$

This map $\mathscr{L}_\xi$ is called the **Lie derivative**. We shall determine the Lie derivative on functions and on vector fields in terms of notions already discussed.

First let φ be a function. Then by the general definition, the Lie derivative of this function with respect to the vector field ξ with flow α is defined to be

$$\mathscr{L}_\xi \varphi(x) = \lim_{t \to 0} \frac{1}{t} [\varphi(\alpha(t, x)) - \varphi(x)],$$

or in other words,

$$\mathscr{L}_\xi \varphi = \frac{d}{dt} (\alpha_t^* \varphi) \bigg|_{t=0}.$$

Our assertion is then that

$$\boxed{\mathscr{L}_\xi \varphi = \xi \varphi.}$$

To prove this, let

$$F(t) = \varphi(\alpha(t, x)).$$

Then

$$F'(t) = \varphi'(\alpha(t, x)) D_1 \alpha(t, x)$$
$$= \varphi'(\alpha(t, x)) \xi(\alpha(t, x)),$$

because α is a flow for ξ. Using the initial condition at $t = 0$, we find that

$$F'(0) = \varphi'(x) \xi(x),$$

which is precisely the value of $\xi \varphi$ at x, thus proving our assertion.

If ξ, η are vector fields, then

$$\boxed{\mathscr{L}_\xi \eta = [\xi, \eta].}$$

As before, let α be a flow for ξ. The Lie derivative is given by

$$\mathscr{L}_\xi \eta = \frac{d}{dt} (\alpha_t^* \eta) \bigg|_{t=0}.$$

Letting ξ and η denote the local representations of the vector fields, we note that the local representation of $(\alpha_t^* \eta)(x)$ is given by

$$(\alpha_t^* \eta)(x) = F(t) = D_2 \alpha(-t, x)\eta(\alpha(t, x)).$$

We must therefore compute $F'(t)$, and then $F'(0)$. Using the chain rule, the formula for the derivative of a product, and the differential equation satisfied by $D_2 \alpha$, we obtain

$$F'(t) = -D_1 D_2 \alpha(-t, x)\eta(\alpha(t, x)) + D_2 \alpha(-t, x)\eta'(\alpha(t, x))D_1 \alpha(t, x)$$
$$= -\xi'(\alpha(-t, x))D_2 \alpha(-t, x)\eta(\alpha(t, x)) + D_2 \alpha(-t, x)\eta'(\alpha(t, x)).$$

Putting $t = 0$ proves our formula, taking into account the initial conditions

$$\alpha(0, x) = x \qquad \text{and} \qquad D_2 \alpha(0, x) = \text{id}.$$

V, §3. EXTERIOR DERIVATIVE

Let X be a manifold. The functor L_a^r (r-multilinear continuous alternating forms) extends to arbitrary vector bundles, and in particular, to the tangent bundle of X. A **differential form** of degree r, or simply an **r-form** on X, is a section of $L_a^r(T(X))$, that is a tensor field of type L_a^r. If X is of class C^p, forms will be assumed to be of a suitable class C^s with $1 \leq s \leq p - 1$. The set of differential forms of degree r will be denoted by $\mathscr{A}^r(X)$ ($\mathscr{A}$ for alternating). It is not only a vector space over $\mathbf{R}$ but a module over the ring of functions on X (of the appropriate order of differentiability). If ω is an r-form, then $\omega(x)$ is an element of $L_a^r(T_x(X))$, and is thus an r-multilinear alternating form of $T_x(X)$ into $\mathbf{R}$. We sometimes denote $\omega(x)$ by ω_x.

Suppose U is open in the Banach space $\mathbf{E}$. Then $L_a^r(T(U))$ is equal to $U \times L_a^r(\mathbf{E})$ and a differential form is entirely described by the projection on the second factor, which we call its **local representation**, following our general system (Chapter III, §4). Such a local representation is therefore a morphism

$$\omega \colon U \to L_a^r(\mathbf{E}).$$

Let ω be in $L_a^r(\mathbf{E})$ and $v_1, \ldots, v_r$ elements of $\mathbf{E}$. We denote the value $\omega(v_1, \ldots, v_r)$ also by

$$\langle \omega, v_1 \times \cdots \times v_r \rangle.$$

Similarly, let $\xi_1, \ldots, \xi_r$ be vector fields on an open set U, and let ω be an *r*-form on *X*. We denote by

$$\langle \omega, \xi_1 \times \cdots \times \xi_r \rangle$$

the mapping from *U* into **R** whose value at a point *x* in *U* is

$$\langle \omega(x), \xi_1(x) \times \cdots \times \xi_r(x) \rangle.$$

Looking at the situation locally on an open set *U* such that *T(U)* is trivial, we see at once that this mapping is a morphism (i.e. a function on *U*) of the same degree of differentiability as ω and the ξ_i.

Proposition 3.1. *Let x_0 be a point of X and ω an r-form on X. If*

$$\langle \omega, \xi_1 \times \cdots \times \xi_r \rangle(x_0)$$

is equal to 0 for all vector fields $\xi_1, \ldots, \xi_r$ at x_0 (i.e. defined on some neighborhood of x_0), then $\omega(x_0) = 0$.

Proof. Considering things locally in terms of their local representations, we see that if $\omega(x_0)$ is not 0, then it does not vanish at some *r*-tuple of vectors $(v_1, \ldots, v_r)$. We can take vector fields at x_0 which take on these values at x_0 and from this our assertion is obvious.

It is convenient to agree that a differential form of degree 0 is a function. In the next proposition, we describe the exterior derivative of an *r*-form, and it is convenient to describe this situation separately in the case of functions.

Therefore let $f: X \to \mathbf{R}$ be a function. For each $x \in X$, the tangent map

$$T_x f: T_x(X) \to T_{f(x)}(\mathbf{R}) = \mathbf{R}$$

is a continuous linear map, and looking at local representations shows at once that the collection of such maps defines a 1-form which will be denoted by *df*. Furthermore, from the definition of the operation of vector fields on functions, it is clear that *df* is the unique 1-form such that for every vector field ξ we have

$$\langle df, \xi \rangle = \xi f.$$

To extend the definition of *d* to forms of higher degree, we recall that if

$$\omega: U \to L_a^r(\mathbf{E})$$

is the local representation of an *r*-form over an open set *U* of **E**, then for each x in U,

$$\omega'(x)\colon \mathbf{E} \to L_a^r(\mathbf{E})$$

is a continuous linear map. Applied to a vector v in $\mathbf{E}$, it therefore gives rise to an r-form on $\mathbf{E}$.

Proposition 3.2. *Let ω be an r-form of class C^{p-1} on X. Then there exists a unique $(r + 1)$-form $d\omega$ on X of class C^{p-2} such that, for any open set U of X and vector fields $\xi_0, \ldots, \xi_r$ on U we have*

$$\langle d\omega, \xi_0 \times \cdots \times \xi_r \rangle$$

$$= \sum_{i=0}^{r} (-1)^i \xi_i \langle \omega, \xi_0 \times \cdots \times \hat{\xi}_i \times \cdots \times \xi_r \rangle$$

$$+ \sum_{i<j} (-1)^{i+j} \langle \omega, [\xi_i, \xi_j] \times \xi_0 \times \cdots \times \hat{\xi}_i \times \cdots \times \hat{\xi}_j \times \cdots \times \xi_r \rangle.$$

If furthermore U is open in $\mathbf{E}$ and $\omega, \xi_0, \ldots, \xi_r$ are the local representations of the form and the vector fields respectively, then at a point x the value of the expression above is equal to

$$\sum_{i=0}^{r} (-1)^i \langle \omega'(x)\xi_i(x), \xi_0(x) \times \cdots \times \widehat{\xi_i(x)} \times \cdots \times \xi_r(x) \rangle.$$

Proof. As before, we observe that the local formula defines a differential form. If we can prove that it gives the same thing as the first formula, which is expressed invariantly, then we can globalize it, and we are done. Let us denote by S_1 and S_2 the two sums occurring in the invariant expression, and let L be the local expression. We must show that $S_1 + S_2 = L$. We consider S_1, and apply the definition of ξ_i operating on a function locally, as in Proposition 1.1, at a point x. We obtain

$$S_1 = \sum_{i=0}^{r} (-1)^i \langle \omega, \xi_0 \times \cdots \times \hat{\xi}_i \times \cdots \times \xi_r \rangle'(x)\xi_i(x).$$

The derivative is perhaps best computed by going back to the definition. Applying this definition directly, and discarding second order terms, we find that S_1 is equal to

$$\sum (-1)^i \langle \omega'(x)\xi_i(x), \xi_0(x) \times \cdots \times \widehat{\xi_i(x)} \times \cdots \times \xi_r(x) \rangle$$

$$+ \sum_i \sum_{i<j} (-1)^i \langle \omega(x), \xi_0(x) \times \cdots \times \xi_j'(x)\xi_i(x) \times \cdots \times \widehat{\xi_i(x)} \times \cdots \times \xi_r(x) \rangle$$

$$+ \sum_i \sum_{j<i} \langle \omega(x), \xi_0(x) \times \cdots \times \widehat{\xi_i(x)} \times \cdots \times \xi_j'(x)\xi_i(x) \times \cdots \times \xi_r(x) \rangle.$$

Of these three sums, the first one is the local formula L. As for the other two, permuting j and i in the first, and moving the term $\xi_j'(x)\xi_i(x)$ to the first position, we see that they combine to give (symbolically)

$$-\sum_i \sum_{i<j} (-1)^{i+j} \langle \omega, (\xi_j'\xi_i - \xi_i'\xi_j) \times \xi_0 \times \cdots \times \overset{\circ}{\xi}_i \times \cdots \times \overset{\circ}{\xi}_j \times \cdots \times \xi_r \rangle$$

(evaluated at x). Using Proposition 1.3, we see that this combination is equal to $-S_2$. This proves that $S_1 + S_2 = L$, as desired.

We call $d\omega$ the **exterior derivative** of ω. Leaving out the order of differentiability for simplicity, we see that d is an **R**-linear map

$$d: \mathscr{A}^r(X) \to \mathscr{A}^{r+1}(X).$$

We now look into the multiplicative properties of d with respect to the wedge product.

Let ω, ψ be continuous multilinear alternating forms of degree r and s respectively on the Banach space **E**. In multilinear algebra, one defines their **wedge product** as an $(r+s)$-continuous multilinear alternating form, by the formula

$$(\omega \wedge \psi)(v_1, \ldots, v_{r+s}) = \frac{1}{r!\, s!} \sum \varepsilon(\sigma)\omega(v_{\sigma 1}, \ldots, v_{\sigma r})\psi(v_{\sigma(r+1)}, \ldots, v_{\sigma(r+s)})$$

the sum being taken over all permutations σ of $(1, \ldots, r+s)$. This definition extends at once to differential forms on a manifold, if we view it as giving the value for $\omega \wedge \psi$ at a point x. The v_i are then elements of the tangent space T_x, and considering the local representations shows at once that the wedge product so defined gives a morphism of the manifold X into $L_a^{r+s}(T(X))$, and is therefore a differential form.

Remark. The coefficient $1/r!\, s!$ is not universally taken to define the wedge product. Some people, e.g. [He 78] and [KoN 63], take $1/(r+s)!$, which causes constants to appear later. I have taken the same factor as [AbM 78] and [GHL 87/93]. I recommend that the reader check out the case with $r = s = 1$ so $r + s = 2$ to see how a factor $\frac{1}{2}$ comes in. With either convention, the wedge product between forms is associative, so with some care, one can carry out a consistent theory with either convention. I leave the proof of associativity to the reader. It follows by induction that if $\omega_1, \ldots, \omega_m$ are forms of degrees $r_1, \ldots, r_m$ respectively, and $t = r_1 + \cdots + r_m$, then

$$(\omega_1 \wedge \cdots \wedge \omega_m)(v_1, \ldots, v_r)$$

$$= \frac{1}{r_1! \cdots r_m!} \sum_\sigma \omega_1(v_{\sigma 1}, \ldots, v_{\sigma r_1})\omega_2(v_{\sigma(r_1+1)}, \ldots, v_{\sigma(r_1+r_2)})\omega_m(v_{\sigma(r-r_m+1)}, \ldots, v_{\sigma r})$$

where the sum is taken over all permutations of $(1, \ldots, r)$.

If we regard functions on X as differential forms of degree 0, then the ordinary product of a function by a differential form can be viewed as the wedge product. Thus if f is a function and ω a differential form, then

$$f\omega = f \wedge \omega.$$

(The form on the left has the value $f(x)\omega(x)$ at x.)

The next proposition gives us more formulas concerning differential forms.

Proposition 3.3. *Let ω, ψ be differential forms on X. Then*

EXD 1. $d(\omega \wedge \psi) = d\omega \wedge \psi + (-1)^{\deg(\omega)}\omega \wedge d\psi.$

EXD 2. $dd\omega = 0$ *(with enough differentiability, say $p \geq 4$).*

Proof. This is a simple formal exercise in the use of the local formula for the local representation of the exterior derivative. We leave it to the reader.

When the manifold is finite dimensional, then one can give a local representation for differential forms and the exterior derivative in terms of local coordinates, which are especially useful in integration which fits the notation better. We shall therefore carry out this local formulation in full. It dates back to Cartan [Ca 28]. There is in addition a theoretical point which needs clarifying. We shall use at first the wedge $\wedge$ in two senses. One sense is defined as above, giving rise to Proposition 3.3. Another sense will come from Theorem A. We shall comment on their relation after Theorem B.

We recall first two simple results from linear (or rather multilinear) algebra. We use the notation $\mathbf{E}^{(r)} = \mathbf{E} \times \mathbf{E} \times \cdots \times \mathbf{E}$, r times.

Theorem A. *Let $\mathbf{E}$ be a finite dimensional vector space over the reals of dimension n. For each positive integer r with $1 \leq r \leq n$ there exists a vector space $\bigwedge^r \mathbf{E}$ and a multilinear alternating map*

$$\mathbf{E}^{(r)} \to \textstyle\bigwedge^r \mathbf{E}$$

denoted by $(u_1, \ldots, u_r) \mapsto u_1 \wedge \cdots \wedge u_r$, having the following property: If $\{v_1, \ldots, v_n\}$ is a basis of $\mathbf{E}$, then the elements

$$\{v_{i_1} \wedge \cdots \wedge v_{i_r}\}, \qquad i_1 < i_2 < \cdots < i_r,$$

form a basis of $\bigwedge^r \mathbf{E}$.

We recall that **alternating** means that $u_1 \wedge \cdots \wedge u_r = 0$ if $u_i = u_j$ for some $i \neq j$. We call $\bigwedge^r \mathbf{E}$ the r-th **alternating** product (or **exterior** product) on **E**. If $r = 0$, we define $\bigwedge^0 \mathbf{E} = \mathbf{R}$. Elements of $\bigwedge^r \mathbf{E}$ which can be written in the form $u_1 \wedge \cdots \wedge u_r$ are called **decomposable**. Such elements generate $\bigwedge^r \mathbf{E}$. If $r > \dim E$, we define $\bigwedge^r \mathbf{E} = \{0\}$.

Theorem B. *For each pair of positive integers (r, s), there exists a unique product (bilinear map)*

$$\bigwedge^r \mathbf{E} \times \bigwedge^s \mathbf{E} \to \bigwedge^{r+s} \mathbf{E}$$

such that if $u_1, \ldots, u_r, w_1, \ldots, w_s \in \mathbf{E}$ then

$$(u_1 \wedge \cdots \wedge u_r) \times (w_1 \wedge \cdots \wedge w_s) \mapsto u_1 \wedge \cdots \wedge u_r \wedge w_1 \wedge \cdots \wedge w_s.$$

This product is associative.

The proofs for these two statements can be found, for instance, in my *Linear Algebra*.

Let $\mathbf{E}^\vee$ be the dual space, $\mathbf{E}^\vee = L(\mathbf{E}, \mathbf{R})$. If $\mathbf{E} = \mathbf{R}^n$ and $\lambda_1, \ldots, \lambda_n$ are the coordinate functions, then each λ_i is an element of the dual space, and in fact $\{\lambda_1, \ldots, \lambda_n\}$ is a basis of this dual space. Let $\mathbf{E} = \mathbf{R}^n$. There is an isomorphism

$$\boxed{\bigwedge^r \mathbf{E}^\vee \xrightarrow{\approx} L_a^r(\mathbf{E}, \mathbf{R})}$$

given in the following manner. If $g_1, \ldots, g_r \in \mathbf{E}^\vee$ and $v_1, \ldots, v_r \in \mathbf{E}$, then the value

$$\det(g_i(v_j))$$

is multilinear alternating both as a function of $(g_1, \ldots, g_r)$ and $(v_1, \ldots, v_r)$. Thus it induces a pairing

$$\bigwedge^r \mathbf{E}^\vee \times \mathbf{E}^r \to \mathbf{R}$$

and a map

$$\bigwedge^r \mathbf{E}^\vee \to L_a^r(\mathbf{E}, \mathbf{R}).$$

This map is the isomorphism mentioned above. Using bases, it is easy to verify that it is an isomorphism (at the level of elementary algebra).

Thus in the finite dimensional case, we may identify $L_a^r(\mathbf{E}, \mathbf{R})$ with the alternating product $\bigwedge^r \mathbf{E}^\vee$, and consequently we may view the local representation of a differential form of degree r to be a map

$$\omega: U \to \bigwedge^r \mathbf{E}^\vee$$

from U into the rth alternating product of $\mathbf{E}^\vee$. We say that the form is of class C^p if the map is of class C^p. (We view $\bigwedge^r \mathbf{E}^\vee$ as a normed vector space, using any norm. It does not matter which, since all norms on a finite dimensional vector space are equivalent.) The wedge product as we gave it, valid in the infinite dimensional case, is compatible with the wedge product and the isomorphism of $\bigwedge^r \mathbf{E}$ with $L_a^r(\mathbf{E}, \mathbf{R})$ given above. If we had taken a different convention for the wedge product of alternating forms, then a constant would have appeared in front of the above determinant to establish the above identification (e.g. the constant $\frac{1}{2}$ in the 2×2 case).

Since $\{\lambda_1, \ldots, \lambda_n\}$ is a basis of $\mathbf{E}^\vee$, we can express each differential form in terms of its coordinate functions with respect to the basis

$$\{\lambda_{i_1} \wedge \cdots \wedge \lambda_{i_r}\}, \qquad (i_1 < \cdots < i_r),$$

namely for each $x \in U$ we have

$$\omega(x) = \sum_{(i)} f_{i_1 \cdots i_r}(x) \lambda_{i_1} \wedge \cdots \wedge \lambda_{i_r},$$

where $f_{(i)} = f_{i_1 \cdots i_r}$ is a function on U. Each such function has the same order of differentiability as ω. We call the preceding expression the **standard form** of ω. We say that a form is **decomposable** if it can be written as just one term $f(x) \lambda_{i_1} \wedge \cdots \wedge \lambda_{i_r}$. Every differential form is a sum of decomposable ones.

We agree to the convention that functions are differential forms of degree 0.

As before, the differential forms on U of given degree r form a vector space, denoted by $\mathscr{A}^r(U)$.

Let $\mathbf{E} = \mathbf{R}^n$. Let f be a function on U. For each $x \in U$ the derivative

$$f'(x): \mathbf{R}^n \to \mathbf{R}$$

is a linear map, and thus an element of the dual space. Thus

$$f': U \to \mathbf{E}^\vee$$

represents a differential form of degree 1, which is usually denoted by df. If f is of class C^p, then df is class C^{p-1}.

Let λ_i be the i-th coordinate function. Then we know that

$$d\lambda_i(x) = \lambda_i'(x) = \lambda_i$$

for each $x \in U$ because $\lambda'(x) = \lambda$ for any continuous linear map λ. Whenever $\{x_1, \ldots, x_n\}$ are used systematically for the coordinates of a point in $\mathbf{R}^n$, it is customary in the literature to use the notation

$$d\lambda_i(x) = dx_i.$$

This is slightly incorrect, but is useful in formal computations. We shall also use it in this book on occasions. Similarly, we also write (incorrectly)

$$\omega = \sum_{(i)} f_{(i)}\, dx_{i_1} \wedge \cdots \wedge dx_{i_r}$$

instead of the correct

$$\omega(x) = \sum_{(i)} f_{(i)}(x)\lambda_{i_1} \wedge \cdots \wedge \lambda_{i_r}.$$

In terms of coordinates, the map df (or f') is given by

$$df(x) = f'(x) = D_1 f(x)\lambda_1 + \cdots + D_n f(x)\lambda_n,$$

where $D_i f(x) = \partial f/\partial x_i$ is the i-th partial derivative. This is simply a restatement of the fact that if $h = (h_1, \ldots, h_n)$ is a vector, then

$$f'(x)h = \frac{\partial f}{\partial x_1} h_1 + \cdots + \frac{\partial f}{\partial x_n} h_n.$$

Thus in old notation, we have

$$df(x) = \frac{\partial f}{\partial x_1} dx_1 + \cdots + \frac{\partial f}{\partial x_n} dx_n.$$

We shall develop the theory of the alternating product and the exterior derivative directly without assuming Propositions 3.2 or 3.3 in the finite dimensional case.

Let ω and ψ be forms of degrees r and s respectively, on the open set U. For each $x \in U$ we can then take the alternating product $\omega(x) \wedge \psi(x)$ and we define the **alternating product** $\omega \wedge \psi$ by

$$(\omega \wedge \psi)(x) = \omega(x) \wedge \psi(x).$$

(It is an exercise to verify that this product corresponds to the product defined previously before Proposition 3.3 under the isomorphism between $L_a^r(\mathbf{E}, \mathbf{R})$ and the r-th alternating product in the finite dimensional case.) If f is a differential form of degree 0, that is a function, then we have again

$$f \wedge \omega = f\omega,$$

where $(f\omega)(x) = f(x)\omega(x)$. By definition, we then have

$$\omega \wedge f\psi = f\omega \wedge \psi.$$

We shall now define the **exterior derivative** $d\omega$ for any differential form ω. We have already done it for functions. We shall do it in general first in terms of coordinates, and then show that there is a characterization independent of these coordinates. If

$$\omega = \sum_{(i)} f_{(i)} \, d\lambda_{i_1} \wedge \cdots \wedge d\lambda_{i_r},$$

we define

$$d\omega = \sum_{(i)} df_{(i)} \wedge d\lambda_{i_1} \wedge \cdots \wedge d\lambda_{i_r}.$$

Example. Suppose $n = 2$ and ω is a 1-form, given in terms of the two coordinates (x, y) by

$$\omega(x, y) = f(x, y) \, dx + g(x, y) \, dy.$$

Then

$$\begin{aligned}
d\omega(x, y) &= df(x, y) \wedge dx + dg(x, y) \wedge dy \\
&= \left(\frac{\partial f}{\partial x} dx + \frac{\partial f}{\partial y} dy \right) \wedge dx + \left(\frac{\partial g}{\partial x} dx + \frac{\partial g}{\partial y} dy \right) \wedge dy \\
&= \frac{\partial f}{\partial y} dy \wedge dx + \frac{\partial g}{\partial x} dx \wedge dy \\
&= \left(\frac{\partial f}{\partial y} - \frac{\partial g}{\partial x} \right) dy \wedge dx
\end{aligned}$$

because the terms involving $dx \wedge dx$ and $dy \wedge dy$ are equal to 0.

Proposition 3.4. *The map d is linear, and satisfies*

$$d(\omega \wedge \psi) = d\omega \wedge \psi + (-1)^r \omega \wedge d\psi$$

if $r = \deg \omega$. The map d is uniquely determined by these properties, and by the fact that for a function f, we have $df = f'$.

Proof. The linearity of d is obvious. Hence it suffices to prove the formula for decomposable forms. We note that for any function f we have

$$d(f\omega) = df \wedge \omega + f \, d\omega.$$

Indeed, if ω is a function g, then from the derivative of a product we get

$d(fg) = f\, dg + g\, df$. If

$$\omega = g\, d\lambda_{i_1} \wedge \cdots \wedge d\lambda_{i_r},$$

where g is a function, then

$$d(f\omega) = d(fg\, d\lambda_{i_1} \wedge \cdots \wedge d\lambda_{i_r}) = d(fg) \wedge d\lambda_{i_1} \wedge \cdots \wedge d\lambda_{i_r}$$
$$= (f\, dg + g\, df) \wedge d\lambda_{i_1} \wedge \cdots \wedge d\lambda_{i_r}$$
$$= f\, d\omega + df \wedge \omega,$$

as desired. Now suppose that

$$\omega = f\, d\lambda_{i_1} \wedge \cdots \wedge d\lambda_{i_r} \quad \text{and} \quad \psi = g\, d\lambda_{j_1} \wedge \cdots \wedge d\lambda_{js}$$
$$= f\tilde{\omega}, \qquad\qquad\qquad\qquad = g\tilde{\psi},$$

with $i_1 < \cdots < i_r$ and $j_1 < \cdots < j_s$ as usual. If some $i_v = j_\mu$, then from the definitions we see that the expressions on both sides of the equality in the theorem are equal to 0. Hence we may assume that the sets of indices $i_i, \ldots, i_r$ and $j_1, \ldots, j_s$ have no element in common. Then $d(\tilde{\omega} \wedge \tilde{\psi}) = 0$ by definition, and

$$d(\omega \wedge \psi) = d(fg\tilde{\omega} \wedge \tilde{\psi}) = d(fg) \wedge \tilde{\omega} \wedge \tilde{\psi}$$
$$= (g\, df + f\, dg) \wedge \tilde{\omega} \wedge \tilde{\psi}$$
$$= d\omega \wedge \psi + f\, dg \wedge \tilde{\omega} \wedge \tilde{\psi}$$
$$= d\omega \wedge \psi + (-1)^r f\tilde{\omega} \wedge dg \wedge \tilde{\psi}$$
$$= d\omega \wedge \psi + (-1)^r \omega \wedge d\psi,$$

thus proving the desired formula, in the present case. (We used the fact that $dg \wedge \tilde{\omega} = (-1)^r \tilde{\omega} \wedge dg$ whose proof is left to the reader.) The formula in the general case follows because any differential form can be expressed as a sum of forms of the type just considered, and one can then use the bilinearity of the product. Finally, d is uniquely determined by the formula, and its effect on functions, because any differential form is a sum of forms of type $f\, d\lambda_i \wedge \cdots \wedge d\lambda_{i_r}$ and the formula gives an expression of d in terms of its effect on forms of lower degree. By induction, if the value of d on functions is known, its value can then be determined on forms of degree ≥ 1. This proves our assertion.

Proposition 3.5. *Let ω be a form of class C^2. Then $dd\omega = 0$.*

Proof. If f is a function, then

$$df(x) = \sum_{j=1}^{n} \frac{\partial f}{\partial x_j} \, dx_j$$

and

$$ddf(x) = \sum_{j=1}^{n} \sum_{k=1}^{n} \frac{\partial^2 f}{\partial x_k \partial x_j} \, dx_k \wedge dx_j.$$

Using the fact that the partials commute, and the fact that for any two positive integers r, s we have $dx_r \wedge dx_s = -dx_s \wedge dx_r$, we see that the preceding double sum is equal to 0. A similar argument shows that the theorem is true for 1-forms, of type $g(x) \, dx_i$ where g is a function, and thus for all 1-forms by linearity. We proceed by induction. It suffices to prove the formula in general for decomposable forms. Let ω be decomposable of degree r, and write

$$\omega = \eta \wedge \psi,$$

where $\deg \psi = 1$. Using the formula for the derivative of an alternating product twice, and the fact that $dd\psi = 0$ and $dd\eta = 0$ by induction, we see at once that $dd\omega = 0$, as was to be shown.

We conclude this section by giving some properties of the pull-back of forms. As we saw at the end of Chapter III, §4, if $f: X \to Y$ is a morphism and if ω is a differential form on Y, then we get a differential form $f^*(\omega)$ on X, which is given at a point $x \in X$ by the formula

$$f^*(\omega)_x = \omega_{f(x)} \circ (T_x f)^r,$$

if ω is of degree r. This holds for $r \geq 1$. The corresponding local representation formula reads

$$\langle f^*\omega(x), \xi_1(x) \times \cdots \times \xi_r(x) \rangle = \langle \omega(f(x)), f'(x)\xi_1(x) \times \cdots \times f'(x)\xi_r(x) \rangle$$

if $\xi_1, \ldots, \xi_r$ are vector fields.

In the case of a 0-form, that is a function, its pull-back is simply the composite function. In other words, if φ is a function on Y, viewed as a form of degree 0, then

$$\boxed{f^*(\varphi) = \varphi \circ f.}$$

It is clear that the pull-back is linear, and satisfies the following properties.

Property 1. *If ω, ψ are two differential forms on Y, then*

$$f^*(\omega \wedge \psi) = f^*(\omega) \wedge f^*(\psi).$$

Property 2. *If ω is a differential form on Y, then*

$$df^*(\omega) = f^*(d\omega).$$

Property 3. *If $f: X \to Y$ and $g: Y \to Z$ are two morphisms, and ω is a differential form on Z, then*

$$f^*(g^*(\omega)) = (g \circ f)^*(\omega).$$

Finally, in the case of forms of degree 0:

Property 4. *If $f: X \to Y$ is a morphism, and g is a function on Y, then*

$$d(g \circ f) = f^*(dg)$$

and at a point $x \in X$, the value of this 1-form is given by

$$T_{f(x)}g \circ T_x f = (dg)_x \circ T_x f.$$

The verifications are all easy, and even trivial, except possibly for **Property 2**. We shall give the proof of **Property 2** in the finite dimensional case and leave the general case to the reader.

For a form of degree 1, say

$$\omega(y) = g(y)\, dy_1,$$

with $y_1 = f_1(x)$, we find

$$(f^*\, d\omega)(x) = (g'(f(x)) \circ f'(x)) \wedge df_1(x).$$

Using the fact that $ddf_1 = 0$, together with Proposition 6 we get

$$(df^*\, \omega)(x) = (d(g \circ f))(x) \wedge df_1(x),$$

which is equal to the preceding expression. Any 1-form can be expressed as a linear combination of form $g_i\, dy_i$, so that our assertion is proved for forms of degree 1.

The general formula can now be proved by induction. Using the linearity of f^*, we may assume that ω is expressed as $\omega = \psi \wedge \eta$ where ψ, η have lower degree. We apply Proposition 3.3 and **Property 1** to

$$f^* d\omega = f^*(d\psi \wedge \eta) + (-1)^r f^*(\psi \wedge d\eta)$$

and we see at once that this is equal to $df^*\omega$, because by induction, $f^*d\psi = df^*\psi$ and $f^*d\eta = df^*\eta$. This proves **Property 2**.

Example 1. Let $y_1, \ldots, y_m$ be the coordinates on V, and let μ_j be the jth coordinate function, $j = 1, \ldots, m$, so that $y_j = \mu_j(y_1, \ldots, y_m)$. Let

$$f: U \to V$$

be the map with coordinate functions

$$y_j = f_j(x) = \mu_j \circ f(x).$$

If

$$\omega(y) = g(y)\, dy_{j_1} \wedge \cdots \wedge dy_{j_s}$$

is a differential form on V, then

$$\boxed{f^*\omega = (g \circ f)\, df_{j_1} \wedge \cdots \wedge df_{j_s}.}$$

Indeed, we have for $x \in U$:

$$(f^*\omega)(x) = g(f(x))(\mu_{j_1} \circ f'(x)) \wedge \cdots \wedge (\mu_{j_s} \circ f'(x))$$

and

$$f_j'(x) = (\mu_j \circ f)'(x) = \mu_j \circ f'(x) = df_j(x).$$

Example 2. Let $f: [a, b] \to \mathbf{R}^2$ be a map from an interval into the plane, and let x, y be the coordinates of the plane. Let t be the coordinate in $[a, b]$. A differential form in the plane can be written in the form

$$\omega(x, y) = g(x, y)\, dx + h(x, y)\, dy,$$

where g, h are functions. Then by definition,

$$f^*\omega(t) = g(x(t), y(t))\frac{dx}{dt}\, dt + h(x(t), y(t))\frac{dy}{dt}\, dt,$$

if we write $f(t) = (x(t), y(t))$. Let $G = (g, h)$ be the vector field whose components are g and h. Then we can write

$$f^*\omega(t) = G(f(t)) \cdot f'(t)\, dt,$$

which is essentially the expression which is integrated when defining the integral of a vector field along a curve.

Example 3. Let U, V be both open sets in n-space, and let $f: U \to V$ be a C^p map. If

$$\omega(y) = g(y)\, dy_1 \wedge \cdots \wedge dy_n,$$

where $y_j = f_j(x)$ is the j-th coordinate of y, then

$$dy_j = D_1 f_j(x)\, dx_1 + \cdots + D_n f_j(x)\, dx_n$$
$$= \frac{\partial y_j}{\partial x_1}\, dx_1 + \cdots + \frac{\partial y_j}{\partial x_n}\, dx_n,$$

and consequently, expanding out the alternating product according to the usual multilinear and alternating rules, we find that

$$f^*\omega(x) = g(f(x))\Delta_f(x)\, dx_1 \wedge \cdots \wedge dx_n,$$

where Δ_f is the determinant of the Jacobian matrix of f.

V, §4. THE POINCARÉ LEMMA

If ω is a differential form on a manifold and is such that $d\omega = 0$, then it is customary to say that ω is **closed**. If there exists a form ψ such that $\omega = d\psi$, then one says that ω is **exact**. We shall now prove that locally, every closed form is exact.

Theorem 4.1 (Poincaré Lemma). *Let U be an open ball in $\mathbf{E}$ and let ω be a differential form of degree ≥ 1 on U such that $d\omega = 0$. Then there exists a differential form ψ on U such that $d\psi = \omega$.*

Proof. We shall construct a linear map k from the r-forms to the $(r-1)$-forms $(r \geq 1)$ such that

$$dk + kd = \mathrm{id}.$$

From this relation, it will follow that whenever $d\omega = 0$, then

$$dk\omega = \omega,$$

thereby proving our proposition. We may assume that the center of the ball is the origin. If ω is an r-form, then we define $k\omega$ by the formula

$$\langle (k\omega)_x, v_1 \times \cdots \times v_{r-1} \rangle = \int_0^1 t^{r-1} \langle \omega(tx), x \times v_1 \times \cdots \times v_{r-1} \rangle\, dt.$$

We can assume that we deal with local representations and that $v_i \in \mathbf{E}$. We have

$$\langle (dk\omega)_x, v_1 \times \cdots \times v_r \rangle$$

$$= \sum_{i=1}^{r} (-1)^{i+1} \langle (k\omega)'(x)v_i, v_1 \times \cdots \times \hat{v}_i \times \cdots \times v_r \rangle$$

$$= \sum (-1)^{i+1} \int_0^1 t^{r-1} \langle \omega(tx), v_i \times v_1 \times \cdots \times \hat{v}_i \times \cdots \times v_r \rangle \, dt$$

$$+ \sum (-1)^{i+1} \int_0^1 t^r \langle \omega'(tx)v_i, x \times v_1 \times \cdots \times \hat{v}_i \times \cdots \times v_r \rangle \, dt.$$

On the other hand, we also have

$$\langle (kd\omega)(x), v_1 \times \cdots \times v_r \rangle$$

$$= \int_0^1 t^r \langle d\omega(x), x \times v_1 \times \cdots \times v_r \rangle \, dt$$

$$= \int_0^1 t^r \langle \omega'(tx)x, v_1 \times \cdots \times v_r \rangle \, dt$$

$$+ \sum (-1)^i \int_0^1 t^r \langle \omega'(tx)v_i, x \times v_1 \times \cdots \times \hat{v}_i \times \cdots \times v_r \rangle \, dt.$$

We observe that the second terms in the expressions for $kd\omega$ and $dk\omega$ occur with opposite signs and cancel when we take the sum. As to the first terms, if we shift v_i to the i-th place in the expression for $dk\omega$, then we get an extra coefficient of $(-1)^{i+1}$. Thus

$$dk\omega + kd\omega = \int_0^1 rt^{r-1} \langle \omega(tx), v_1 \times \cdots \times v_r \rangle \, dt$$

$$+ \int_0^1 t^r \langle \omega'(tx)x, v_1 \times \cdots \times v_r \rangle \, dt.$$

This last integral is simply the integral of the derivative with respect to t of

$$\langle t^r \omega(tx), v_1 \times \cdots \times v_r \rangle.$$

Evaluating this expression between $t = 0$ and $t = 1$ yields

$$\langle \omega(x), v_1 \times \cdots \times v_r \rangle$$

which proves the theorem.

We observe that we could have taken our open set U to be star-shaped instead of an open ball. For more information on the relationship between closed and exact forms, see Chapter XIII, §1.

V, §5. CONTRACTIONS AND LIE DERIVATIVE

Let ξ be a vector field and let ω be an r-form on a manifold X, $r \geq 1$. Then we can define an $(r-1)$-form $C_\xi \omega$ by the formula

$$(C_\xi \omega)(x)(v_2, \ldots, v_r) = \omega(\xi(x), v_2, \ldots, v_r),$$

for $v_2, \ldots, v_r \in T_x$. Using local representations shows at once that $C_\xi \omega$ has the appropriate order of differentiability (the minimum of ω and ξ). We call $C_\xi \omega$ the **contraction** of ω by ξ, and also denote $C_\xi \omega$ by

$$\omega \circ \xi.$$

If f is a function, we define $C_\xi f = 0$. Leaving out the order of differentiability, we see that contraction gives an **R**-linear map

$$C: \mathscr{A}^r(X) \to \mathscr{A}^{r-1}(X).$$

This operation of contraction satisfies the following properties.

CON 1. $C_\xi \circ C_\xi = 0.$

CON 2. *The association* $(\xi, \omega) \mapsto C_\xi \omega = \omega \circ \xi$ *is bilinear. It is in fact bilinear with respect to functions, that is if* φ *is a function, then*

$$C_{\varphi \xi} = \varphi C_\xi \quad and \quad C_\xi(\varphi \omega) = \varphi C_\xi \omega.$$

CON 3. *If* ω, ψ *are differential forms and* $r = \deg \omega$, *then*

$$C_\xi(\omega \wedge \psi) = (C_\xi \omega) \wedge \psi + (-1)^r \omega \wedge C_\xi \psi.$$

These three properties follow at once from the definitions.

Example. Let $X = \mathbf{R}^n$, and let

$$\omega(x) = dx_1 \wedge \cdots \wedge dx_n.$$

If ξ is a vector field on $\mathbf{R}^n$, then we have the local representation

$$(\omega \circ \xi)(x) = \sum_{i=1}^{n} (-1)^{i+1} \xi_i(x) \, dx_1 \wedge \cdots \wedge \widehat{dx_i} \wedge \cdots \wedge dx_n.$$

We also have immediately from the definition of the exterior derivative,

$$d(\omega \circ \xi) = \sum_{i=1}^{n} \frac{\partial \xi_i(x)}{\partial x_i} dx_1 \wedge \cdots \wedge dx_n,$$

letting $\xi = (\xi_1, \ldots, \xi_n)$ in terms of its components ξ_i.

We can define the **Lie derivative** of an r-form as we did before for vector fields. Namely, we shall evaluate the following limit:

$$(\mathscr{L}_\xi \omega)(x) = \lim_{t \to 0} \frac{1}{t} [(\alpha_t^* \omega)(x) - \omega(x)],$$

or in other words,

$$\mathscr{L}_\xi \omega = \frac{d}{dt} (\alpha_t^* \omega) \Big|_{t=0}$$

where α is the flow of the vector field ξ, and we call $\mathscr{L}_\xi$ the **Lie derivative** again, applied to the differential form ω. We may rewrite this definition in terms of the value on vector fields $\xi_1, \ldots, \xi_r$ as follows:

$$(\mathscr{L}_\xi \omega)(\xi_1, \ldots, \xi_r) = \frac{d}{dt} \langle \omega \circ \alpha_t, \alpha_{t*} \xi_1 \times \cdots \times \alpha_{t*} \xi_r \rangle \Big|_{t=0}$$

Proposition 5.1. *Let ξ be a vector field and ω a differential form of degree $r \geq 1$. The Lie derivative $\mathscr{L}_\xi$ is a derivation, in the sense that*

$$\mathscr{L}_\xi \big(\omega(\xi_1, \ldots, \xi_r)\big) = (\mathscr{L}_\xi \omega)(\xi_1, \ldots, \xi_r) + \sum_{i=1}^{r} \omega(\xi_1, \ldots, \mathscr{L}_\xi \xi_i, \ldots, \xi_r)$$

where of course $\mathscr{L}_\xi \xi_i = [\xi, \xi_i]$.

If ξ, ξ_i, ω denote the local representations of the vector fields and the form respectively, then the Lie derivative $\mathscr{L}_\xi \omega$ has the local representation

$$\langle (\mathscr{L}_\xi \omega)(x), \xi_1(x) \times \cdots \times \xi_r(x) \rangle$$
$$= \langle \omega'(x)\xi(x), \xi_1(x) \times \cdots \times \xi_r(x) \rangle$$
$$+ \sum_{i=1}^{r} \langle \omega(x), \xi_1(x) \times \cdots \times \xi'(x)\xi_i(x) \times \cdots \times \xi_r(x) \rangle.$$

Proof. The proof is routine using the definitions. The first assertion is obvious by the definition of the pull back of a form. For the local expression we actually derive more, namely we derive a local expression for $\alpha_t^*\omega$ and $\dfrac{d}{dt}\alpha_t^*\omega$ which are characterized by their values at $(\xi_1, \ldots, \xi_r)$. So we let

(1) $F(t) = \langle (\alpha_t^*\omega)(x), \xi_1(x) \times \cdots \times \xi_r(x) \rangle$

$\qquad\qquad = \langle \omega(\alpha(t, x)), D_2\alpha(t, x)\xi_1(x) \times \cdots \times D_2\alpha(t, x)\xi_r(x) \rangle.$

Then the Lie derivative $(\mathscr{L}_\xi\omega)(x)$ is precisely $F'(0)$, but we obtain also the local representation for $\dfrac{d}{dt}\alpha_t^*\omega$:

(2) $F'(t) = \left\langle \dfrac{d}{dt}\alpha_t^*\omega(x), \xi_1(x) \times \cdots \times \xi_r(x) \right\rangle =$

(3) $\langle \omega'(\alpha(t, x))D_1\alpha(t, x), D_2\alpha(t, x)\xi_1(x) \times \cdots \times D_2\alpha(t, x)\xi_r(x) \rangle$

$\displaystyle + \sum_{i=1}^{r} \langle \omega(\alpha(t, x)), D_2\alpha(t, x)\xi_1(x) \times \cdots \times D_1 D_2\alpha(t, x)\xi_i(x) \times \cdots \times D_2\alpha(t, x)\xi_r(x) \rangle$

by the rule for the derivative of a product. Putting $t = 0$ and using the differential equation satisfied by $D_2\alpha(t, x)$, we get precisely the local expression as stated in the proposition. Remember the initial condition $D_2\alpha(0, x) = \mathrm{id}$.

From Proposition 5.1, we conclude that the Lie derivative gives an **R**-linear map

$$\mathscr{L}_\xi: \mathscr{A}^r(X) \to \mathscr{A}^r(X).$$

We may use expressions (1) and (3) in the above proof to derive a formula which holds even more generally for time-dependent vector fields.

Proposition 5.2. *Let ξ_t be a time-dependent vector field, α its flow, and let ω be a differential form. Then*

$$\frac{d}{dt}(\alpha_t^*\omega) = \alpha_t^*(\mathscr{L}_{\xi_t}\omega) \qquad or \qquad \frac{d}{dt}(\alpha_t^*\omega) = \alpha_t^*(\mathscr{L}_\xi\omega)$$

for a time-independent vector field.

Proof. Proposition 5.1 gives us a local expression for $(\mathscr{L}_{\xi_t}\omega)(y)$, replacing x by y because we shall now put $y = \alpha(t, x)$. On the other hand, from (1) in the proof of Proposition 5.1, we obtain

$$\alpha_t^*(\mathscr{L}_{\xi_t}\omega)(x) = \langle (\mathscr{L}_{\xi_t}\omega)(y), D_2\alpha(t, x)\xi_1(x) \times \cdots \times D_2\alpha(t, x)\xi_r(x) \rangle.$$

Substituting the local expression for $(\mathscr{L}_{\xi_t}\omega)(y)$, we get expression (3) from the proof of Proposition 5.1, thereby proving Proposition 5.2.

Proposition 5.3. *As a map on differential forms, the Lie derivative satisfies the following properties.*

LIE 1. $\mathscr{L}_\xi = d \circ C_\xi + C_\xi \circ d$, *so* $\mathscr{L}_\xi = C_\xi \circ d$ *on functions.*

LIE 2. $\mathscr{L}_\xi(\omega \wedge \psi) = \mathscr{L}_\xi\omega \wedge \psi + \omega \wedge \mathscr{L}_\xi\psi$.

LIE 3. $\mathscr{L}_\xi$ *commutes with* d *and* C_ξ.

LIE 4. $\mathscr{L}_{[\xi,\eta]} = \mathscr{L}_\xi \circ \mathscr{L}_\eta - \mathscr{L}_\eta \circ \mathscr{L}_\xi$.

LIE 5. $C_{[\xi,\eta]} = \mathscr{L}_\xi \circ C_\eta - C_\eta \circ \mathscr{L}_\xi$.

LIE 6. $\mathscr{L}_{f\xi}\omega = f\mathscr{L}_\xi\omega + df \wedge C_\xi\omega$ *for all forms ω and functions f.*

Proof. Let $\xi_1, \ldots, \xi_r$ be vector fields, and ω an r-form. Using the definition of the contraction and the local formula of Proposition 5.1, we find that $C_\xi d\omega$ is given locally by

$$\langle C_\xi d\omega(x), \xi_1(x) \times \cdots \times \xi_r(x)\rangle$$

$$= \langle \omega'(x)\xi(x), \xi_1(x) \times \cdots \times \xi_r(x)\rangle$$

$$+ \sum_{i=1}^{r} (-1)^i \langle \omega'(x)\xi_i(x), \xi(x) \times \xi_1(x) \times \cdots \times \widehat{\xi_i(x)} \times \cdots \times \xi_r(x)\rangle.$$

On the other hand, $dC_\xi\omega$ is given by

$$\langle dC_\xi\omega(x), \xi_1(x) \times \cdots \times \xi_r(x)\rangle$$

$$= \sum_{i=1}^{r} (-1)^{i+1} \langle (C_\xi\omega)'(x)\xi_i(x), \xi_1(x) \times \cdots \times \widehat{\xi_i(x)} \times \cdots \times \xi_r(x)\rangle.$$

To compute $(C_\xi\omega)'(x)$ is easy, going back to the definition of the derivative. At vectors $v_1, \ldots, v_{r-1}$, the form $C_\xi\omega(x)$ has the value

$$\langle \omega(x), \xi(x) \times v_1 \times \cdots \times v_{r-1}\rangle.$$

Differentiating this last expression with respect to x and evaluating at a vector h we get

$$\langle \omega'(x)h, \xi(x) \times v_1 \times \cdots \times v_{r-1}\rangle + \langle \omega(x), \xi'(x)h \times v_1 \times \cdots \times v_{r-1}\rangle.$$

Hence $\langle dC_\xi\omega(x), \xi_1(x) \times \cdots \times \xi_r(x)\rangle$ is equal to

$$\sum_{i=1}^{r} (-1)^{i+1} \langle \omega'(x)\xi_i(x), \xi(x) \times \xi_1(x) \times \cdots \times \widehat{\xi_i(x)} \times \cdots \times \xi_r(x)\rangle$$

$$+ \sum_{i=1}^{r} (-1)^{i+1} \langle \omega(x), \xi'(x)\xi_i(x) \times \xi_1(x) \times \cdots \times \widehat{\xi_i(x)} \times \cdots \times \xi_r(x)\rangle.$$

Shifting $\zeta'(x)\xi_i(x)$ to the i-th place in the second sum contributes a sign of $(-1)^{i-1}$ which gives 1 when multiplied by $(-1)^{i+1}$. Adding the two local representations for $dC_\xi\omega$ and $C_\xi\,d\omega$, we find precisely the expression of Proposition 5.1, thus proving **LIE 1**.

As for **LIE 2**, it consists in using the derivation rule for d and C_ξ in Proposition 3.3, **EXD 1**, and **CON 3**. The corresponding rule for $\mathscr{L}_\xi$ follows at once. (Terms will cancel just the right way.) The other properties are then clear.

V, §6. VECTOR FIELDS AND 1-FORMS UNDER SELF DUALITY

Let **E** be a Banach space and let

$$(v, w) \mapsto \langle v, w \rangle$$

be a continuous bilinear function of $\mathbf{E} \times \mathbf{E} \to \mathbf{R}$. We call such a function a **bilinear form**. This form induced a linear map

$$\lambda: \mathbf{E} \to \mathbf{E}^\vee$$

which to each $v \in \mathbf{E}$ associates the functional λ_v such that

$$\lambda_v(w) = \langle v, w \rangle.$$

We have a similar map on the other side. If both these mappings are toplinear isomorphisms of $\mathbf{E}$ and $\mathbf{E}^\vee$ then we say that the bilinear form is **non-singular**. If such a non-singular form exists, then we say that $\mathbf{E}$ is **self-dual**. For instance, a Hilbert space is self-dual.

If **E** is finite dimensional, it suffices for a bilinear form to be non-singular that its kernels on the right and on the left be 0. (The kernels are the kernels of the associated maps λ as above.) However, in the infinite dimensional case, this condition on the kernels is not sufficient any more.

Let **E** be a self dual Banach space with respect to the non-singular form $(v, w) \mapsto \langle v, w \rangle$, and let

$$\Omega: \mathbf{E} \times \mathbf{E} \to \mathbf{R}$$

be a continuous bilinear map. There exists a unique operator A such that
$$\Omega(v, w) = \langle Av, w \rangle$$

for all $v, w \in \mathbf{E}$. (An **operator** is a continuous linear map by definition.)

Remarks. *Suppose that the form* $(v, w) \mapsto \langle v, w \rangle$ *is symmetric,* i.e.

$$\langle v, w \rangle = \langle w, v \rangle$$

for all v, $w \in \mathbf{E}$. Then Ω is **symmetric** (resp. **alternating**) if and only if A is symmetric (resp. skew-symmetric). Recall that A symmetric (with respect to $\langle \, , \, \rangle$) means that

$$\langle Av, w \rangle = \langle v, Aw \rangle \qquad \text{for all} \quad v, w \in \mathbf{E}.$$

That A is skew-symmetric means that $\langle Av, w \rangle = -\langle Aw, v \rangle$ for all v, $w \in \mathbf{E}$. For any operator $A: \mathbf{E} \to \mathbf{E}$ there is another operator ${}^t A$ (the transpose of A with respect to the non-singular form $\langle \, , \, \rangle$) such that for all v, $w \in \mathbf{E}$ we have

$$\langle Av, w \rangle = \langle v, {}^t Aw \rangle.$$

Thus A is symmetric (resp. skew-symmetric) if and only if ${}^t A = A$ (resp. ${}^t A = -A$).

The above remarks apply to any continuous bilinear form Ω. For invertibility, we have the criterion:

The form Ω is non-singular if and only if the operator A representing the form with respect to $\langle \, , \, \rangle$ is invertible.

The easy verification is left to the reader. Of course, in the finite dimensional case, invertibility or non-singularity can be checked by verifying that the matrix representing the linear map with respect to bases has non-zero determinant. Similarly, the form is also represented by a matrix with respect to a choice of bases, and its being non-singular is equivalent to the matrix representing the form being invertible.

We recall that the set of invertible operators in Laut($\mathbf{E}$) is an open subset. Alternatively, the set of non-singular bilinear forms on $\mathbf{E}$ is an open subset of $L^2(\mathbf{E})$.

We may now globalize these notions to a vector bundle (and eventually especially to the tangent bundle) as follows.

Let X be a manifold, and $\pi: E \to X$ a vector bundle over X with fibers which are toplinearly isomorphic to $\mathbf{E}$, or as we shall also say, **modeled** on $\mathbf{E}$. Let Ω be a tensor field of type L^2 on E, that is to say, a section of the bundle $L^2(E)$ (or $L^2(\pi)$), or as we shall also say, a **bilinear tensor field** on E. Then for each $x \in X$, we have a continuous bilinear form Ω_x on E_x.

If Ω_x is non-singular for each $x \in X$ then we say that Ω is **non-singular**. If π is trivial, and we have a trivialisation $X \times E$, then the local representation of Ω can be described by a morphism of X into the Banach space of operators. If Ω is non-singular, then the image of this morphism is contained in the open set of invertible operators. (If Ω is a 2-form, this image is contained in the submanifold of skew-symmetric operators.) For example, in a chart U, we can represent Ω over U by a morphism

$$A: U \to L(\mathbf{E}, \mathbf{E}) \qquad \text{such that} \qquad \Omega_x(v, w) = \langle A_x v, w \rangle$$

for all v, $w \in \mathbf{E}$. Here we wrote A_x instead of $A(x)$ to simplify the typography.

A non-singular Ω as above can be used to establish a linear isomorphism

$$\Gamma(E) \to \Gamma L^1(E), \qquad \text{also denoted by} \quad \Gamma L(E) \text{ or } \Gamma E^\vee,$$

between the **R**-vector spaces of sections $\Gamma(E)$ of E and the 1-forms on E in the following manner. Let ξ be a section of E. For each $x \in X$ we define a continuous linear map

$$(\Omega \circ \xi)_x: E_x \to \mathbf{R}$$

by the formula

$$(\Omega \circ \xi)_x(w) = \Omega_x(\xi(x), w).$$

Looking at local trivialisations of π, we see at once that $\Omega \circ \xi$ is a 1-form on E.

Conversely, let ω be a given 1-form on E. For each $x \in X$, ω_x is therefore a 1-form on E_x and since Ω is non-singular, there exists a unique element $\xi(x)$ of E_x such that

$$\Omega_x(\xi(x), \omega) = \omega_x(w)$$

for all $w \in E_x$. In this fashion, we obtain a mapping ξ of x into E and we contend that ξ is a morphism (and therefore a section).

To prove our contention we can look at the local representations. We use Ω and ω to denote these. They are represented over a suitable open set U by two morphisms

$$A: U \to \operatorname{Aut}(\mathbf{E}) \qquad \text{and} \qquad \eta: U \to \mathbf{E}$$

such that

$$\Omega_x(v, w) = \langle A_x v, w \rangle \qquad \text{and} \qquad \omega_x(w) = \langle \eta(x), w \rangle.$$

From this we see that

$$\xi(x) = A_x^{-1} \eta(x),$$

from which it is clear that ξ is a morphism. We may summarize our discussion as follows.

Proposition 6.1. *Let X be a manifold and $\pi: E \to X$ a vector bundle over X modeled on $\mathbf{E}$. Let Ω be a non-singular bilinear tensor field on $\mathbf{E}$. Then Ω induces an isomorphism of $\mathfrak{F}(X)$-modules*

$$\Gamma E \to \Gamma E^{\vee}.$$

A section ξ corresponds to a 1-form ω if and only if $\Omega \circ \xi = \omega$.

In many applications, one takes the differential form to be df for some function f. The vector field corresponding to df is then called the **gradient of f with respect to Ω**.

Remark. There is no universally accepted notation to denote the correspondence between a 1-form and a vector field under Ω as above. Some authors use sharps and flats, which have two disadvantages. First, they do not provide a symbols for the mapping, and second they do not contain the Ω in the notation. I would propose the check sign $\bigvee_{\Omega}$ to denote either isomorphism

$$\bigvee_{\Omega}: \Gamma L(E) \to \Gamma E \qquad \text{denoted on elements by} \qquad \omega \mapsto \bigvee_{\Omega} \omega = \omega^{\vee} = \xi_{\omega}$$

and also

$$\bigvee_{\Omega}: \Gamma E \to \Gamma L(E) \qquad \text{denoted on elements by} \qquad \xi \mapsto \bigvee_{\Omega} \xi = \xi^{\vee} = \omega_{\xi}.$$

If Ω is fixed throughout a discussion and need not be referred to, then it is useful to write $\xi^{\vee}$ or $\lambda^{\vee}$ in some formulas. We have $\bigvee_{\Omega} \circ \bigvee_{\Omega} = \mathrm{id}$. Instead of the sharp and flat superscript, I prefer the single $^{\vee}$ sign.

Many important applications of the above duality occur when Ω is a non-singular symmetric bilinear tensor field on the tangent bundle TX. Such a tensor field is then usually denoted by g. If ξ, η are vector fields, we may then define their scalar product to be the function

$$\langle \xi, \eta \rangle_g = g(\xi, \eta).$$

On the other hand, by the duality of Proposition 6.1, if i.e. ω, λ are 1-forms, i.e. sections of the dual bundle $T^{\vee}X$, then $\omega^{\vee}$ and $\lambda^{\vee}$ are vector fields, and we define the scalar product of the 1-forms to be

$$\langle \omega, \lambda \rangle_g = \langle \omega^{\vee}, \lambda^{\vee} \rangle_g.$$

This duality is especially important for Riemannian metrics, as in Chapter X.

The rest of this section will not be used in the book.

In Proposition 6.1, we dealt with a quite general non-singular bilinear tensor field on E. We now specialize to the case when $E = TX$ is the tangent bundle of X, and Ω is a 2-form, i.e. Ω is alternating. A pair (X, Ω) consisting of a manifold and a non-singular closed 2-form is called a **symplectic manifold**. (Recall that **closed** means $d\Omega = 0$.)

We denote by ξ, η vector fields over X, and by f, h functions on X, so that df, dh are 1-forms. We let ξ_{df} be the vector field on X which corresponds to df under the 2-forms Ω, according to Proposition 6.1. Vector fields on X which are of type $\xi \, df$ are called **Hamiltonian** (with respect to the 2-form). More generally, we denote by ξ_ω the vector field corresponding to a 1-form ω. By definition we have the formula

Ω 1. $\Omega \circ \xi_\omega = \omega$ so in particular $\Omega \circ \xi_{df} = df.$

In Chapter VII, §6 we shall consider a particularly important example, when the base manifold is the cotangent bundle; the function is the **kinetic energy**

$$K(v) = \tfrac{1}{2}\langle v, v\rangle_g$$

with respect to the scalar product g of a Riemannian or pseudo Riemannian metric, and the 2-form Ω arises canonically from the pseudo Riemannian metric.

In general, by **LIE 1** of Proposition 5.3 formula **Ω 1**, and the fact that $d\Omega = 0$, we find for any 1-form ω that:

Ω 2. $\mathscr{L}_{\xi_\omega}\Omega = d\omega.$

The next proposition reinterprets this formula in terms of the flow when $d\omega = 0$.

Proposition 6.2. *Let ω be such that $d\omega = 0$. Let α be the flow of ξ_ω. Then $\alpha_t^*\Omega = \Omega$ for all t (in the domain of the flow).*

Proof. By Proposition 5.2,

$$\frac{d}{dt}\alpha_t^*\Omega = \alpha_t^*\mathscr{L}_{\xi_\omega}\Omega = 0 \quad \text{by } \Omega \text{ 2.}$$

Hence $\alpha_t^*\Omega$ is constant, equal to $\alpha_0^*\Omega = \Omega$, as was to be shown.

A special case of Proposition 6.2 in Hamiltonian mechanics is when $\omega = dh$ for some function h. Next by **LIE 5**, we obtain for any vector fields ξ, η:

$$\mathscr{L}_\xi(\Omega \circ \eta) = (\mathscr{L}_\xi\Omega) \circ \eta + \Omega \circ [\xi, \eta].$$

In particular, since $ddf = 0$, we get

Ω 3. $\mathscr{L}_{\xi_{df}}(\Omega \circ \xi_{dh}) = \Omega \circ [\xi_{df}, \xi_{dh}]$.

One defines the **Poisson bracket** between two functions f, h to be

$$\{f, h\} = \xi_{df} \cdot h.$$

Then the preceding formula may be rewritten in the form

Ω 4. $[\xi_{df}, \xi_{dh}] = \xi_{d\{f,h\}}$.

It follows immediately from the definitions and the antisymmetry of the ordinary bracket between vector fields that the Poisson bracket is also antisymmetric, namely

$$\{f, h\} = -\{h, f\}.$$

In particular, we find that

$$\xi_{df} \cdot f = 0.$$

In the case of the cotangent bundle with a symplectic 2-form as in the next section, physicists think of f as an energy function, and interpret this formula as a law of conservation of energy. The formula expresses the property that f is constant on the integral curves of the vector field ξ_{df}. This property follows at once from the definition of the Lie derivative of a function. Furthermore:

Proposition 6.3. *If* $\xi_{df} \cdot h = 0$ *then* $\xi_{dh} \cdot f = 0$.

This is immediate from the antisymmetry of the Poisson bracket. It is interpreted as conservation of momentum in the physical theory of Hamiltonian mechanics, when one deals with the canonical 2-form on the cotangent bundle, to be defined in the next section.

V, §7. THE CANONICAL 2-FORM

Consider the functor $\mathbf{E} \mapsto L(\mathbf{E})$ (continuous linear forms). If $E \to X$ is a vector bundle, then $L(E)$ will be called the **dual bundle**, and will be denoted by $\mathbf{E}^\vee$. For each $x \in X$, the fiber of the dual bundle is simply $L(E_x)$.

If $E = T(X)$ is the tangent bundle, then its dual is denoted by $T^\vee(X)$ and is called the **cotangent bundle**. Its elements are called **cotangent**

vectors. The fiber of $T^\vee(X)$ over a point x of X is denoted by $T_x^\vee(X)$. For each $x \in X$ we have a pairing

$$T_x^\vee \times T_x \to \mathbf{R}$$

given by

$$\langle \lambda, u \rangle = \lambda(u)$$

for $\lambda \in T_x^\vee$ and $u \in T_x$ (it is the value of the linear form λ at u).

We shall now describe how to construct a canonical 1-form on the cotangent bundle $T^\vee(X)$. For each $\lambda \in T^\vee(X)$ we must define a 1-form on $T_v(T^\vee(X))$.

Let $\pi: T^\vee(X) \to X$ be the canonical projection. Then the induced tangent map

$$T\pi = \pi_*: T(T^\vee(X)) \to T(X)$$

can be applied to an element z of $T_v(T^\vee(X))$ and one sees at once that $\pi_* z$ lies in $T_x(X)$ if λ lies in $T_x^\vee(X)$. Thus we can take the pairing

$$\langle \lambda, \pi_* z \rangle = \theta_\lambda(z)$$

to define a map (which is obviously continuous linear):

$$\theta_\lambda: T_\lambda(T^\vee(X)) \to \mathbf{R}.$$

Proposition 7.1. *This map defines a 1-form on $T^\vee(X)$. Let $X = U$ be open in* $\mathbf{E}$ *and*

$$T^\vee(U) = U \times \mathbf{E}^\vee, \qquad T(T^\vee(U)) = (U \times \mathbf{E}^\vee) \times (\mathbf{E} \times \mathbf{E}^\vee).$$

If $(x, \lambda) \in U \times \mathbf{E}^\vee$ and $(u, \omega) \in \mathbf{E} \times \mathbf{E}^\vee$, then the local representation $\omega_{(x,v)}$ is given by

$$\langle \theta_{(x,\lambda)}, (u, w) \rangle = \lambda(u).$$

Proof. We observe that the projection $\pi: U \times \mathbf{E}^\vee \to U$ is linear, and hence that its derivative at each point is constant, equal to the projection on the first factor. Our formula is then an immediate consequence of the definition. The local formula shows that θ is in fact a 1-form locally, and therefore globally since it has an invariant description.

Our 1-form is called the **canonical 1-form on the cotangent bundle**. We define the **canonical 2-form** Ω on the cotangent bundle $T^\vee X$ to be

$$\Omega = -d\theta.$$

The next proposition gives a local description of Ω.

Proposition 7.2. *Let U be open in $\mathbf{E}$, and let Ω be the local representation of the canonical 2-form on $T^{\vee}U = U \times \mathbf{E}^{\vee}$. Let $(x, \lambda) \in U \times \mathbf{E}^{\vee}$. Let (u_1, w_1) and (u_2, w_2) be elements of $\mathbf{E} \times \mathbf{E}^{\vee}$. Then*

$$\langle \Omega_{(x, \lambda)}, (u_1, w_1) \times (u_2, w_2) \rangle = \langle u_1, w_2 \rangle - \langle u_2, w_1 \rangle$$
$$= w_2(u_1) - w_1(u_2).$$

Proof. We observe that θ is linear, and thus that θ' is constant. We then apply the local formula for the exterior derivative, given in Proposition 3.2. Our assertion becomes obvious.

The canonical 2-form plays a fundamental role in Lagrangian and Hamiltonian mechanics, cf. [AbM 78], Chapter 3, §3. I have taken the sign of the canonical 2-form both so that its value is a 2×2 determinant, and so that it fits with, for instance, [LoS 68] and [AbM 78]. We observe that Ω is closed, that is $d\Omega = 0$, because $\Omega = -d\theta$. Thus $(T^{\vee}X, \Omega)$ is a symplectic manifold, to which the properties listed at the end of the last section apply.

In particular, let ξ be a vector field on X. Then to ξ is **associated a function** called the **momentum function**

$$f_{\xi}: T^{\vee}X \to \mathbf{R} \qquad \text{such that} \qquad f_{\xi}(\lambda_x) = \lambda_x(\xi(x))$$

for $\lambda_x \in T_x^{\vee}X$. Then df_{ξ} is a 1-form on $T^{\vee}X$. Classical Hamiltonian mechanics then applies Propositions 6.2 and 6.3 to this situation. We refer the interested reader to [LoS 68] and [AbM 78] for further information on this topic. For an important theorem of Marsden–Weinstein [MaW 74] and applications to vector bundles, see [Ko 87].

V, §8. DARBOUX'S THEOREM

If $\mathbf{E} = \mathbf{R}^n$ then the usual scalar product establishes the self-duality of $\mathbf{R}^n$. This self-duality arises from other forms, and in this section we are especially interested in the self-duality arising from alternating forms. If $\mathbf{E}$ is finite dimensional and ω is an element of $L_a^2(\mathbf{E})$, that is an alternating 2-form, which is non-singular, then one sees easily that the dimension of $\mathbf{E}$ is even.

Example. An example of such a form on $\mathbf{R}^{2n}$ is the following. Let

$$v = (v_1, \ldots, v_n, v'_1, \ldots, v'_n),$$
$$w = (w_1, \ldots, w_n, w'_1, \ldots, w'_n),$$

be elements of $\mathbf{R}^{2n}$, with components v_i, v_i', w_i, w_i'. Letting

$$\omega(v, w) = \sum_{i=1}^{n} (v_i w_i' - v_i' w_i)$$

defines a non-singular 2-form ω on $\mathbf{R}^{2n}$. It is an exercise of linear algebra to prove that any non-singular 2-form on $\mathbf{R}^{2n}$ is linearly isomorphic to this particular one in the following sense. If

$$f: E \to F$$

is a linear isomorphism between two finite dimensional spaces, then it induces an isomorphism

$$f^*: L_a^2(F) \to L_a^2(E).$$

We call forms ω on E and ψ on F **linearly isomorphic** if there exists a linear isomorphism f such that $f^*\psi = \omega$. Thus up to a linear isomorphism, there is only one non-singular 2-form on $\mathbf{R}^{2n}$. (For a proof, cf. for instance my book *Algebra*.)

We are interested in the same question on a manifold locally. Let U be open in the Banach space $\mathbf{E}$ and let $x_0 \in U$. A 2-form

$$\omega: U \to L_a^2(\mathbf{E})$$

is said to be **non-singular** if each form $\omega(x)$ is non-singular. If ξ is a vector field on U, then $\omega \circ \xi$ is a 1-form, whose value at (x, ω) is given

$$(\omega \circ \xi)(x)(w) = \omega(x)(\xi(x), w).$$

As a special case of Proposition 6.1, we have:

Let ω be a non-singular 2-form on an open set U in $\mathbf{E}$. The association

$$\xi \mapsto \omega \circ \xi$$

is a linear isomorphism between the space of vector fields on U and the space of 1-forms on U.

Let

$$\omega: U \to L_a^2(U)$$

be a 2-form on an open set U in $\mathbf{E}$. If there exists a local isomorphism f at a point $x_0 \in U$, say

$$f: U_1 \to V_1,$$

and a 2-form ψ on V_1 such that $f^*\psi = \omega$ (or more accurately, ω restricted to U_1), then we say that ω is **locally isomorphic** to ψ at x_0. Observe that in the case of an isomorphism we can take a direct image of forms, and we shall also write

$$f_*\omega = \psi$$

instead of $\omega = f^*\psi$. In other words, $f_* = (f^{-1})^*$.

Example. On $\mathbf{R}^{2n}$ we have the constant form of the previous example. In terms of local coordinates $(x_1, \ldots, x_n, y_1, \ldots, y_n)$, this form has the local expression

$$\omega(x, y) = \sum_{i=1}^{n} dx_i \wedge dy_i.$$

This 2-form will be called the **standard 2-form** on $\mathbf{R}^{2n}$.

The Darboux theorem states that any non-singular closed 2-form in $\mathbf{R}^{2n}$ is locally isomorphic to the standard form, that is that in a suitable chart at a point, it has the standard expression of the above example. A technique to show that certain forms are isomorphic was used by Moser [Mo 65], who pointed out that his arguments also prove the classical Darboux theorem. Moser's theorem will be given in Chapter XI, §6.

Alan Weinstein observed that Moser's proof applies to the infinite dimensional case, whose statement is as follows.

Theorem 8.1 (Darboux Theorem). *Let* $\mathbf{E}$ *be a self-dual Banach space. Let*

$$\omega: U \to L_a^2(\mathbf{E})$$

be a non-singular closed 2-form on an open set of $\mathbf{E}$, *and let* $x_0 \in U$. *Then* ω *is locally isomorphic at* x_0 *to the constant form* $\omega(x_0)$.

Proof. Let $\omega_0 = \omega(x_0)$, and let

$$\omega_t = \omega_0 + t(\omega - \omega_0), \qquad 0 \leqq t \leqq 1.$$

We wish to find a time-dependent vector field ξ_t locally at 0 such that if α denotes its flow, then

$$\alpha_t^*\omega_t = \omega_0.$$

Then the local isomorphism α_1 satisfies the requirements of the theorem. By the Poincaré lemma, there exists a 1-form θ locally at 0 such that

$$\omega - \omega_0 = d\theta,$$

and without loss of generality, we may assume that $\theta(x_0) = 0$. We contend that the time-dependent vector field ξ_t, such that

$$\omega_t \circ \xi_t = -\theta,$$

has the desired property. Let α be its flow. If we shrink the domain of the vector field near x_0 sufficiently, and use the fact that $\theta(x_0) = 0$, then we can use the local existence theorem (Proposition 1.1 of Chapter IV) to see that the flow can be integrated at least to $t = 1$ for all points x in this small domain. We shall now verify that

$$\frac{d}{dt}(\alpha_t^* \omega_t) = 0.$$

This will prove that $\alpha_t^* \omega_t$ is constant. Since we have $\alpha_0^* \omega_0 = \omega_0$ because

$$\alpha(0, x) = x \qquad \text{and} \qquad D_2 \alpha(0, x) = \text{id},$$

it will conclude the proof of the theorem.

We compute locally. We use the local formula of Proposition 5.2, and formula **LIE 1**, which reduces to

$$\mathscr{L}_{\xi_t} \omega_t = d(\omega_t \circ \xi_t),$$

because $d\omega_t = 0$. We find

$$\frac{d}{dt}(\alpha_t^* \omega_t) = \alpha_t^* \left(\frac{d}{dt} \omega_t \right) + \alpha_t^* (\mathscr{L}_{\xi_t} \omega_t)$$

$$= \alpha_t^* \left(\frac{d}{dt} \omega_t + d(\omega_t \circ \xi_t) \right)$$

$$= \alpha_t^* (\omega - \omega_0 - d\theta)$$

$$= 0.$$

This proves Darboux's theorem.

Remark 1. For the analogous uniqueness statement in the case of a non-singular symmetric form, see the Morse–Palais lemma of Chapter VII, §5. Compare also with Theorem 2.2 of Chapter XIII.

Remark 2. The proof of the Poincaré lemma can also be cast in the above style. For instance, let $\phi_t(x) = tx$ be a retraction of a star shaped open set around 0. Let ξ_t be the vector field whose flow is ϕ_t, and let ω

be a closed form. Then

$$\frac{d}{dt}\phi_t^*\omega = \phi_t^*\mathcal{L}_{\xi_t}\omega = \phi_t^*\, dC_{\xi_t}\,\omega = d\phi_t^*\, C_{\xi_t}\omega.$$

Since $\phi_0^*\omega = 0$ and ϕ_1 is the identity, we see that

$$\omega = \phi_1^*\omega - \phi_0^*\omega = \int_0^1 \frac{d}{dt}\phi_t^*\omega\, dt = d\int_0^1 \phi_t^* C_{\xi_t}\omega\, dt$$

is exact, thus concluding a proof of Poincaré's theorem.

CHAPTER VI

The Theorem of Frobenius

Having acquired the language of vector fields, we return to differential equations and give a generalization of the local existence theorem known as the Frobenius theorem, whose proof will be reduced to the standard case discussed in Chapter IV. We state the theorem in §1. Readers should note that one needs only to know the definition of the bracket of two vector fields in order to understand the proof. It is convenient to insert also a formulation in terms of differential forms, for which the reader needs to know the local definition of the exterior derivative. However, the condition involving differential forms is proved to be equivalent to the vector field condition at the very beginning, and does not reappear explicitly afterwards.

We shall follow essentially the proof given by Dieudonné in his *Foundations of Modern Analysis*, allowing for the fact that we use freely the geometric language of vector bundles, which is easier to grasp.

It is convenient to recall in §2 the statements concerning the existence theorems for differential equations depending on parameters. The proof of the Frobenius theorem proper is given in §3. An important application to Lie groups is given in §5, after formulating the theorem globally.

The present chapter will not be used in the rest of this book.

VI, §1. STATEMENT OF THE THEOREM

Let X be a manifold of class C^p ($p \geq 2$). A subbundle E of its tangent bundle will also be called a **tangent subbundle** over X. We contend that the following two conditions concerning such a subbundle are equivalent.

FR 1. *For each point $z \in X$ and vector fields ξ, η at z (i.e. defined on an open neighborhood of z) which lie in E (i.e. such that the image of each point of X under ξ, η lies in E), the bracket $[\xi, \eta]$ also lies in E.*

FR 2. *For each point $z \in X$ and differential form ω of degree 1 at z which vanishes on E, the form $d\omega$ vanishes on $\xi \times \eta$ whenever ξ, η are two vector fields at z which lie in E.*

The equivalence is essentially a triviality. Indeed, assume **FR 1**. Let ω vanish on E. Then

$$\langle d\omega, \xi \times \eta \rangle = -\langle \omega, [\xi, \eta] \rangle - \eta \langle \omega, \xi \rangle + \xi \langle \omega, \eta \rangle.$$

By assumption the right-hand side is 0 when evaluated at z. Conversely, assume **FR 2**. Let ξ, η be two vector fields at z lying in E. If $[\xi, \eta](z)$ is not in E, then we see immediately from a local product representation and the Hahn-Banach theorem that there exists a differential form ω of degree 1 defined on a neighborhood of z which is 0 on E_z and non-zero on $[\xi, \eta](z)$, thereby contradicting the above formula.

We shall now give a third condition equivalent to the above two, and actually, we shall not refer to **FR 2** any more. We remark merely that in the finite dimensional case, it is easy to prove that when a differential form ω satisfies condition **FR 2**, then $d\omega$ can be expressed locally in a neighborhood of each point as a finite sum

$$d\omega = \sum \gamma_i \wedge \omega_i$$

where γ_i and ω_i are of degree 1 and each ω_i vanishes on E. We leave this as an exercise to the reader.

Let E be a tangent subbundle over X. We shall say that E is **integrable** at a point x_0 if there exists a submanifold Y of X containing x_0 such that the tangent map of the inclusion

$$j: Y \to X$$

induces a VB-isomorphism of TY with the subbundle E restricted to Y. Equivalently, we could say that for each point $y \in Y$, the tangent map

$$T_y j: T_y Y \to T_y X$$

induces a toplinear isomorphism of $T_y Y$ on E_y. Note that our condition defining integrability is local at x_0. We say that E is **integrable** if it is integrable at every point.

Using the functoriality of vector fields, and their relations under tangent maps and the bracket product, we see at once that if E is integrable, then it satisfies **FR 1**. Indeed, locally, vector fields having their values in E are related to vector fields over Y under the inclusion mapping.

Frobenius' theorem asserts the converse.

Theorem 1.1. *Let X be a manifold of class C^p ($p \geq 2$) and let E be a tangent subbundle over X. Then E is integrable if and only if E satisfies condition* **FR 1**.

The proof of Frobenius' theorem will be carried out by analyzing the situation locally and reducing it to the standard theorem for ordinary differential equations. Thus we now analyze the condition **FR 1** in terms of its local representation.

Suppose that we work locally, over a product $U \times V$ of open subsets of Banach spaces **E** and **F**. Then the tangent bundle $T(U \times V)$ can be written in a natural way as a direct sum. Indeed, for each point (x, y) in $U \times V$ we have

$$T_{(x,y)}(U \times V) = T_x(U) \times T_y(V).$$

One sees at once that the collection of fibers $T_x(U) \times 0$ (contained in $T_x(U) \times T_y(V)$) forms a subbundle which will be denoted by $T_1(U \times V)$ and will be called the **first factor** of the tangent bundle. One could define $T_2(U \times V)$ similarly, and

$$T(U \times V) = T_1(U \times V) \oplus T_2(U \times V).$$

A subbundle E of $T(X)$ is integrable at a point $z \in X$ if and only if there exists an open neighborhood W of z and an isomorphism

$$\varphi \colon U \times V \to W$$

of a product onto W such that the composition of maps

$$T_1(U \times V) \xrightarrow{\text{inc.}} T(U \times V) \xrightarrow{T\varphi} T(W)$$

induces a VB-isomorphism of $T_1(U \times V)$ onto $E|W$ (over φ). Denoting by φ_y the map of U into W given by $\varphi_y(x) = \varphi(x, y)$, we can also express the integrability condition by saying that $T_x\varphi_y$ should induce a toplinear isomorphism of **E** onto $E_{\varphi(x,y)}$ for all (x, y) in $U \times V$. We note that in terms of our local product structure, $T_x\varphi_y$ is nothing but the partial derivative $D_1\varphi(x, y)$.

Given a subbundle of $T(X)$, and a point in the base space X, we know from the definition of a subbundle in terms of a local product decomposition that we can find a product decomposition of an open neighborhood of this point, say $U \times V$, such that the point has coordinates (x_0, y_0) and such that the subbundle can be written in the form of an exact sequence

$$0 \to U \times V \times \mathbf{E} \xrightarrow{\ \tilde{f}\ } U \times V \times \mathbf{E} \times \mathbf{F}$$

with the map

$$f(x_0, y_0): \mathbf{E} \to \mathbf{E} \times \mathbf{F}$$

equal to the canonical embedding of $\mathbf{E}$ on $\mathbf{E} \times 0$. For a point (x, y) in $U \times V$ the map $f(x, y)$ has two components $f_1(x, y)$ and $f_2(x, y)$ into $\mathbf{E}$ and $\mathbf{F}$ respectively. Taking a suitable VB-automorphism of $U \times V \times \mathbf{E}$ if necessary, we may assume without loss of generality that $f_1(x, y)$ is the identity. We now write $f(x, y) = f_2(x, y)$. Then

$$f: U \times V \to L(\mathbf{E}, \mathbf{F})$$

is a morphism (of class C^{p-1}) which describes our subbundle completely.

We shall interpret condition **FR 1** in terms of the present situation. If

$$\xi: U \times V \to \mathbf{E} \times \mathbf{F}$$

is the local representation of a vector field over $U \times V$, we let ξ_1 and ξ_2 be its projections on $\mathbf{E}$ and $\mathbf{F}$ respectively. Then ξ lies in the image of $\tilde{f}$ if and only if

$$\xi_2(x, y) = f(x, y)\xi_1(x, y)$$

for all (x, y) in $U \times V$, or in other words, if and only if ξ is of the form

$$\xi(x, y) = \big(\xi_1(x, y), f(x, y)\xi_1(x, y)\big)$$

for some morphism (of class C^{p-1})

$$\xi_1: U \times V \to \mathbf{E}.$$

We shall also write the above condition symbolically, namely

(1) $\xi = (\xi_1, f \cdot \xi_1).$

If ξ, η are the local representations of vector fields over $U \times V$, then the reader will verify at once from the local definition of the bracket (Proposition 1.3 of Chapter V) that $[\xi, \eta]$ lies in the image of $\tilde{f}$ if and only if

$$Df(x, y) \cdot \xi(x, y) \cdot \eta_1(x, y) = Df(x, y) \cdot \eta(x, y) \cdot \xi_1(x, y)$$

or symbolically,

$$(2) \qquad\qquad Df \cdot \xi \cdot \eta_1 = Df \cdot \eta \cdot \xi_1.$$

We have now expressed all the hypotheses of Theorem 1.1 in terms of local data, and the heart of the proof will consist in proving the following result.

Theorem 1.2. *Let* U, V *be open subsets of Banach spaces* **E**, **F** *respectively. Let*

$$f \colon U \times V \to L(\mathbf{E}, \mathbf{F})$$

be a C^r*-morphism* $(r \geq 1)$. *Assume that if*

$$\xi_1, \eta_1 \colon U \times V \to \mathbf{E}$$

are two morphisms, and if we let

$$\xi = (\xi_1, f \cdot \xi_1) \qquad and \qquad \eta = (\eta_1, f \cdot \eta_1)$$

then relation (2) *above is satisfied. Let* (x_0, y_0) *be a point of* $U \times V$. *Then there exists open neighborhoods* U_0, V_0 *of* x_0, y_0 *respectively, contained in* U, V, *and a unique morphism* $\alpha \colon U_0 \times V_0 \to V$ *such that*

$$D_1 \alpha(x, y) = f(x, \alpha(x, y))$$

and $\alpha(x_0, y) = y$ *for all* (x, y) *in* $U_0 \times V_0$.

We shall prove Theorem 1.2 in §3. We now indicate how Theorem 1.1 follows from it. We denote by α_y the map $\alpha_y(x) = \alpha(x, y)$, viewed as a map of U_0 into V. Then our differential equation can be written

$$D\alpha_y(x) = f(x, \alpha_y(x)).$$

We let

$$\varphi \colon U_0 \times V_0 \to U \times V$$

be the map $\varphi(x, y) = (x, \alpha_y(x))$. It is obvious that $D\varphi(x_0, y_0)$ is a top-linear isomorphism, so that φ is a local isomorphism at (x_0, y_0). Furthermore, for $(u, v) \in \mathbf{E} \times \mathbf{F}$ we have

$$D_1 \varphi(x, y) \cdot (u, v) = (u, D\alpha_y(x) \cdot u) = (u, f(x, \alpha_y(x)) \cdot u)$$

which shows that our subbundle is integrable.

VI, §2. DIFFERENTIAL EQUATIONS DEPENDING ON A PARAMETER

Proposition 2.1. *Let* U, V *be open sets in Banach spaces* **E**, **F** *respectively. Let* J *be an open interval of* **R** *containing* 0, *and let*

$$g: J \times U \times V \to \mathbf{F}$$

be a morphism of class C^r $(r \geq 1)$. *Let* (x_0, y_0) *be a point in* $U \times V$. *Then there exists open balls* J_0, U_0, V_0 *centered at* 0, x_0, y_0 *and contained* J, U, V *respectively, and a unique morphism of class* C^r

$$\beta: J_0 \times U_0 \times V_0 \to V$$

such that $\beta(0, x, y) = y$ *and*

$$D_1 \beta(t, x, y) = g(t, x, \beta(t, x, y))$$

for all $(t, x, y) \in J_0 \times U_0 \times V_0$.

Proof. This follows from the existence and uniqueness of local flows, by considering the ordinary vector field on $U \times V$

$$G: J \times U \times V \to \mathbf{E} \times \mathbf{F}$$

given by $G(t, x, y) = (0, g(t, x, y))$. If $B(t, x, y)$ is the local flow for G, then we let $\beta(t, x, y)$ be the projection on the second factor of $B(t, x, y)$. The reader will verify at once that β satisfies the desired conditions. The uniqueness is clear.

Let us keep the initial condition y fixed, and write

$$\beta(t, x) = \beta(t, x, y).$$

From Chapter IV, §1, we obtain also the differential equation satisfied by β in its second variable:

Proposition 2.2. *Let notation be as in Proposition 2.1, and with* y *fixed, let* $\beta(t, x) = \beta(t, x, y)$. *Then* $D_2 \beta(t, x)$ *satisfies the differential equation*

$$D_1 D_2 \beta(t, x) \cdot v = D_2 g(t, x, \beta(t, x)) \cdot v + D_3 g(t, x, \beta(t, x)) \cdot D_2 \beta(t, x) \cdot v,$$

for every $v \in \mathbf{E}$.

Proof. Here again, we consider the vector field as in the proof of Proposition 2.1, and apply the formula for the differential equation satisfied by $D_2\beta$ as in Chapter IV, §1.

VI, §3. PROOF OF THE THEOREM

In the application of Proposition 2.1 to the proof of Theorem 1.2, we take our morphism g to be

$$g(t, z, y) = f(x_0 + tz, y) \cdot z$$

with z in a small ball $\mathbf{E}_0$ around the origin in $\mathbf{E}$, and y in V. It is convenient to make a translation, and without loss of generality we can assume that $x_0 = 0$ and $y_0 = 0$. From Proposition 2.1 we then obtain

$$\beta: J_0 \times \mathbf{E}_0 \times V_0 \to V$$

with initial condition $\beta(0, z, y) = y$ for all $z \in \mathbf{E}_0$, satisfying the differential equation

$$D_1\beta(t, z, y) = f(tz, \beta(t, z, y)) \cdot z.$$

Making a change of variables of type $t = as$ and $z = a^{-1}x$ for a small positive number a, we see at once that we may assume that J_0 contains 1, provided we take $\mathbf{E}_0$ sufficiently small. As we shall keep y fixed from now on, we omit it from the notation, and write $\beta(t, z)$ instead of $\beta(t, z, y)$. Then our differential equation is

$$(3) \qquad\qquad D_1\beta(t, z) = f(tz, \beta(t, z)) \cdot z.$$

We observe that if we knew the existence of α in the statement of Theorem 2, then letting $\beta(t, z) = \alpha(x_0 + tz)$ would yield a solution of our differential equation. Thus the uniqueness of α follows. To prove its existence, we start with β and contend that the map

$$\alpha(x) = \beta(1, x)$$

has the required properties for small $|x|$. To prove our contention it will suffice to prove that

$$(4) \qquad\qquad D_2\beta(t, z) = tf(tz, \beta(t, z))$$

because if that relation holds, then

$$D\alpha(x) = D_2\beta(1, x) = f(x, \beta(1, x)) = f(x, \alpha(x))$$

which is precisely what we want.

From Proposition 2.2, we obtain for any vector $v \in \mathbf{E}$,

$$D_1 D_2 \beta(t, z) \cdot v = t D_1 f(tz, \beta(t, z)) \cdot v \cdot z$$

$$+ D_2 f(tz, \beta(t, z)) \cdot D_2 \beta(t, z) \cdot v \cdot z + f(tz, \beta(t, z)) \cdot v.$$

We now let $k(t) = D_2\beta(t, z) \cdot v - tf(tz, \beta(t, z)) \cdot v$. Then one sees at once that $k(0) = 0$ and we contend that

$$(5) \qquad Dk(t) = D_2 f(tz, \beta(t, z)) \cdot k(t) \cdot z.$$

We use the main hypothesis of our theorem, namely relation (2), in which we take ξ_1 and η_1 to be the fields v and z respectively. We compute Df using the formula for the partial derivatives, and apply it to this special case. Then (5) follows immediately. It is a linear differential equation satisfied by $k(t)$, and by Corollary 1.7 of Chapter IV, we know that the solution 0 is the unique solution. Thus $k(t) = 0$ and relation (4) is proved. The theorem also.

VI, §4. THE GLOBAL FORMULATION

Let X be a manifold. Let F be a tangent subbundle. By an **integral manifold** for F, we shall mean an injective immersion

$$f: Y \to X$$

such that at every point $y \in Y$, the tangent map

$$T_y f: T_y Y \to T_{f(y)} X$$

induces a toplinear isomorphism of $T_y Y$ on the subspace $F_{f(y)}$ of $T_{f(y)} X$. Thus Tf induces locally an isomorphism of the tangent bundle of Y with the bundle F over $f(Y)$.

Observe that the image $f(Y)$ itself may not be a submanifold of X. For instance, if F has dimension 1 (i.e. the fibers of F have dimension 1), an integral manifold for F is nothing but an integral curve from the theory of differential equations, and this curve may wind around X in such a way that its image is dense. A special case of this occurs if we consider the torus as the quotient of the plane by the subgroup generated by the two unit vectors. A straight line with irrational slope in the plane gets mapped on a dense integral curve on the torus.

If Y is a submanifold of X, then of course the inclusion $j: Y \to X$ is an injective immersion, and in this case, the condition that it be an integral manifold for F simply means that $T(Y) = F|Y$ (F restricted to Y).

We now have the local uniqueness of integral manifolds, corresponding to the local uniqueness of integral curves.

Theorem 4.1. *Let Y, Z be integral submanifolds of X for the subbundle F of TX, passing through a point x_0. Then there exists an open neighborhood U of x_0 in X, such that*

$$Y \cap U = Z \cap U.$$

Proof. Let U be an open neighborhood of x_0 in X such that we have a chart

$$U \to V \times W$$

with

$$x_0 \mapsto (y_0, w_0),$$

and Y corresponds to all points (y, w_0), $y \in V$. In other words, Y corresponds to a factor in the product in the chart. If V is open in $\mathbf{F}_1$ and W open in $\mathbf{F}_2$, with $\mathbf{F}_1 \times \mathbf{F}_2 = \mathbf{E}$, then the subbundle F is represented by the projection

$$V \times W \times \mathbf{F}_1$$
$$\downarrow$$
$$V \times W$$

Shrinking Z, we may assume that $Z \subset U$. Let $h: Z \to V \times W$ be the restriction of the chart to Z, and let $h = (h_1, h_2)$ be represented by its two components. By assumption, $h'(x)$ maps $\mathbf{E}$ into $\mathbf{F}_1$ for every $x \in Z$. Hence h_2 is constant, so that $h(Z)$ is contained in the factor $V \times \{w_0\}$. It follows at once that $h(Z) = V_1 \times \{w_0\}$ for some open V_1 in V, and we can shrink U to a product $V_1 \times W_1$ (where W_1 is a small open set in W containing w_0) to conclude the proof.

We wish to get a maximal connected integral manifold for an integrable subbundle F of TX passing through a given point, just as we obtained a maximal integral curve. For this, it is just as easy to deal with the nonconnected case, following Chevalley's treatment in his book on *Lie Groups*. (Note the historical curiosity that vector bundles were invented about a year after Chevalley published his book, so that the language of vector bundles, or the tangent bundle, is absent from

Chevalley's presentation. In fact, Chevalley used a terminology which now appears terribly confusing for the notion of a tangent subbundle, and it will not be repeated here!)

We give a new manifold structure to X, depending on the integrable tangent subbundle F, and the manifold thus obtained will be denoted by X_F. This manifold has the same set of points as X. Let $x \in X$. We know from the local uniqueness theorem that a submanifold Y of X which is at the same time an integral manifold for F is locally uniquely determined. A chart for this submanifold locally at x is taken to be a chart for X_F. It is immediately verified that the collection of such charts is an atlas, which defines our manifold X_F. (We lose one order of differentiability.) The identity mapping

$$j: X_F \to X$$

is then obviously an injective immersion, satisfying the following universal properties.

Theorem 4.2. *Let F be an integrable tangent subbundle over X. If*

$$f: Y \to X$$

is a morphism such that $Tf: TY \to TX$ maps TY into F, then the induced map

$$f_F: Y \to X_F$$

(same values as f but viewed as a map into the new manifold X_F) is also a morphism. Furthermore, if f is an injective immersion, then f_F induces an isomorphism of Y onto an open subset of X_F.

Proof. Using the local product structure as in the proof of the local uniqueness Theorem 4.1, we see at once that f_F is a morphism. In other words, locally, f maps a neighborhood of each point of Y into a submanifold of X which is tangent to F. If in addition f is an injective immersion, then from the definition of the charts on X_F, we see that f_F maps Y bijectively onto an open subset of X_F, and is a local isomorphism at each point. Hence f_F induces an isomorphism of Y with an open subset of X_F, as was to be shown.

Corollary 4.3. *Let $X_F(x_0)$ be the connected component of X_F containing a point x_0. If $f: Y \to X$ is an integral manifold for F passing through x_0, and Y is connected, then there exists a unique morphism*

$$h: Y \to X_F(x_0)$$

making the following diagram commutative:

$$Y \xrightarrow{\ h\ } X_F(x_0)$$

with f from Y to X and j from $X_F(x_0)$ to X.

and h induces an isomorphism of Y onto an open subset of $X_F(x_0)$.

Proof. Clear from the preceding discussion.

Note the general functorial behavior of the integral manifold. If

$$g: X \to X'$$

is an isomorphism, and F is an integrable tangent subbundle over X, then $F' = (Tg)(F) = g_* F$ is an integrable bundle over X'. Then the following diagram is commutative:

$$
\begin{array}{ccc}
X_F & \xrightarrow{\ g_F\ } & X'_{F'} \\
\downarrow{\scriptstyle j} & & \downarrow{\scriptstyle j'} \\
X & \xrightarrow{\ g\ } & X'
\end{array}
$$

The map g_F is, of course, the map having the same values as g, but viewed as a map on the manifold X_F.

VI, §5. LIE GROUPS AND SUBGROUPS

It is not our purpose here to delve extensively into Lie groups, but to lay the groundwork for their theory. For more results, we refer the reader to texts on Lie groups, differential geometry, and also to the paper by W. Graeub [Gr 61]. Although seemingly written to apply only to the finite dimensional case, this paper holds essentially in its entirety for the Banach case (and Hilbert case when dealing with Riemannian metrics), and is written on foundations corresponding to those of the present book.

By a **group manifold**, or a **Lie group** G, we mean a manifold with a group structure, that is a law of composition and inverse,

$$\tau: G \times G \to G \qquad \text{and} \qquad G \to G$$

which are morphisms. Thus each $x \in G$ gives rise to a left translation

$$\tau^x: G \to G$$

such that $\tau^x(y) = xy$.

When dealing with groups, we shall have to distinguish between isomorphisms in the category of manifolds, and isomorphisms in the category of group manifolds, which are also group homomorphisms. Thus we shall use prefixes, and speak of group manifold isomorphism, or manifold isomorphism as the case may be. We abbreviate these by GM-isomorphism or M-isomorphism. We see that left translation is an M-isomorphism, but not a GM-isomorphism.

Let e denote the origin (unit element) of G. If $v \in T_eG$ is a tangent vector at the origin, then we can translate it, and we obtain a map

$$(x, v) \mapsto \tau^x_* v = \xi_v(x)$$

which is easily verified to be a VB-isomorphism

$$G \times T_eG \to TG$$

from the product bundle to the tangent bundle of G. This is done at once using charts. Recall that T_eG can be viewed as a Banachable space, using any local trivialization of G at e to get a toplinear isomorphism of T_eG with the standard Banachable space on which G is modeled. Thus we see that the tangent bundle of a Lie group is trivializable.

A vector field ξ over G is called **left invariant** if $\tau^x_* \xi = \xi$ for all $x \in G$. Note that the map

$$x \mapsto \xi_v(x)$$

described above is a left invariant vector field, and that the association

$$v \mapsto \xi_v$$

obviously establishes a linear isomorphism between T_eG and the vector space of left invariant vector fields on G. The space of such vector fields will be denoted by $\mathfrak{g}$ or $\mathfrak{l}(G)$, and will be called the **Lie algebra** of G, because of the following result.

Proposition 5.1. *Let ξ, η be left invariant vector fields on G. Then $[\xi, \eta]$ is also left invariant.*

Proof. This follows from the general functorial formula

$$\tau^x_* [\xi, \eta] = [\tau^x_* \xi, \tau^x_* \eta] = [\xi, \eta].$$

Under the linear isomorphism of $T_e G$ with $\mathfrak{l}(G)$, we can view $\mathfrak{l}(G)$ as a Banachable space. By a **Lie subalgebra** of $\mathfrak{l}(G)$ we shall mean a closed subspace $\mathfrak{h}$ which splits, and having the property that if ξ, $\eta \in \mathfrak{h}$, then $[\xi, \eta] \in \mathfrak{h}$ also.

Note. In the finite dimensional case, every subspace is closed and splits, so that only this last condition about the bracket product need be mentioned explicitly.

Let G, H be Lie groups. A map

$$f : H \to G$$

will be called a **homomorphism** if it is a group homomorphism and a morphism in the category of manifolds. Such a homomorphism induces a continuous linear map

$$T_e f = f_* : T_e H \to T_e G,$$

and it is clear that it also induces a corresponding linear map

$$\mathfrak{l}(H) \to \mathfrak{l}(G),$$

also denoted by f_*. Namely, if $v \in T_e H$ and ξ_v is the left invariant vector field on H induced by v, then

$$f_* \xi_v = \xi_{f_* v}.$$

The general functorial property of related vector fields applies to this case, and shows that the induced map

$$f_* : \mathfrak{l}(H) \to \mathfrak{l}(G)$$

is also a Lie algebra homomorphism, namely for ξ, $\eta \in \mathfrak{l}(H)$ we have

$$f_*[\xi, \eta] = [f_* \xi, f_* \eta].$$

Now suppose that the homomorphism $f : H \to G$ is also an immersion at the origin of H. Then by translation, one sees that it is an immersion at every point. If in addition it is an injective immersion, then we shall say that f is a **Lie subgroup** of G. We see that in this case, f induces a splitting injection

$$f_* : \mathfrak{l}(H) \to \mathfrak{l}(G).$$

The image of $\mathfrak{l}(H)$ and $\mathfrak{l}(G)$ is a Lie subalgebra of $\mathfrak{l}(G)$.

In general, let $\mathfrak{h}$ be a Lie subalgebra of $\mathfrak{l}(G)$ and let F_e be the corresponding subspace of $T_e G$. For each $x \in G$, let

$$F_x = \tau_*^x F_e.$$

Then F_x is a split subspace of $T_x G$, and using local charts, it is clear that the collection $F = \{F_x\}$ is a subbundle of TG, which is left invariant. Furthermore, if

$$f: H \to G$$

is a homomorphism which is an injective immersion, and if $\mathfrak{h}$ is the image of $\mathfrak{l}(H)$, then we also see that f is an integral manifold for the subbundle F. We shall now see that the converse holds, using Frobenius' theorem.

Theorem 5.2. *Let G be a Lie group, $\mathfrak{h}$ a Lie subalgebra of $\mathfrak{l}(G)$, and let F be the corresponding left invariant subbundle of TG. Then F is integrable.*

Proof. I owe the proof to Alan Weinstein. It is based on the following lemma.

Lemma 5.3. *Let X be a manifold, let ξ, η be vector fields at a point x_0, and let F be a subbundle of TX. If $\xi(x_0) = 0$ and ξ is contained in F, then $[\xi, \eta](x_0) \in F$.*

Proof. We can deal with the local representations, such that $X = U$ is open in $\mathbf{E}$, and F corresponds to a factor, that is

$$TX = U \times \mathbf{F}_1 \times \mathbf{F}_2 \quad \text{and} \quad F = U \times \mathbf{F}_1.$$

We may also assume without loss of generality that $x_0 = 0$. Then $\xi(0) = 0$, and $\xi: U \to \mathbf{F}_1$ may be viewed as a map into $\mathbf{F}_1$. We may write

$$\xi(x) = A(x)x,$$

with a morphism $A: U \to L(\mathbf{E}, \mathbf{F}_1)$. Indeed,

$$\xi(x) = \int_0^1 \xi'(tx)\, dt \cdot x,$$

and $A(x) = \mathrm{pr}_1 \circ \int_0^1 \xi'(tx)\, dt$, where pr_1 is the projection on $\mathbf{F}_1$. Then

$$[\xi, \eta](x) = \eta'(x)\xi(x) - \xi'(x)\eta(x)$$
$$= \eta'(x)A(x)x - A'(x) \cdot x \cdot \eta(x) - A(x) \cdot \eta(x),$$

whence

$$[\xi, \eta](0) = A(0)\eta(0).$$

Since $A(0)$ maps $\mathbf{E}$ into $\mathbf{F}_1$, we have proved our lemma.

Back to the proof of the proposition. Let ξ, η be vector fields at a point x_0 in G, both contained in the invariant subbundle F. There exist invariant vector fields ξ_0 and η_0 at x_0 such that

$$\xi(x_0) = \xi_0(x_0) \quad \text{and} \quad \eta(x_0) = \eta_0(x_0).$$

Let

$$\xi_1 = \xi - \xi_0 \quad \text{and} \quad \eta_1 = \eta - \eta_0.$$

Then ξ_1, η_1 vanish at x_0 and lie in F. We get:

$$[\xi, \eta] = \sum_{i,j} [\xi_i, \eta_j].$$

The proposition now follows at once from the lemma.

Theorem 5.4. *Let G be a Lie group, let $\mathfrak{h}$ be a Lie subalgebra of $\mathfrak{l}(G)$, and let F be its associated invariant subbundle. Let*

$$j: H \to G$$

be the maximal connected integral manifold of F passing through e. Then H is a subgroup of G, and $j: H \to G$ is a Lie subgroup of G. The association between $\mathfrak{h}$ and $j: H \to G$ establishes a bijection between Lie subalgebras of $\mathfrak{l}(G)$ and Lie subgroups of G.

Proof. Let $x \in H$. The M-isomorphism τ^x induces a VB-isomorphism of F onto itself, in other words, F is invariant under τ^x_*. Furthermore, since H passes through e, and xe lies in H, it follows that $j: H \to G$ is also the maximal connected integral manifold of F passing through x. Hence x maps H onto itself. From this we conclude that if $y \in H$, then $xy \in H$, and there exists some $y \in H$ such that $xy = e$, whence $x^{-1} \in H$. Hence H is a subgroup. The other assertions are then clear.

If H is a Lie subgroup of G, belonging to the Lie algebra $\mathfrak{h}$, and F is the associated integrable left invariant tangent subbundle, then the integral manifold for F passing through a given point x is simply the translation xH, as one sees from first functorial principles.

When $\mathfrak{h}$ is 1-dimensional, then it is easy to see that the Lie subgroup is in fact a homomorphic image of an integral curve

$$\alpha: \mathbf{R} \to G$$

which is a homomorphism, and such that $\alpha'(0) = v$ is any vector in T_eG which is the value at e of a non-zero element of $\mathfrak{h}$. Changing this vector merely reparametrizes the curve. The integral curve may coincide with the subgroup, or it comes back on itself, and then the subgroup is essentially a circle. Thus the integral curve need not be equal to the subgroup. However, locally near $t = 0$, they do coincide. Such an integral curve is called a **one-parameter subgroup** of G.

Using Theorem 1.5 of Chapter V, it is then easy to see that if the Lie algebra of a connected Lie group G is commutative, then G itself is commutative. One first proves this for elements in a neighborhood of the origin, using 1-parameter subgroups, and then one gets the statement globally by expressing G as a union of products

$$UU \cdots U,$$

where U is a symmetric connected open neighborhood of the unit element. All of these statements are easy to prove, and belong to the first chapter of a book on Lie groups. Our purpose here is merely to lay the general foundations essentially belonging to general manifold theory.

Warning. The group of differential automorphisms of a finite dimensional manifold is "infinite dimensional" but usually not a Lie group, because multiplication is usually continuous only in each variable separately. For an analysis of this, also in the context of H^p (Sobolev) spaces, cf. Ebin and Marsden [EbM 70].

CHAPTER VII

Metrics

In our discussion of vector bundles, we put no greater structure on the fibers than that of topological vector space (of the same category as those used to build up manifolds). One can strengthen the notion so as to include the metric structure, and we are thus led to consider Hilbert bundles, whose fibers are Hilbert spaces.

Aside from the definitions, and basic properties, we deal with two special topics. On the one hand, we complete our uniqueness theorem on tubular neighborhoods by showing that when a Riemannian metric is given, a tubular neighborhood can be straightened out to a metric one. Secondly, we show how a Riemannian metric gives rise in a natural way to a spray, and thus how one recovers geodesics. The fundamental 2-form is used to identify the vector fields and 1-forms on the tangent bundle, identified with the cotangent bundle by the Riemannian metric.

We assume throughout that our manifolds are Hausdorff and are sufficiently differentiable so that all our statements make sense. (For instance, when dealing with sprays, we take $p \geq 3$.)

Of necessity, we shall use the standard spectral theorem for (bounded) symmetric operators. A self-contained treatment will be given in the appendix.

VII, §1. DEFINITION AND FUNCTORIALITY

For Riemannian geometry, we shall deal with a Hilbertable vector space, that is a topological vector space which is complete, and whose topology can be defined by the norm associated with a bilinear form, which is symmetric and positive definite. All facts needed in the sequel concerning Hilbert spaces can be found in the Appendix.

It turns out that some basic properties have only to do with a weaker property of the space **E** on which a manifold is modeled, namely that the Banach space **E** is self dual, via a symmetric non-singular bilinear form. Thus we only assume this property until more is needed. We recall that such a form is a continuous bilinear map

$$(v, w) \mapsto \langle v, w \rangle \qquad \text{of} \quad \mathbf{E} \times \mathbf{E} \to \mathbf{R}$$

such that $\langle v, w \rangle = \langle w, v \rangle$ for all v, $w \in \mathbf{E}$, and the corresponding map of **E** into the dual space $L(\mathbf{E})$ is a topological isomorphism.

Examples. Of course, the standard positive definite scalar product on Euclidean space provides the easiest (in some sense) example of a self dual vector space. But the physicists are interested in $\mathbf{R}^4$ with the scalar product such that the square of a vector (x, y, z, t) is $x^2 + y^2 + z^2 - t^2$. This scalar product is non-singular. For one among many nice applications of the indefinite case, cf. for instance [He 84] and [Gu 91], dealing with Huygens' principle.

We consider $L_{\mathrm{sym}}^2(\mathbf{E})$, the vector space of continuous bilinear forms

$$\lambda \colon \mathbf{E} \times \mathbf{E} \to \mathbf{R}$$

which are symmetric. If x is fixed in **E**, then the continuous linear form $\lambda_x(y) = \lambda(x, y)$ is represented by an element of **E** which we denote by Ax, where A is a continuous linear map of **E** into itself. The symmetry of λ implies that A is symmetric, that is we have

$$\lambda(x, y) = \langle Ax, y \rangle = \langle x, Ay \rangle$$

for all $x, y \in \mathbf{E}$. Conversely, given a symmetric continuous linear map $A \colon \mathbf{E} \to \mathbf{E}$ we can define a continuous bilinear form on **E** by this formula. Thus $L_{\mathrm{sym}}^2(\mathbf{E})$ is in bijection with the set of such operators, and is itself a Banach space, the norm being the usual operator norm. Suppose **E** is a Hilbert space, and in particular, **E** is self dual.

The subset of $L_{\mathrm{sym}}^2(\mathbf{E})$ consisting of those forms corresponding to symmetric **positive definite operators** (by definition such that $A \geqq \varepsilon I$ for some $\varepsilon > 0$) will be called the **Riemannian** of **E** and be denoted by Ri(**E**). Forms λ in Ri(**E**) are called positive definite. The associated operator A of such a form is invertible, because its spectrum does not contain 0 and the continuous function $1/t$ is invertible on the spectrum.

In general, suppose only that **E** is self dual. The space $L_{\mathrm{sym}}^2(\mathbf{E})$ contains as an open subset the set of non-singular symmetric bilinear forms, which we denote by Met(**E**), and which we call the set of **metrics or pseudo Riemannian metrics**. In view of the operations on vector bundles (Chapter III, §4) we can apply the functor L_{sym}^2 to any bundle whose fibers are self dual. Thus if $\pi: E \to X$ is such a bundle, then we can form $L^2_{\text{sym}}(\pi)$. A section of $L^2_{\text{sym}}(\pi)$ will be called by definition a **symmetric bilinear form** on π. A **(pseudo Riemannian) metric** on π (or on E) is defined to be a symmetric bilinear form on π, whose image lies in the open set of metrics at each point. We let $\text{Met}(\pi)$ be the set of metrics on π, which we also call the set of metrics on E, and may denote by $\text{Met}(E)$.

If E is a Hilbert space and the image of the section of $L^2_{\text{sym}}(\pi)$ lies in the Riemannian space $\text{Ri}(\pi_x)$ at each point x, in order words, if on the fiber at each point the non-singular symmetric bilinear form is actually positive definite, then we call the metric **Riemannian**. Let us denote a metric by g, so that $g(x) \in \text{Met}(E_x)$ for each $x \in X$, and lies in $\text{Ri}(E_x)$ if the metric is Riemanian. Then $g(x)$ is a non-singular symmetric bilinear form in general, and in the Riemannian case, it is positive definite in addition.

A pair (X, g) consisting of a manifold X and a (pseudo Riemannian) metric g will be called a **pseudo Riemannian manifold**. It will be called a **Riemannian manifold** if the manifold is modeled on a Hilbert space, and the metric is Riemannian.

Observe that the sections of $L^2_{\text{sym}}(\pi)$ form a vector space (abstract) but that the Riemannian metrics do not. They form a convex cone. Indeed, if a, $b > 0$ and g_1, g_2 are two Riemannian metrics, then $ag_1 + bg_2$ is also a Riemannian metric.

Suppose we are given a VB-trivialization of π over an open subset U of X, say

$$\tau: \pi^{-1}(U) \to U \times \mathbf{E}.$$

We can transport a given pseudo Riemannian metric g (or rather its restriction to $\pi^{-1}(U)$) to $U \times \mathbf{E}$. In the local representation, this means that for each $x \in U$ we can identify $g(x)$ with a symmetric invertible operator A_x giving rise to the metric. The operator A_x is positive definite in the Riemannian case. Furthermore, the map

$$x \mapsto A_x$$

from U into the Banach space $L(\mathbf{E}, \mathbf{E})$ is a morphism.

As a matter of notation, we sometimes write g_x instead of $g(x)$. Thus if v, w are two vectors in E_x, then $g_x(v, w)$ is a number, and is more convenient to write than $g(x)(v, w)$. We shall also write $\langle v, w \rangle_x$ if the metric g is fixed once for all.

Proposition 1.1. *Let X be a manifold admitting partitions of unity. Let $\pi: E \to X$ be a vector bundle whose fibers are Hilbertable vector spaces. Then π admits a Riemannian metric.*

Proof. Find a partition of unity $\{U_i, \varphi_i\}$ such that $\pi|U_i$ is trivial, that is such that we have a trivialization

$$\pi_i \colon \pi^{-1}(U_i) \to U_i \times \mathbf{E}$$

(working over a connected component of X, so that we may assume the fibers toplinearly isomorphic to a fixed Hilbert space $\mathbf{E}$). We can then find a Riemannian metric on $U_i \times \mathbf{E}$ in a trivial way. By transport of structure, there exists a Riemannian metric g_i on $\pi|U_i$ and we let

$$g = \sum \varphi_i g_i.$$

Then g is a Riemannian metric on x.

Let us investigate the functorial behavior of metrics.
Consider a VB-morphism

$$
\begin{array}{ccc}
E' & \xrightarrow{\ f\ } & E \\
{\scriptstyle \pi'}\downarrow & & \downarrow{\scriptstyle \pi} \\
X & \xrightarrow[\ f_0\ ]{} & Y
\end{array}
$$

with vector bundles E' and E over X and Y respectively, whose fibers are self dual spaces. Let g be a symmetric bilinear form on π, so that for each $y \in Y$ we have a continuous, bilinear, symmetric map

$$g(y)\colon E_y \times E_y \to \mathbf{R}.$$

Then the composite map

$$E_x' \times E_x' \to E_y \times E_y \to \mathbf{R}$$

with $y = f(x)$ is a symmetric bilinear form on E_x' and one verifies immediately that it gives rise to such a form, on the vector bundle π', which will be denoted by $f^*(g)$. Then f induces a map

$$L_{\text{sym}}^2(f) = f^*\colon L_{\text{sym}}^2(\pi) \to L_{\text{sym}}^2(\pi').$$

Furthermore, if f_x is injective and splits for each $x \in X$, and g is a metric (resp. g is a Riemannian metric in the Hilbert case), then obviously so is $f^*(g)$, and we can view f^* as mapping $\mathrm{Met}(\pi)$ into $\mathrm{Met}(\pi')$ (resp. $\mathrm{Ri}(\pi)$ into $\mathrm{Ri}(\pi')$ in the Riemannian case).

Let X be a manifold modeled on a Hilbertable space and let $T(X)$ be its tangent bundle. By abuse of language, we call a metric on $T(X)$ also a metric on X and write Met(X) instead of Met$(T(X))$. Similarly, we write Ri(X) instead of Ri$(T(X))$.

Let $f: X \to Y$ be an immersion. Then for each $x \in X$, the linear map

$$T_x f: T_x(X) \to T_{f(x)}(Y)$$

is injective, and splits, and thus we obtain a contravariant map

$$f^*: \text{Ri}(Y) \to \text{Ri}(X),$$

each Riemannian metric on Y inducing a Riemannian metric on X.

A similar result applies in the pseudo Riemannian case. If (Y, g) is Riemannian, and f is merely of class C^1 but not necessarily an immersion, then the pull back $f^*(g)$ is not necessarily positive definite, but is merely what we call **semipositive**. In general, if (X, h) is pseudo Riemannian and $h(v, v) \geqq 0$ for all $v \in T_x X$, all x, then (X, h) is called **semi Riemannian**. Thus the pull back of a semi Riemannian metric is semi Riemannian.

For a major result concerning Riemannian embeddings of manifolds in Euclidean space, see Nash [Na 56], followed by Moser [Mo 61], as well as the exposition I gave in [La 61]. Even though dealing a priori with finite dimensional manifolds, the imbedding problem is essentially concerned with the infinite dimensional manifold of Riemannian metrics. The problem partly amounts to obtaining an inverse mapping theorem in a context more complicated than that of Banach spaces, namely Frechet spaces, when all C^p norms intervene, for $p = 1, 2, \ldots$. Newton approximation is used instead of the shrinking lemma to solve the local isomorphism problem in this case.

The next five sections will be devoted to considerations which apply specifically to the Riemannian case, where positivity plays a central role.

VII, §2. THE HILBERT GROUP

Let **E** be a Hilbert space. The group of toplinear automorphisms Laut(**E**) contains the group Hilb(**E**) of Hilbert automorphisms, that is those toplinear automorphisms which preserve the inner product:

$$\langle Av, Aw \rangle = \langle v, w \rangle$$

for all $v, w \in$ **E**. We note that A is Hilbertian if and only if $A^*A = I$.

As usual, we say that a linear continuous map $A: \mathbf{E} \to \mathbf{E}$ is **symmetric** if $A^* = A$ and that it is **skew-symmetric** if $A^* = -A$. We have a direct sum decomposition of the Banach space $L(\mathbf{E}, \mathbf{E})$ in terms of the two closed subspaces of symmetric and skew-symmetric operators:

$$A = \tfrac{1}{2}(A + A^*) + \tfrac{1}{2}(A - A^*).$$

We denote by Sym(**E**) and Sk(**E**) the Banach spaces of symmetric and skew-symmetric maps respectively. The word **operator** will always mean continuous linear map of **E** into itself.

Proposition 2.1. *For all operators A, the series*

$$\exp(A) = I + A + \frac{A^2}{2!} + \cdots$$

converges. If A commutes with B, then

$$\exp(A + B) = \exp(A)\exp(B).$$

For all operators sufficiently close to the identity I, the series

$$\log(A) = \frac{(A - I)}{1} + \frac{(A - I)^2}{2} + \cdots$$

converges, and if A commutes with B, then

$$\log(AB) = \log(A) + \log(B).$$

Proof. Standard.

We leave it as an exercise to the reader to show that the exponential function gives a C^∞-morphism of $L(\mathbf{E}, \mathbf{E})$ into itself. Similarly, a function admitting a development in power series say around 0 can be applied to the set of operators whose bound is smaller than the radius of convergence of the series, and gives a C^∞-morphism.

Proposition 2.2. *If A is symmetric (resp. skew-symmetric), then $\exp(A)$ is symmetric positive definite (resp. Hilbertian). If A is toplinear automorphism sufficiently close to I and is positive definite symmetric (resp. Hilbertian), then $\log(A)$ is symmetric (resp. skew-symmetric).*

Proof. The proofs are straightforward. As an example, let us carry out the proof of the last statement. Suppose A is Hilbertian and sufficiently close to I. Then $A^*A = I$ and $A^* = A^{-1}$. Then

$$\log(A)^* = \frac{(A^* - I)}{1} + \cdots$$

$$= \log(A^{-1}).$$

If A is close to I, so is A^{-1}, so that these statements make sense. We now conclude by noting that $\log(A^{-1}) = -\log(A)$. All the other proofs are carried out in a similar fashion, taking a star operator in series term by term, under conditions which insure convergence.

The exponential and logarithm functions give inverse C^∞ mappings between neighborhoods of 0 in $L(\mathbf{E}, \mathbf{E})$ and neighborhoods of I in Laut($\mathbf{E}$). Furthermore, the direct sum decomposition of $L(\mathbf{E}, \mathbf{E})$ into symmetric and skew-symmetric subspaces is reflected locally in a neighborhood of I by a C^∞ direct product decomposition into positive definite and Hilbertian automorphisms. This direct product decomposition can be translated multiplicatively to any toplinear automorphism, because if $A \in$ Laut($\mathbf{E}$) and B is close to A, then

$$B = AA^{-1}B = A\bigl(I - (I - A^{-1}B)\bigr)$$

and $(I - A^{-1}B)$ is small. This proves:

Proposition 2.3. *The Hilbert group of automorphisms of* $\mathbf{E}$ *is a closed submanifold of* Laut($\mathbf{E}$).

In addition to this local result, we get a global one also:

Proposition 2.4. *The exponential map gives a* C^∞*-isomorphism from the space* Sym($\mathbf{E}$) *of symmetric endomorphisms of* $\mathbf{E}$ *and the space* Pos($\mathbf{E}$) *of symmetric positive definite automorphisms of* $\mathbf{E}$.

Proof. We must construct its inverse, and for this we use the spectral theorem. Given A, symmetric positive definite, the analytic function $\log t$ is defined on the spectrum of A, and thus $\log A$ is symmetric. One verifies immediately that it is the inverse of the exponential function (which can be viewed in the same way). We can expand $\log t$ around a large positive number c, in a power series uniformly and absolutely convergent in an interval $0 < \varepsilon \leqq t \leqq 2c - \varepsilon$, to achieve our purposes.

Proposition 2.5. *The manifold of toplinear automorphisms of the Hilbert space* $\mathbf{E}$ *is* C^∞*-isomorphic to the product of the Hilbert automorphisms and the positive definite symmetric automorphisms, under the mapping*

$$\text{Hilb}(\mathbf{E}) \times \text{Pos}(\mathbf{E}) \to \text{Laut}(\mathbf{E})$$

given by

$$(H, P) \to HP.$$

Proof. Our map is induced by a continuous bilinear map of

$$L(\mathbf{E}, \mathbf{E}) \times L(\mathbf{E}, \mathbf{E})$$

into $L(\mathbf{E}, \mathbf{E})$ and so is C^∞. We must construct an inverse, or in other words express any given toplinear automorphism A in a unique way as a product $A = HP$ where H is Hilbertian, P is symmetric positive definite, and both H, P depend C^∞ on A. This is done as follows. First we note that A^*A is symmetric positive definite (because $\langle A^*Av, v \rangle = \langle Av, Av \rangle$, and furthermore, A^*A is a toplinear automorphism, so that 0 cannot be in its spectrum, and hence $A^*A \geq \varepsilon I > O$ since the spectrum is closed). We let

$$P = (A^*A)^{1/2}$$

and let $H = AP^{-1}$. Then H is Hilbertian, because

$$H^*H = (P^{-1})^*A^*AP^{-1} = I.$$

Both P and H depend differentiably on A since all constructions involved are differentiable.

There remains to be shown that the expression as a product is unique. If $A = H_1P_1$ where H_1, P_1 are Hilbertian and symmetric positive definite respectively, then

$$H^{-1}H_1 = PP_1^{-1},$$

and we get $H_2 = PP_1^{-1}$ for some Hilbertian automorphism H_2. By definition,

$$I = H_2^*H_2 = (PP_1^{-1})^*PP_1^{-1}$$

and from the fact that $P^* = P$ and $P_1^* = P_1$, we find

$$P^2 = P_1^2.$$

Taking the log, we find $2 \log P = 2 \log P_1$. We now divide by 2 and take the exponential, thus giving $P = P_1$ and finally $H = H_1$. This proves our proposition.

VII, §3. REDUCTION TO THE HILBERT GROUP

We define a new category of bundles, namely the **Hilbert bundles** over X, denoted by $\mathrm{HB}(X)$. As before, we would denote by $\mathrm{HB}(X, \mathbf{E})$ or $\mathrm{HB}(X, \mathfrak{A})$ those Hilbert bundles whose fiber is a Hilbert space $\mathbf{E}$ or lies in a category $\mathfrak{A}$.

Let $\pi: E \to X$ be a vector bundle over X, and assume that it has a trivialization $\{(U_i, \tau_i)\}$ with trivializing maps

$$\tau_i: \pi^{-1}(U_i) \to U_i \times \mathbf{E}$$

where $\mathbf{E}$ is a Hilbert space, such that each toplinear automorphism $(\tau_j\tau_i^{-1})_x$ is a Hilbert automorphism. Equivalently, we could also say that τ_{ix} is a Hilbert isomorphism. Such a trivialization will be called a **Hilbert trivialization**. Two such trivializations are called **Hilbert-compatible** if their union is again a Hilbert trivialization. An equivalence class of such compatible trivializations constitutes what we call a **Hilbert bundle** over X. Any such Hilbert bundle determines a unique vector bundle, simply by taking the VB-equivalence class determined by the trivialization.

Given a Hilbert trivialization $\{(U_i, \tau_i)\}$ of a vector bundle π over X, we can define on each fiber π_x a Hilbert space structure. Indeed, for each x we select an open set U_i in which x lies, and then transport to π_x the scalar product in $\mathbf{E}$ by means of τ_{ix}. By assumption, this is independent of the choice of U_i in which x lies. Thus in a Hilbert bundle, we can assume that the fibers are Hilbert spaces, not only Hilbertable.

It is perfectly possible that several distinct Hilbert bundles determine the same vector bundle.

Any Hilbert bundle determining a given vector bundle π will be said to be a **reduction of π to the Hilbert group**.

We can make Hilbert bundles into a category, if we take for the **HB-morphisms** the VB-morphisms which are injective and split at each point, and which preserve the metric, again at each point.

Each reduction of a vector bundle to the Hilbert group determines a Riemannian metric on the bundle. Indeed, defining for each $z \in X$ and v, $w \in \pi_x$ the scalar product

$$g_x(v, w) = \langle \tau_{ix} v, \tau_{ix} w \rangle$$

with any Hilbert-trivializing map τ_{ix} such that $x \in U_i$, we get a morphism

$$x \mapsto g_x$$

of X into the sections of $L^2_{\text{sym}}(\pi)$ which are positive definite. We also have the converse.

Theorem 3.1. *Let π be a vector bundle over a manifold X, and assume that the fibers of π are all toplinearly isomorphic to a Hilbert space $\mathbf{E}$. Then the above map, from reductions of π to the Hilbert group, into the Riemannian metrics, is a bijection.*

Proof. Suppose that we are given an ordinary VB-trivialization $\{(U_i, \tau_i)\}$ of π. We must construct an HB-trivialization. For each i, let g_i be the Riemannian metric on $U_i \times \mathbf{E}$ transported from $\pi^{-1}(U_i)$ by means of τ_i. Then for each $x \in U_i$, we have a positive definite symmetric operator A_{ix} such that

$$g_{ix}(v, w) = \langle A_{ix}v, w \rangle$$

for all v, $w \in \mathbf{E}$. Let B_{ix} be the square root of A_{ix}. We define the trivialization σ_i by the formula

$$\sigma_{ix} = B_{ix}\tau_{ix}$$

and contend that $\{(U_i, \sigma_i)\}$ is a Hilbert trivialization. Indeed, from the definition of g_{ix}, it suffices to verify that the VB-isomorphism

$$B_i \colon U_i \times \mathbf{E} \to U_i \times \mathbf{E}$$

given by B_{ix} on each fiber, carries g_i on the usual metric. But we have, for v, $w \in E$:

$$\langle B_{ix}v, B_{ix}w \rangle = \langle A_{ix}v, w \rangle$$

since B_{ix} is symmetric, and equal to the square root of A_{ix}. This proves what we want.

At this point, it is convenient to make an additional comment on normal bundles.

Let α, β be two Hilbert bundles over the manifold X, and let $f \colon \alpha \to \beta$ be an HB-morphism. Assume that

$$0 \to \alpha \xrightarrow{\ f\ } \beta$$

is exact. Then by using the Riemannian metric, there is a natural way of constructing a splitting for this sequence (cf. Chapter III, §5).

Using Theorem 1.2 of the Appendix, we see at once that if $\mathbf{F}$ is a (closed) subspace of a Hilbert space, then $\mathbf{E}$ is the direct sum

$$\mathbf{E} = \mathbf{F} \oplus \mathbf{F}^{\perp}$$

of $\mathbf{F}$ and its orthogonal complement, consisting of all vectors perpendicular to $\mathbf{F}$.

In our exact sequence, we may view f as an injection. For each x we let $\alpha_x^{\perp}$ be the orthogonal complement of α_x in β_x. Then we shall find an exact sequence of VB-morphisms

$$\beta \xrightarrow{\ h\ } \alpha \to 0$$

whose kernel is $\alpha^{\perp}$ (set theoretically). In this manner, the collection of orthogonal complements $\alpha_x^{\perp}$ can be given the structure of a Hilbert bundle.

For each x we can write $\beta_x = \alpha_x \oplus \alpha_x^{\perp}$ and we define h_x to be the projection in this direct sum decomposition. This gives us a mapping $h \colon \beta \to \alpha$, and it will suffice to prove that h is a VB-morphism. In order

to do this, we may work locally. In that case, after taking suitable
VB-automorphisms over a small open set U of X, we can assume that
we deal with the following situation.

Our vector bundle β is equal to $U \times \mathbf{E}$ and α is equal to $U \times \mathbf{F}$ for
some subspace $\mathbf{F}$ of $\mathbf{E}$, so that we can write $\mathbf{E} = \mathbf{F} \times \mathbf{F}^{\perp}$. Our HB-
morphism is then represented for each x by an injection $f_x \colon \mathbf{F} \to \mathbf{E}$:

$$U \times \mathbf{F} \xrightarrow{\;f\;} U \times \mathbf{E}.$$

By the definition of exact sequences, we can find two VB-isomorphisms τ
and σ such that the following diagram is commutative:

$$
\begin{array}{ccc}
U \times \mathbf{F} & \xrightarrow{\;f\;} & U \times \mathbf{E} \\
{\scriptstyle \sigma}\downarrow & & \downarrow{\scriptstyle \tau} \\
U \times \mathbf{F} & \longrightarrow & U \times \mathbf{E}
\end{array}
$$

and such that the bottom map is simply given by the ordinary inclusion
of $\mathbf{F}$ in $\mathbf{E}$. We can transport the Riemannian structure of the bundles on
top to the bundles on the bottom by means of σ^{-1} and τ^{-1} respectively.
We are therefore reduced to the situation where f is given by the simple
inclusion, and the Riemannian metric on $U \times \mathbf{E}$ is given by a family A_x
of symmetric positive definite operators on $\mathbf{E}$ ($x \in U$). At each point x,
we have $\langle v, w \rangle_x = \langle A_x v, w \rangle$. We observe that the map

$$A \colon U \times \mathbf{E} \to U \times \mathbf{E}$$

given by A_x on each fiber is a VB-automorphism of $U \times \mathbf{E}$. Let $\mathrm{pr}_{\mathbf{F}}$ be
the projection of $U \times \mathbf{E}$ on $U \times \mathbf{F}$. It is a VB-morphism. Then the
composite

$$h = \mathrm{pr}_{\mathbf{F}} \circ A$$

gives us a VB-morphism of $U \times \mathbf{E}$ on $U \times \mathbf{F}$, and the sequence

$$U \times \mathbf{E} \xrightarrow{\;h\;} U \times \mathbf{F} \to 0$$

is exact. Finally, we note that the kernel of h consists precisely of the
orthogonal complement of $U \times \mathbf{F}$ in each fiber. This proves what we
wanted.

VII, §4. HILBERTIAN TUBULAR NEIGHBORHOODS

Let $\mathbf{E}$ be a Hilbert space. Then the open ball of radius 1 is isomorphic
to $\mathbf{E}$ itself under the mapping

$$v \mapsto \frac{v}{(1 - |v|^2)^{1/2}},$$

the inverse mapping being

$$w \mapsto \frac{w}{(1 + |w|^2)^{1/2}}.$$

If $a > 0$, then any ball of radius a is isomorphic to the unit ball under multiplication by the scalar a (or a^{-1}).

Let X be a manifold, and $\sigma: X \to \mathbf{R}$ a function (morphism) such that $\sigma(x) > 0$ for all $x \in X$. Let $\pi: E \to X$ be a Hilbert bundle over X. We denote by $E(\sigma)$ the subset of E consisting of those vectors v such that, if v lies in E_x, then

$$|v|_x < \sigma(x).$$

Then $E(\sigma)$ is an open neighborhood of the zero section.

Proposition 4.1. *Let X be a manifold and $\pi: E \to X$ a Hilbert bundle. Let $\sigma: X \to \mathbf{R}$ be a morphism such that $\sigma(x) > 0$ for all x. Then the mapping*

$$w \to \frac{\sigma(\pi w)w}{(1 + |w|^2)^{1/2}}$$

gives an isomorphism of E onto $E(\sigma)$.

Proof. Obvious. The invese mapping is constructed in the obvious way.

Corollary 4.2. *Let X be a manifold admitting partitions of unity, and let $\pi: E \to X$ be a Hilbert bundle over X. Then E is compressible.*

Proof. Let Z be an open neighborhood of the zero section. For each $x \in X$, there exists an open neighborhood V_x and a number $a_x > 0$ such that the vectors in $\pi^{-1}(V_x)$ which are of length $< a_x$ lie in Z. We can find a partition of unity $\{(U_i, \varphi_i)\}$ on X such that each U_i is contained in some $V_{x(i)}$. We let σ be the function

$$\sum a_{x(i)} \varphi_i.$$

Then $E(\sigma)$ is contained in Z, and our assertion follows from the proposition.

Proposition 4.3. *Let X be a manifold. Let $\pi: E \to X$ and $\pi_1: E_1 \to X$ be two Hilbert bundles over X. Let*

$$\lambda: E \to E_1$$

be a VB-*isomorphism. Then there exists an isotopy of* VB-*isomorphisms*

$$\lambda_t \colon E \to E_1$$

with proper domain $[0, 1]$ *such that* $\lambda_1 = \lambda$ *and* λ_0 *is an* HB-*isomorphism.*

Proof. We find reductions of E and E_1 to the Hilbert group, with Hilbert trivializations $\{(U_i, \tau_i)\}$ for E and $\{(U_i, \rho_i)\}$ for E_1. We can then factor $\rho_i \lambda \tau_i^{-1}$ as in Proposition 2.5, applied to each fiber map:

$$
\begin{array}{ccccc}
U_i \times \mathbf{E} & \longrightarrow & U_i \times \mathbf{E} & \longrightarrow & U_i \times \mathbf{E} \\
\tau_i \uparrow & & \tau_i \uparrow & & \uparrow \rho_i \\
\pi^{-1}(U_i) & \xrightarrow[\lambda_P]{} & \pi(U_i^{-1}) & \xrightarrow[\lambda_H]{} & \pi_1^{-1}(U_i)
\end{array}
$$

and obtain a factorization of λ into $\lambda = \lambda_H \lambda_P$ where λ_H is a HB-isomorphism and λ_P is a positive definite symmetric VB-automorphism. The latter form a convex set, and our isotopy is simply

$$\lambda_t = \lambda_H \circ \bigl(tI + (1 + t)\lambda_P\bigr).$$

(Smooth out the end points if you wish.)

Theorem 4.4. *Let X be a submanifold of Y. Let $\pi \colon E \to X$ and $\pi_1 \colon E_1 \to X$ be two Hilbert bundles. Assume that E is compressible. Let $f \colon E \to Y$ and $g \colon E_1 \to Y$ be two tubular neighborhoods of X in Y. Then there exists an isotopy*

$$f_t \colon E \to Y$$

of tubular neighborhoods with proper domain $[0, 1]$ and there exists an HB-*isomorphism $\mu \colon E \to E_1$ such that $f_1 = f$ and $f_0 = g\mu$.*

Proof. From Theorem 6.2 of Chapter IV, we know already that there exists a VB-isomorphism λ such that $f \approx g\lambda$. Using the preceding proposition, we know that $\lambda \approx \mu$ where μ is a HB-isomorphism. Thus $g\lambda \approx g\mu$ and by transitivity, $f \approx \mu$, as was to be shown.

Remark. In view of Proposition 4.1, we could of course replace the condition that E be compressible by the more useful condition (in practice) that X admit partitions of unity.

VII, §5. THE MORSE–PALAIS LEMMA

Let U be an open set in some (real) Hilbert space **E**, and let f be a C^{p+2} function on U, with $p \geq 1$. We say that x_0 is a **critical point** for f if $Df(x_0) = 0$. We wish to investigate the behavior of f at a critical point. After translations, we can assume that $x_0 = 0$ and that $f(x_0) = 0$. We observe that the second derivative $D^2f(0)$ is a continuous bilinear form on **E**. Let $\lambda = D^2f(0)$, and for each $x \in \mathbf{E}$ let λ_x be the functional such that $y \mapsto \lambda(x, y)$. If the map $x \mapsto \lambda_x$ is a toplinear isomorphism of **E** with its dual space $\mathbf{E}^{\vee}$, then we say that λ is **non-singular**, and we say that the critical point is **non-degenerate**.

We recall that a local C^p-isomorphism φ at 0 is a C^p-invertible map defined on an open set containing 0.

Theorem 5.1. *Let f be a C^{p+2} function defined on an open neighborhood of 0 in the Hilbert space **E**, with $p \geq 1$. Assume that $f(0) = 0$, and that 0 is a non-degenerate critical point of f. Then there exists a local C^p-isomorphism at 0, say φ, and an invertible symmetric operator A such that*

$$f(x) = \langle A\varphi(x), \varphi(x) \rangle.$$

Proof. We may assume that U is a ball around 0. We have

$$f(x) = f(x) - f(0) = \int_0^1 Df(tx)x \, dt,$$

and applying the same formula to Df instead of f, we get

$$f(x) = \int_0^1 \int_0^1 D^2f(stx)tx \cdot x \, ds \, dt = g(x)(x, x)$$

where

$$g(x) = \int_0^1 \int_0^1 D^2f(stx)t \, ds \, dt.$$

Then g is a C^p map into the Banach space of continuous bilinear maps on **E**, and even the space of symmetric such maps. We know that this Banach space is toplinearly isomorphic to the space of symmetric operators on **E**, and thus we can write

$$f(x) = \langle A(x)x, x \rangle$$

where $A: U \to \mathrm{Sym}(\mathbf{E})$ is a C^p map of U into the space of symmetric operators on **E**. A straightforward computation shows that

$$D^2f(0)(v, w) = \langle A(0)v, w \rangle.$$

Since we assumed that $D^2 f(0)$ is non-singular, this means that $A(0)$ is invertible, and hence $A(x)$ is invertible for all x sufficiently near 0.

Theorem 5.1 is then a consequence of the following result, which expresses locally the uniqueness of a non-singular symmetric form.

Theorem 5.2. *Let $A: U \to \mathrm{Sym}(\mathbf{E})$ be a C^p map of U into the open set of invertible symmetric operators on $\mathbf{E}$. Then there exists a C^p isomorphism of an open subset U_1 containing 0, of the form*

$$\varphi(x) = C(x)x, \qquad \text{with a } C^p \text{ map} \quad C: U_1 \to \mathrm{Laut}(\mathbf{E})$$

such that

$$\langle A(x)x, x \rangle = \langle A(0)\varphi(x), \varphi(x) \rangle = \langle A(0)C(x)x, C(x)x \rangle.$$

Proof. We seek a map C such that

$$C(x)^* A(0) C(x) = A(x).$$

If we let $B(x) = A(0)^{-1} A(x)$, then $B(x)$ is close to the identity I for small x. The square root function has a power series expansion near 1, which is a uniform limit of polynomials, and is C^∞ on a neighborhood of I, and we can therefore take the square root of $B(x)$, so that we let

$$C(x) = B(x)^{1/2}.$$

We contend that this $C(x)$ does what we want. Indeed, since both $A(0)$ and $A(x)$ $\left(\text{or } A(x)^{-1}\right)$ are self-adjoint, we find that

$$B(x)^* = A(x)A(0)^{-1},$$

whence

$$B(x)^* A(0) = A(0)B(x).$$

But $C(x)$ is a power series in $I - B(x)$, and $C(x)^*$ is the same power series in $I - B(x)^*$. The preceding relation holds if we replace $B(x)$ by any power of $B(x)$ (by induction), hence it holds if we replace $B(x)$ by any polynomial in $I - B(x)$, and hence finally, it holds if we replace $B(x)$ by $C(x)$, and thus

$$C(x)^* A(0) C(x) = A(0)C(x)C(x) = A(0)B(x) = A(x).$$

which is the desired relation.

All that remains to be shown is that φ is a local C^p-isomorphism at 0. But one verifies that in fact, $D\varphi(0) = C(0)$, so that what we need follows from the inverse mapping theorem. This concludes the proof of Theorems 5.1 and 5.2.

Corollary 5.3. *Let f be a C^{p+2} function near 0 on the Hilbert space* **E**, *such that 0 is a non-degenerate critical point. Then there exists a local C^p-isomorphism ψ at 0, and an orthogonal decomposition* **E** $=$ **F** $+$ **F**$^\perp$, *such that if we write $\psi(x) = y + z$ with $y \in$ **F** and $z \in$ **F**$^\perp$, then*

$$f(\psi(x)) = \langle y, y \rangle - \langle z, z \rangle.$$

Proof. On a space where A is positive definite, we can always make the toplinear isomorphism $x \mapsto A^{1/2}x$ to get the quadratic form to become the given hermitian product $\langle \ , \ \rangle$, and similarly on a space where A is negative definite. In general, we use the spectral theorem to decompose **E** into a direct orthogonal sum such that the restriction of A to the factors is positive definite and negative definite respectively.

Note. The Morse–Palais lemma was proved originally by Morse in the finite dimensional case, using the Gram–Schmidt orthogonalization process. The elegant generalization and its proof in the Hilbert space case is due to Palais [Pa 69]. It shows (in the language of coordinate systems) that a function near a critical point can be expressed as a quadratic form after a suitable change of coordinate system (satisfying requirements of differentiability). It comes up naturally in the calculus of variations. For instance, one considers a space of paths (of various smoothness) $\sigma: [a, b] \to$ **E** where **E** is a Hilbert space. One then defines a length function (see next section) or the **energy function**

$$f(\sigma) = \int_a^b \langle \sigma'(t), \sigma'(t) \rangle \, dt,$$

and one investigates the critical points of this function, especially its minimum values. These turn out to be the solutions of the variational problem, by definition of what one means by a variational problem. Even if **E** is finite dimensional, so a Euclidean space, the space of paths is infinite dimensional. Cf. [Mi 63] and [Pa 63].

VII, §6. THE RIEMANNIAN DISTANCE

Let (X, g) be a Riemannian manifold. For each C^1 curve

$$\gamma: [a, b] \to X$$

we define its **length**

$$L_g(\gamma) = L(\gamma) = \int_a^b \langle \gamma'(t), \gamma'(t) \rangle_g^{1/2} \, dt = \int_a^b \|\gamma'(t)\|_g \, dt.$$

The norm is the one associated with the positive definite scalar product, i.e. the Hilbert space norm at each point. We can extend the length to piecewise C^1 paths by taking the sum over the C^1 curves constituting the path. *We assume that X is connected, which is equivalent to the property that any two points can be joined by a piecewise C^1 path.* (If X is connected, then the set of points which can be joined to a given point x_0 by a piecewise C^1 path is immediately verified to be open and closed, so equal to X. The converse, that pathwise connectedness implies connectedness, is even more obvious.)

We define the **g-distance** on X for any two points $x, y \in X$ by:

dist$_g(x, y)$ = greatest lower bound of $L(\gamma)$ for paths γ in X
          joining x and y.

When g is fixed throughout, we may omit g from the notation and write simply dist(x, y). It is clear that dist$_g$ is a semidistance, namely it is symmetric in (x, y) and satisfies the triangle inequality. To prove that it is a distance, we have to show that if $x \neq y$ then dist$_g(x, y) > 0$. In a chart, there is a neighborhood U of x which contains a closed ball $\overline{\mathbf{B}}(x, r)$ with $r > 0$, and such that y lies outside this closed ball. Then any path between x and y has to cross the sphere $\mathbf{S}(x, r)$. *Here we are using the Hilbert space norm in the chart.* We can also take r so small that the norm in the chart is given by

$$\langle v, w \rangle_{g(x)} = \langle v, A(x)w \rangle,$$

for $v, w \in \mathbf{E}$, and $x \mapsto A(x)$ is a morphism from U into the set of invertible symmetric positive definite operators, such that there exist a number $C_1 > 0$ for which

$$A(x) \geqq C_1 I \qquad \text{for all } x \in \overline{\mathbf{B}}(x, r).$$

We then claim that there exists a constant $C > 0$ depending only on r, such that for any piecewise C^1 path γ between x and a point on the sphere $\mathbf{S}(x, r)$ we have

$$L(\gamma) \geqq Cr.$$

This will prove that dist$_g(x, y) \geqq Cr > 0$, and will conclude the proof that dist$_g$ is a distance.

By breaking up the path into a sum of C^1 curves, we may assume without loss of generality that our path is such a curve. Furthermore, we may take the interval $[a, b]$ on which γ is defined to be such that $\gamma(b)$ is the first point such that $\gamma(t)$ lies on $\mathbf{S}(x, r)$, and otherwise $\gamma(t) \in \overline{\mathbf{B}}(x, r)$ for $t \in [a, b]$. Let $\gamma(b) = ru$, where u is a unit vector. Write $\mathbf{E}$ as an orthogonal direct sum

$$\mathbf{E} = \mathbf{R}u \perp \mathbf{F},$$

where $\mathbf{F}$ is a closed subspace. Then $\gamma(t) = s(t)u + w(t)$ with $|s(t)| \leq r$, $s(a) = 0$, $s(b) = r$ and $w(t) \in \mathbf{F}$. Then

$$L(\gamma) = \int_a^b \|\gamma'(t)\|_g \, dt = \int_a^b \langle \gamma'(t), A(\gamma(t))\gamma'(t)\rangle^{1/2} \, dt$$

$$\geq C_1^{1/2} \int_a^b \langle \gamma'(t), \gamma'(t)\rangle^{1/2} \, dt$$

$$\geq C_1^{1/2} \int_a^b |s'(t)| \, dt \quad \text{by Pythagoras}$$

$$\geq C_1^{1/2} r$$

as was to be shown.

In addition, the above local argument also proves:

Proposition 6.1. *The distance* dist$_g$ *defines the given topology on* X. *Equivalently, a sequence* $\{x_n\}$ *in* X *converges to a point* x *in the given topology if and only if* dist$_g(x_n, x)$ *converges to* 0.

We conclude this section with some remarks on reparametrization. Let

$$\gamma \colon [a, b] \to X$$

be a piecewise C^1 path in X. To reparametrize γ, we may do so on each subinterval where γ is actually C^1, so assume γ is C^1. Let

$$\varphi \colon [c, d] \to [a, b]$$

be a C^1 map such that $\varphi(c) = a$ and $\varphi(d) = b$. Then $\gamma \circ \varphi$ is C^1, and is called a **reparametrization** of γ. The chain rule shows that

$$L(\gamma \circ \varphi) = L(\gamma).$$

Define the function $s \colon [a, b] \to \mathbf{R}$ by

$$s(t) = \int_a^t \|\gamma(t)\|_g \, dt, \quad \text{so} \quad s(b) = L = L(\gamma).$$

Then s is monotone and $s(a) = 0$, while $s(b) = L(\gamma)$. Suppose that there is only a finite number of values $t \in [a, b]$ such that $\gamma'(t) = 0$. We may then break up $[a, b]$ into subintervals where $\gamma'(t) \neq 0$ except at the end points of the subintervals. Consider each subinterval separately, and say

$$a < a_1 < b_1 < b$$

with $\gamma'(t) \neq 0$ for $t \in (a_1, b_1)$. Let $s(a_1)$ be the length of the curve over the interval $[a, a_1]$. Define

$$s(t) = s(a_1) + \int_{a_1}^{t} \|\gamma'(t)\|_g \, dt \qquad \text{for} \quad a_1 \leqq t \leqq b_1.$$

Then s is strictly increasing, and therefore the inverse function $t = \varphi(s)$ is defined over the interval. Thus we can reparametrize the curve by the variable s over the interval $a_1 \leqq t \leqq b_1$, with the variable s satisfying

$$s(a_1) \leqq s \leqq s(b_1).$$

Thus the whole path γ on $[a, b]$ is reparametrized by another path

$$\gamma \circ \varphi \colon [0, L] \to X$$

via a piecewise map $f \colon [0, L] \to [a, b]$, such that

$$\|(\gamma \circ \varphi)'(s)\|_g = 1 \qquad \text{and} \qquad L_0^s(\gamma \circ \varphi) = s.$$

We now define a path $\gamma \colon [a, b] \to X$ to be **parametrized by arc length** if $\|\gamma'(t)\|_g = 1$ for all $t \in [a, b]$. We see that starting with any path γ, with the condition that there is only a finite number of points where $\gamma'(t) = 0$ for convenience, there is a reparametrization of the path by arc length.

Let $f \colon Y \to X$ be a C^p map with $p \geqq 1$. We shall deal with several notions of isomorphisms in different categories, so in the C^p category, we may call f a differential morphism. Suppose (X, g) and (Y, h) are Riemannian manifolds. We say that f is an **isometry**, or a **differential metric isomorphism** if f is a differential isomorphism and $f^*(g) = h$. If f is an isometry, then it is immediate that f preserves distances, i.e. that

$$\text{dist}_g\big(f(y_1), f(y_2)\big) = \text{dist}_h(y_1, y_2) \qquad \text{for all} \quad y_1, y_2 \in Y.$$

Note that there is another circumstance of interest with somewhat weaker conditions when $f \colon Y \to X$ is an immersion, so induces an injection $df(y)$: $T_y Y \to T_{f(y)} X$ for every $y \in Y$, and we can speak of f being a metric immersion if $f^*(g) = h$. It may even happen that f is a local differential isomorphism at each point of y, as for instance if f is covering map. In such a case, f may be a local isometry, but not a global one, whereby f may not preserve distances on all of Y, possibly because two points $y_1 \neq y_2$ may have the same image $f(y_1) = f(y_2)$.

VII, §7. THE CANONICAL SPRAY

We now come back to the pseudo Riemannian case.

Let X be a pseudo Riemannian manifold, modeled on the self dual space $\mathbf{E}$. The scalar product $\langle \ , \ \rangle$ in $\mathbf{E}$ identifies $\mathbf{E}$ with its dual $\mathbf{E}^\vee$. The metric on X gives a toplinear isomorphism of each tangent space $T_x(X)$ with $T_x^\vee(X)$. If we work locally with $X = U$ open in $\mathbf{E}$ and we make the identification

$$T(U) = U \times \mathbf{E} \quad \text{and} \quad T^\vee(U) = U \times \mathbf{E}^\vee \approx T(U)$$

then the metric gives a VB-isomorphism

$$h: T(U) \to T(U)$$

by means of a morphism

$$g: U \to L(\mathbf{E}, \mathbf{E})$$

such that $h(x, v) = (x, g(x)v)$. (In the finite dimensional case, with respect to an orthonormal basis, $g(x)$ is represented by a symmetric matrix $(g_{ij}(x))$, so the notation here fits what's in other books with their g_{ij}.) The scalar product of the metric at each point x is then given by the formula

$$\langle v, w \rangle_x = \langle v, g(x)w \rangle = \langle g(x)v, w \rangle \quad \text{for} \quad v, w \in \mathbf{E}.$$

For each $x \in U$ we note that $g'(x)$ maps $\mathbf{E}$ into $L(\mathbf{E}, \mathbf{E})$. For $x \in U$ and $u, v \in E$ we write

$$\big(g'(x)u\big)(v) = g'(x)u \cdot v = g'(x)(u, v).$$

From the symmetry of g, differentiating the symmetry relation of the scalar product, we find that for all $u, v, w \in \mathbf{E}$,

$$\langle g'(x)u \cdot w, v \rangle = \langle g'(x)u \cdot v, w \rangle.$$

So we can interchange the last two arguments in the scalar product without changing the value.

Observe that locally, the tangent linear map

$$T(h): T\big(T(U)\big) \to T\big(T(U)\big)$$

is then given by

$$T(h): (x, v, u_1, u_2) \mapsto \big(x, g(x)v, u_1, g'(x)u_1 \cdot v + g(x)u_2\big).$$

If we pull back the canonical 2-form described in Proposition 7.2 of Chapter V from $T^{\vee}(U) \approx T(U)$ to $T(U)$ by means of h then its description locally can be written on $U \times \mathbf{E}$ in the following manner.

(1) $\langle \Omega_{(x, v)}, (u_1, u_2) \times (w_1, w_2) \rangle$

$$= \langle u_1, g(x)w_2 \rangle - \langle u_2, g(x)w_1 \rangle - \langle g'(x)u_1 \cdot v, w_1 \rangle + \langle g'(x)w_1 \cdot v, u_1 \rangle.$$

From the simple formula giving our fundamental 2-form on the cotangent bundle in Chapter V, we see at once that it is nonsingular on $T(U)$. Since h is a VB-isomorphism, it follows that the pull-back of this 2-form to the tangent bundle is also non-singular.

We shall now apply the results of the preceding section. To do so, we construct a 1-form on $T(X)$. Indeed, we have a function (**kinetic energy!**)

$$K: T(X) \to \mathbf{R}$$

given by $K(v) = \frac{1}{2}\langle v, v \rangle_x$ if v is in T_x. Then dK is a 1-form. By Proposition 6.1 of Chapter V, it corresponds to a vector field on $T(X)$, and we contend:

Theorem 7.1. *The vector field F on $T(X)$ corresponding to $-dK$ under the fundamental 2-form is a spray over X, called the* **canonical spray.**

Proof. We work locally. We take U open in $\mathbf{E}$ and have the double tangent bundle

$$(U \times \mathbf{E}) \times (\mathbf{E} \times \mathbf{E})$$

$$\downarrow$$

$$U \times \mathbf{E}$$

$$\downarrow$$

$$U.$$

Our function K can be written

$$K(x, v) = \frac{1}{2}\langle v, v \rangle_x = \frac{1}{2}\langle v, g(x)v \rangle,$$

and dK at a point (x, v) is simply the ordinary derivative

$$DK(x, v): \mathbf{E} \times \mathbf{E} \to \mathbf{R}.$$

The derivative DK is completely described by the two partial derivatives, and we have

$$DK(x, v) \cdot (w_1, w_2) = D_1 K(x, v) \cdot w_1 + D_2 K(x, v) \cdot w_2.$$

From the definition of derivative, we find

$$D_1 K(x, v) \cdot w_1 = \tfrac{1}{2} \langle v, g'(x)w_1 \cdot v \rangle$$

$$D_2 K(x, v) \cdot w_2 = \langle w_2, g(x)v \rangle = \langle v, g(x)w_2 \rangle.$$

We use the notation of Proposition 3.2 of Chapter IV. We can represent the vector field F corresponding to dK under the canonical 2-form Ω by a morphism $f: U \times E \to E \times E$, which we write in terms of its two components:

$$f(x, v) = \big(f_1(x, v), f_2(x, v) \big) = (u_1, u_2).$$

Then by definition:

$$(2) \quad \langle \Omega_{(x,v)}, \big(f_1(x, v), f_2(x, v) \big) \times (w_1, w_2) \rangle = \langle DK(x, v), (w_1, w_2) \rangle$$

$$= D_1 K(x, v) \cdot w_1 + \langle v, g(x)w_2 \rangle.$$

Comparing expressions (1) to (2), we find that as functions of w_2 they have only one term on the right side depending on w_2. From the equality of the two expressions, we conclude that

$$\langle f_1(x, v), g(x)w_2 \rangle = \langle v, g(x)w_2 \rangle$$

for all w_2, and hence that $f_1(x, v) = v$, whence our vector field F is a second order vector field on X.

Again we compare expression (1) and (2), using the fact just proved that $u_1 = f_1(x, v) = v$. Setting the right sides of the two expressions equal to each other, and using $u_2 = f_2(u, v)$, we obtain:

Proposition 7.2. *In the chart U, let $f = (f_1, f_2): U \times E \to E \times E$ represent F. Then $f_2(x, v)$ is the unique vector such that for all $w_1 \in E$ we have:*

$$\langle f_2(x, v), g(x)w_1 \rangle = \tfrac{1}{2} \langle g'(x)w_1 \cdot v, v \rangle - \langle g'(x) \cdot v \cdot v, w_1 \rangle.$$

From this one sees that f_2 is homogeneous of degree 2 in the second variable v, in other words that it represents a spray. This concludes the proof of Theorem 7.1.

Remark. Having represented $f_{U,2}(x, v)$ in the chart, we could also represent the associated bilinear map B_U. We shall give the formula for B_U in the context of Theorem 4.2 of Chapter VIII.

CHAPTER VIII

Covariant Derivatives and Geodesics

Throughout this chapter, by a manifold, we shall mean a C^∞ manifold, for simplicity of language. Vector fields, forms and other objects will also be assumed to be C^∞ unless otherwise specified. We let X be a manifold. We denote the $\mathbf{R}$-vector space of vector fields by $\Gamma T(X)$. Observe that $\Gamma T(X)$ is also a module over the ring of functions $\mathfrak{F} = \mathfrak{F}^\infty(X)$. We let

$$\pi \colon TX \to X$$

be the natural map of the tangent bundle onto X.

VIII, §1. BASIC PROPERTIES

By a **covariant derivative** D we mean an $\mathbf{R}$-bilinear map

$$D \colon \Gamma T(X) \times \Gamma T(X) \to \Gamma T(X),$$

denoted by $(\xi, \eta) \mapsto D_\xi \eta$, satisfying the two conditions:

COVD 1. (a) In the first variable ξ, $D_\xi \eta$ is $\mathfrak{F}$-linear.

(b) For a function φ, define $D_\xi \varphi = \xi \varphi = \mathscr{L}_\xi \varphi$ to be the Lie derivative of the function. Then in the second variable η, $D_\xi \eta$ is a derivation. Thus (a) and (b) can be written in the form:

$$D_{\varphi \xi} \eta = \varphi D_\xi \eta \qquad \text{and} \qquad D_\xi(\varphi \eta) = (D_\xi \varphi)\eta + \varphi D_\xi \eta.$$

COVD 2. $D_\xi \eta - D_\eta \xi = [\xi, \eta]$.

Remark. This second condition can be eliminated to give rise to a more general notion, following the ideas of a connection as described at the end of Chapter IV, §3. However, we concentrate here on what we need for some basic results, rather than develop systematically the general theory of connections.

Having defined D_ξ on functions and vector fields, we may extend the definition to all differential forms, or even to multilinear tensor fields. Let ω be in $\Gamma L'(T(X))$, i.e. ω is a multilinear tensor field on X, *not necessarily alternating*. We define $D_\xi\omega$ by giving its value on vector fields $\eta_1, \ldots, \eta_r$, namely

$$(D_\xi\omega)(\eta_1, \ldots, \eta_r) = \mathscr{L}_\xi(\omega(\eta_1, \ldots, \eta_r)) - \sum_{j=1}^{r} \omega(\eta_1, \ldots, D_\xi\eta_j, \ldots, \eta_r).$$

The definition of D_ξ is such that D_ξ satisfies the derivation property with respect to the $r+1$ variables $\omega, \eta_1, \ldots, \eta_r$, that is

$$D_\xi(\omega(\eta_1, \ldots, \eta_r)) = (D_\xi\omega)(\eta_1, \ldots, \eta_r) + \sum_{j=1}^{r} \omega(\eta_1, \ldots, D_\xi\eta_j, \ldots, \eta_r).$$

Recall that $D_\xi = \mathscr{L}_\xi$ on function, as on the left side of this equation. Looking in a local chart shows that $D_\xi\omega$ is again a multilinear tensor field. It is immediate from the definition that if ω is alternating, then so is $D_\xi\omega$. In particular, D_ξ is a derivation with respect to contractions and it is also a derivation with respect to the wedge product, that is:

COVD 3. $D_\xi(\omega \circ \eta_1) = (D_\xi\omega) \circ \eta_1 + \omega \circ D_\xi\eta_1$.

COVD 4. On the algebra of alternating forms, the covariant derivative D_ξ is a derivation, in the sense that for two forms ω and γ, we have

$$D_\xi(\omega \wedge \gamma) = D_\xi\omega \wedge \gamma + \omega \wedge D_\xi\gamma.$$

The proof comes directly from the definition of the wedge product in Chapter V, §3. In the finite dimensional case, when a form is a sum of decomposable forms, i.e. wedge products of forms of degree 0 and 1, it follows that the above definition is the unique extension of D_ξ to the algebra of differential forms. Furthermore, similarly to the formula of Proposition 5.1 of Chapter V, for the Lie derivative of a form, one has:

COVD 5. $(\mathscr{L}_\xi\omega)(\eta_1, \ldots, \eta_r) = (D_\xi\omega)(\eta_1, \ldots, \eta_r) + \sum_{i=1}^{r} \omega(\eta_1, \ldots, D_{\eta_i}\xi, \ldots, \eta_r),$

which is an alternative to

$$\mathscr{L}_\xi(\omega(\eta_1, \ldots, \eta_r)) = (D_\xi\omega)(\eta_1, \ldots, \eta_r) + \sum_{i=1}^{r} \omega(\eta_1, \ldots, D_\xi\eta_i, \ldots, \eta_r).$$

COVD 6. $(d\omega)(\xi_0, \xi_1, \ldots, \xi_r) = \sum_{i=0}^{r} (-1)^i (D_{\xi_i}\omega)(\xi_1, \ldots, \xi_{i-1}, \xi_0, \xi_{i+1}, \ldots, \xi_r).$

Proof. One uses the formulas given in propositions of Chapter V, Proposition 3.2, for $d\omega$, and Proposition 5.1 for the Lie derivative. One replaces brackets $[\beta, \gamma]$ by $D_\beta \gamma - D_\gamma \beta$. The desired formulas drop out. Note that only **COVD 2** has been used in the proof.

Next we give a finite dimensional formula. Recall that a frame of vector fields $\xi_1, \dots, \xi_n$ is such that for each x, $\xi_1(x), \dots, \xi_n(x)$ are a basis of $T_x X$.

Proposition 1.1. *Let* $\{\xi_1, \dots, \xi_n\}$ *be a frame of vector fields. Let* λ_1, $\dots, \lambda_n$ *be the dual frame of 1-forms (so* $\lambda_i(\xi_j) = \delta_{ij}$). *For any form* $\omega \in \mathscr{A}^r(X)$ *we have*

$$d\omega = \sum_{i=1}^n \lambda_i \wedge D_{\xi_i} \omega.$$

Proof. Let $d'\omega = \sum \lambda_i \wedge D_{\xi_i}\omega$. Then d' defines an anti-derivation of the alternating algebra of forms, that is if $\psi \in \mathscr{A}^q(X)$ for any q, then

$$d'(\omega \wedge \psi) = (d'\omega) \wedge \psi + (-1)^r \omega \wedge d'\psi.$$

Furthermore, $d' = d$ on functions (as is immediately verified), and we verify that $d' = d$ on $\mathscr{A}^1(X)$ as follows:

$$
\begin{aligned}
(d'\omega)(\xi, \eta) &= \sum (\lambda_i \wedge D_{\xi_i}\omega)(\xi, \eta) \\
&= \sum [\lambda_i(\xi)\langle D_{\xi_i}\omega, \eta \rangle - \lambda_i(\eta)\langle D_{\xi_i}\omega, \xi \rangle] \\
&= \sum [\langle D_{\lambda_i(\xi)}\omega, \eta \rangle - \langle D_{\lambda_i(\eta)}\omega, \xi \rangle] \\
&= \langle D_\xi \omega, \eta \rangle - \langle D_\eta \omega, \xi \rangle \\
&= (d\omega)(\xi, \eta) \quad \text{by } \textbf{COVD 6},
\end{aligned}
$$

which concludes the proof for 1-forms. Since 1-forms generate the algebra of forms in the finite dimensional case, the proposition is proved in general.

The above finite dimensional formula won't be used until we meet strictly finite dimensional results, in connection with volume forms and integration. We included it here for completeness of the general formalism. We now return to the general case which may be infinite dimensional.

We can extend the covariant derivative to TX-valued forms i.e. sections of the bundle $L^r(TX, TX)$. If ω is such a section, we define $D_\xi \omega$ by its values on vector fields $\eta_1, \dots, \eta_r$ to be

$$(D_\xi \omega)(\eta_1, \dots, \eta_r) = D_\xi(\omega(\eta_1, \dots, \eta_r)) - \sum_{j=1}^r \omega(\eta_1, \dots, D_\xi \eta_j, \dots, \eta_r),$$

so D_ξ satisfies the derivation property with respect to the $r + 1$ variables ω, η_1, ..., η_r. We note that $\omega(\eta_1, ..., \eta_r) \in \Gamma T X$ is a vector field, so we know how to apply the covariant derivative $D_\xi(\omega(\eta_1, ..., \eta_r))$ instead of $\mathscr{L}_\xi(\omega(\eta_1, ..., \eta_r))$ for ordinary **R**-valued forms, in which case $\omega(\eta_1, ..., \eta_r)$ is a function on X. When ω is TX-valued, we have on the other hand

$$\mathscr{L}_\xi(\omega(\eta_1, ..., \eta_r)) = [\xi, \omega(\eta_1, ..., \eta_r)].$$

A local formula will be given in Proposition 2.2.

VIII, §2. SPRAYS AND COVARIANT DERIVATIVES

Let F be a spray over a manifold X. In a chart U, we index geometric objects by U to indicate their representatives in the chart. Thus the representative ξ_U of a vector field over U is a morphism

$$\xi_U : U \to \mathbf{E}.$$

Similarly, we have the symmetric bilinear map associated with the spray, and its representative

$$B_U(x) = \tfrac{1}{2} D_2^2 f_{U,2}(x, 0),$$

where $f_{U,2}$ is the second component of the representative for the spray, as described in Chapter IV, §3.

Theorem 2.1. *Given a spray F over X, there exists a unique covariant derivative D such that in a chart U, the derivative is given by the local formula*

$$(D_\xi \eta)_U(x) = \eta_U'(x)\xi_U(x) - B_U(x; \xi_U(x), \eta_U(x)).$$

Or, suppressing the index U for simplicity, and thus using ξ, η to denote the local representatives of the vector fields in the chart, we have

$$(D_\xi \eta)(x) = \eta'(x)\xi(x) - B(x; \xi(x), \eta(x))$$

or simply

$$D_\xi \eta = \eta' \cdot \xi - B(\xi, \eta).$$

Proof. Let us define $D_\xi \eta$ over U by the formula of the theorem. It is immediately verified that $D_\xi \eta$ is a vector field over U, and that the association $(\xi, \eta) \mapsto D_\xi \eta$ is a covariant derivative over U: It is $\mathfrak{F}(U)$-linear in the variable ξ, it is a derivation in the variable η with respect to multiplication by functions, and we have

$$D_\xi \eta - D_\eta \xi = [\xi, \eta].$$

This last property follows from the representation of the bracket in a chart given by Proposition 1.3 of Chapter V. Thus a spray gives rise to a covariant derivative in a chart, in a natural fashion.

We now claim that when the spray is given globally, there exists a unique covariant derivative on the manifold X which has the above representation in a chart. For this we must verify how the local representation changes under a change of chart. Let

$$h: U \to V$$

be a C^∞-isomorphism, i.e. a change of chart. Then we claim that the natural image of $D_{\xi_U}\eta_U$ under the change of chart is $D_{\xi_V}\eta_V$, so that we may define $D_\xi\eta$ for any two vector fields on the manifold via the local representations.

In other words, we have to verify that

$$(D_{\xi_V}\eta_V)(h(x)) = h'(x)(D_{\xi_U}\eta_U)(x).$$

But we have

$$\eta_V(h(x)) = h'(x)\eta_U(x),$$

whence by the rule for the derivative of a product, we obtain

$$(\eta_V \circ h)'(x) = h''(x)\eta_U(x) + h'(x)\eta_U'(x).$$

Hence putting $v = \xi_U(x)$, $w = \eta_U(x)$, we get by using the change of variable formula for a spray in a chart, Proposition 3.3 of Chapter IV, together with the fact that $h''(x)$ is a symmetric bilinear map:

$$
\begin{aligned}
(D_{\xi_V}\eta_V)(h(x)) &= \eta_V'(h(x))h'(x)\xi_U(x) - B_V(h(x); h'(x)v, h'(x)w) \\
&= (\eta_V \circ h)'(x)\xi_U(x) - h''(x)(v, w) - h'(x)B_U(x; v, w) \\
&= h''(x)(w, v) + h'(x)\eta_U'(x)\xi_U(x) \\
&\quad - h''(x)(v, w) - h'(x)B_U(x; v, w) \\
&= h'(x)\big(\eta_U'(x)\xi_U(x) - B_U(x; v, w)\big)
\end{aligned}
$$

(appreciate the cancellation!)

$$= h'(x)(D_\xi\eta)_U(x),$$

which proves the change of variable formula, and therefore concludes the proof of Theorem 2.1.

The covariant derivative defined in Theorem 2.1 will be called the **covariant derivative determined by the spray**, or **associated with the spray**. As mentioned previously, one could give a similar definition of a covariant derivative associated to any connection (even without the symmetry condition on the bilinear map).

There is of course an analogous local representation for differential forms as follows.

Proposition 2.2. *Let* $\omega \in \Gamma L'(TX, \mathbf{R})$ *or* $\Gamma L'(TX, TX)$. *Let* $\xi, \eta_1, \ldots, \eta_r$ *be vector fields over* X. *If* $\omega \in \Gamma L'(TX, \mathbf{R})$, *then in a chart* U *we have the formula*

$$(D_\xi \omega)_U(\eta_{1U}, \ldots, \eta_{rU})$$

$$= \omega'_U(\xi_U)(\eta_{1U}, \ldots, \eta_{rU}) + \sum_{j=1}^{r} \omega_U(\eta_{1U}, \ldots, B_U(\xi_U, \eta_{jU}), \ldots, \eta_{rU}).$$

If $\omega \in \Gamma L'(TX, TX)$, *then*

$$(D_\xi \omega)_U(\eta_{1U}, \ldots, \eta_{rU}) = same\ expression - B_U(\xi_U, \omega_U(\eta_{1U}, \ldots, \eta_{rU})).$$

Proof. This comes directly from the definitions in §1. Observe that in applying the definitions, the sum

$$\sum_{j=1}^{r} \omega_U(\eta_{1U}, \ldots, \eta'_{1U} \cdot \xi, \ldots, \eta_{rU})$$

occurs twice, once with a $+$ sign and once with a $-$ sign, so cancels in the end.

For the limited purposes of this book, we will not need the proposition. It has an analogue for lifts of curves, which we shall discuss briefly at the end of §3.

Converse, from covariant derivatives to sprays

We now wish to discuss the converse of Theorem 2.1, and for this purpose, we have to make general remarks on localization. Let $\mathbf{E}$ be a Banach space. We say that $\mathbf{E}$ **admits cut off functions** if given two positive real numbers $0 < r < s$, there exists a C^∞-function (simply called function) φ such that $\varphi = 1$ on the ball $B_r(0)$ and $\varphi = 0$ on the complement of $B_s(0)$. Given any point $x_0 \in \mathbf{E}$, we may then find similarly a function which is 1 in the ball $B_r(x_0)$ and 0 outside $B_s(x_0)$. If X is a manifold modeled on $\mathbf{E}$, then one can then find such cut off functions equal to 1 in a given neighborhood of a point, and 0 outside a slightly larger neighborhood. Manifolds modelled on a Hilbert space, and especially finite dimensional manifolds, admit cut off functions.

Assume that X admits cut off functions. Let E be a vector bundle over X, and let ξ be a section of E. Let $x_0 \in X$. Let φ a cut off function near x_0. Then $\varphi\xi$ is a section of E, having the same values as ξ in a neighborhood of x_0. Suppose that $E = TX$ and that D is a connection. Then

$$(D_\xi \eta)(x) = (D_{\varphi\xi} \eta)(x)$$

for all x in a sufficiently small neighborhood of x_0, because D is $\mathfrak{F}$-linear in the first variable. Since φ is constant near x_0, it follows that

$$(\mathscr{L}_\xi \varphi)(x) = 0 \qquad \text{for } x \text{ near } x_0,$$

and it therefore follows also that

$$(D_\xi(\varphi\eta))(x) = (D_\xi \eta)(x)$$

for all x sufficiently close to x_0.

Now given an open neighborhood U_0 of x_0 corresponding to a chart, we pick cut off functions φ, ψ near x_0 such that the supports of φ, ψ are contained in U_0, and φ, $\psi = 1$ on an open neighborhood U of x whose closure is contained in U_0. Then U also corresponds to a chart, and we may compute

$$(D_\xi \eta)(x) = (D_{\varphi\xi}(\psi\eta))(x) \qquad \text{for} \quad x \in U.$$

Thus the determination of the values of a covariant derivative can be carried out locally in a chart. We still need a criterion when the value of the covariant derivative at a given point depends only on the value of ξ at the given point.

Lemma 2.3. *Let E, F be vector bundles over X, with E finite dimensional and X admitting cut off functions. Let*

$$H : \Gamma E \to \Gamma F$$

be a linear map which is $\mathfrak{F}(X)$-linear, that is $H(\varphi\xi) = \varphi H(\xi)$ for $\varphi \in \mathfrak{F}$. Given a point $x \in X$, the value $H(\xi)(x)$ depends only on the value $\xi(x)$.

Proof. It suffices to prove that if $\xi(x_0) = 0$ then $H(\xi)(x_0) = 0$. There exists a cut off function φ near x_0 by assumption, so we may give the proof locally. By assumption, there exists a finite number of sections e_1, ...,e_r of E which form a basis for the sections locally, so there exist functions $\varphi_1, \ldots, \varphi_r$ such that

$$\xi = \varphi_1 e_1 + \cdots + \varphi_r e_r$$

locally. Then

$$H(\xi) = \varphi_1 H(e_1) + \cdots + \varphi_r H(e_r).$$

The condition $\xi(x_0) = 0$ is equivalent with the conditions $\varphi_i(x_0) = 0$ for all i. Hence $H(\xi)(x_0) = 0$, thus proving the lemma.

Observe that when we obtain a covariant derivative from a spray, the value of the covariant derivative at a point x depends only on the value of the vector field $\xi(x)$ (a derivative of η however enters). This was clear from the local formula in Theorem 2.1, because for instance $B_U(x; u, w)$ is defined for arbitrary vectors u, w which can then be taken to be the values $\xi_U(x)$ and $\eta_U(x)$ respectively.

Conversely, we are now interested in reversing the procedure. Specifically, let D be a covariant derivative. We assume the existence of cut off functions throughout. In a chart over an open set U in $\mathbf{E}$, define

$(\mathbf{B}_U)$ $\qquad\qquad B_U(x; \xi, \eta) = \eta'(x)\xi(x) - (D_{\xi_U}\eta_U)(x).$

It is immediately verified from the two properties of a covariant derivative that $B_U(x)$ is symmetric in ξ_U, η_U by **COVD 2**, and then $B_U(x)$ is $\mathfrak{F}(U)$-bilinear in ξ_U, η_U. Given vectors u, $w \in \mathbf{E}$ one wants to define

$$B_U(x)(u, w) = B_U\big(x; \xi(x), \eta(x)\big)$$

for any vector fields ξ, η such that $\xi(x) = u$ and $\eta(x) = w$. At this point, we need to know that the value on the right of $(\mathbf{B}_U)$ is independent of the vector fields ξ, η chosen so that $\xi(x) = v$ and $\eta(x) = w$. By Lemma 2.3 we can certainly achieve this in the finite dimensional case, and in that case we obtain:

Theorem 2.4. *Assume X finite dimensional. Then the association of a covariant derivative to a spray establishes a bijection between sprays over X and covariant derivatives.*

In practice, Theorem 2.4 is not that useful (and it will **NOT** be used in this book) because one either starts from a spray to get a covariant derivative, or if one starts from some natural covariant derivative, and one needs the spray, the situation provides the tools to show that a spray can indeed be defined in a natural manner to give the covariant derivative. We shall see an example of this in §4, when we discuss the Riemannian covariant derivative. Furthermore, the finite dimensional device used in Lemma 2.3 has had historically the unfortunate effect of obscuring the natural bilinear map B, thus obscuring a fundamental structure in expositions of differential geometry. Quite generally, connections on any vector bundle give rise to covariant derivatives. These are applicable to many contexts of topology and analysis, see for example [BGV 92], Chapter I, and also for instance [MokSY 93] for an entirely different direction.

VIII, §3. DERIVATIVE ALONG A CURVE AND PARALLELISM

Instead of using vector fields ξ, η we may carry out a similar construction of a differentiation dealing only with curves, as follows.

We continue to denote by F a spray over X. Let $\pi: TX \to X$ be the tangent bundle, and let

$$\alpha: J \to X$$

be a C^1 curve. By a **lift** γ of α to TX we mean a C^1 curve $\gamma: J \to TX$ such that $\pi\gamma = \alpha$. We then also say that γ **lies above** α. We denote the set of lifts of α by $\mathrm{Lift}(\alpha)$. It is clear that $\mathrm{Lift}(\alpha)$ is a vector space over $\mathbf{R}$, and a module over the ring of functions on J. We wish to define $D_{\alpha'}\gamma$ in a way analogous to the way we defined $D_\xi\eta$ for vector fields ξ, η. This is done by the next theorem. As in §2, we let B_U denote the bilinear map associated to the spray in a chart U.

Theorem 3.1. *There exists a unique linear map*

$$D_{\alpha'}: \mathrm{Lift}(\alpha) \to \mathrm{Lift}(\alpha)$$

which in a chart U has the expression

$$(D_{\alpha'}\gamma)_U(t) = \gamma'_U(t) - B_U\big(\alpha(t);\, \alpha'_U(t),\, \gamma_U(t)\big).$$

The map $D_{\alpha'}$ satisfies the derivation property for a C^1 function φ on J:

$$\big(D_{\alpha'}(\varphi\gamma)\big)(t) = \varphi'(t)(D_{\alpha'}\gamma)(t) + \varphi(t)(D_{\alpha'}\gamma)(t).$$

Remark. In the present context, the local representation γ_U of a curve in $TU = U \times \mathbf{E}$ is taken to be the map on the second component, i.e.

$$\gamma_U: J \to \mathbf{E}.$$

Thus $\gamma'_U(t)$ is the ordinary derivative, with values $\gamma'_U(t) \in \mathbf{E}$. Note that in the case of the representation $\alpha_U: J \to U$, we have $\alpha'_U(t) \in \mathbf{E}$ also. Thus $\alpha'_U(t)$, $\gamma_U(t)$ and $\gamma'_U(t)$ are "vectors."

Proof of Theorem 3.1. The proof is entirely analogous to the proof for Theorem 2.1, using the local representation of the bilinear map B_U associated with a spray in charts. We have to verify that the formula of Theorem 3.1 transforms in the proper way under a change of charts, i.e. under an isomorphism

$$h: U \to V.$$

Note that the local representation γ_V of the curve by definition is given by

$$\gamma_V(t) = h'(\alpha_U(t))\gamma_U(t).$$

Therefore by the rule for the derivative of a product, we find:

$$\gamma'_V(t) = h''(\alpha_U(t))(\alpha'_U(t), \gamma_U(t)) + h'(\alpha_U(t), \gamma'_U(t)).$$

Hence using the transformation rule from B_U to B_V, Proposition 3.3 of Chapter IV, we get

$$
\begin{aligned}
(D_{\alpha'}\gamma)_V(t) &= \gamma'_V(t) - B_V(\alpha(t); \alpha'_V(t), \gamma_V(t)) \\
&= h''(\alpha_U(t))(\alpha'_U(t), \gamma_U(t)) + h'(\alpha_U(t))\gamma'_U(t) - h''(\alpha_U(t))(\alpha'_U(t), \gamma_U(t)) \\
&\quad - h'(\alpha_U(t))B_U(\alpha(t), \alpha'_U(t), \gamma_U(t)) \\
&= h'(\alpha_U(t))(D_{\alpha'}\gamma)_U(t) \quad \text{(because the } h'' \text{ term cancels!)},
\end{aligned}
$$

which proves the desired transformation formula for $(D_{\alpha'}\gamma)_U$ in charts. Thus we have proved the existence of $D_{\alpha'}\gamma$ as asserted. Its being a derivation is immediate from the local representation in charts. This concludes the proof of Theorem 3.1.

Corollary 3.2. *Let η be a vector field and suppose $\gamma(t) = \eta(\alpha(t))$, $t \in J$. Let ξ be a vector field on X such that $\alpha'(t_0) = \xi(\alpha(t_0))$ for some $t_0 \in J$. Then*

$$(D_{\alpha'}\gamma)(t_0) = (D_\xi\eta)(\alpha(t_0)).$$

Proof. Immediate from the chain rule and the local representation of Theorem 3.1.

Let $\alpha: J \to X$ be a C^2-morphism. We say that a lift $\gamma: J \to TX$ of α is α-**parallel** if $D_{\alpha'}\gamma = 0$. In the chart U, this is equivalent to the condition that

$$\gamma'_U(t) = B_U(\alpha_U(t); \alpha'_U(t), \gamma_U(t)),$$

which defines a first-order linear differential equation for γ_U. From the basic properties of linear differential equations, we obtain:

Theorem 3.3. *Let $\alpha: J \to X$ be a C^2 curve in X. Let $t_0 \in J$. Given $v \in T_{\alpha(t_0)}X$, there exists a unique lift $\gamma_v: J \to TX$ which is α-parallel and such that $\gamma_v(t_0) = v$. Let $\mathrm{Par}(\alpha)$ denote the set of α-parallel lifts of α. The map $v \mapsto \gamma_v$ is a linear isomorphism of $T_{\alpha(t_0)}X$ with $\mathrm{Par}(\alpha)$.*

Proof. The existence and uniqueness simply comes from the existence and uniqueness of solutions of differential equations. Note that from the linearity of the equation, the integral curve γ is defined on the whole interval of definition J by Proposition 1.9 of Chapter IV.

Of course, the notion of parallelism is *with respect to the given spray*, which has been left out of the notation. We express the linearity of Theorem 3.3 another way in the next theorem.

Theorem 3.4. *Fix $t_0 \in J$. For $t \in J$ define the map*

$$P_{\alpha, t_0, t} = P_t: T_{\alpha(t_0)}X \to T_{\alpha(t)}X \qquad by \qquad P_t(v) = \gamma(t, v),$$

where $t \mapsto \gamma(t, v)$ is the unique curve in TX which is α-parallel and $\gamma(t_0, v) = v$. Then P_t is a linear isomorphism.

Proof. We must verify that $P_t(sv) = sP_t(v)$ and $P_t(v + w) = P_t(v) + P_t(w)$ for $s \in \mathbf{R}$ and $v, w \in T_xX$. But these properties follow at once from the linearity of the differential equation satisfied by γ, and the uniqueness theorem for its solutions with given initial conditions.

The map P_t is called **parallel translation** along α.

Multilinear tensor fields

Instead of dealing with vector fields, we may deal with TX-valued multilinear tensor fields, or **R**-valued multilinear tensor fields at essentially no extra cost. *Let E denote either TX or* **R**. We extend $D_{\alpha'}$ to a linear map

$$D_{\alpha'}: \text{Lift}(\alpha, L^r(TX, E)) \to \text{Lift}(\alpha, L^r(TX, E))$$

as follows. Let $\omega: J \to L^r(TX, E)$ be a lift of $\alpha: J \to X$. Let $\eta_1, \ldots, \eta_r$ be lifts of α in TX (sometimes called **vector fields along the curve** α). We **define** $D_{\alpha'}\omega$ by its values on $(\eta_1, \ldots, \eta_r)$ to be

$$(D_{\alpha'}\omega)(\eta_1, \ldots, \eta_r) = D_{\alpha'}(\omega(\eta_1, \ldots, \eta_r)) - \sum_{j=1}^{r} \omega(\eta_1, \ldots, D_{\alpha'}\eta_j, \ldots, \eta_r).$$

Thus $D_{\alpha'}$ satisfies the Leibniz rule for the derivative of a multifold product with the $r + 1$ variables $\omega, \eta_1, \ldots, \eta_r$. Note that *if $\eta_1, \ldots, \eta_r$ are* α-parallel, so $D_{\alpha'}\eta_j = 0$, then the formula simplifies to

$$(D_{\alpha'}\omega)(\eta_1, \ldots, \eta_r) = D_{\alpha'}(\omega(\eta_1, \ldots, \eta_r)).$$

We shall obtain a local formula as usual. Given an index j, we define a linear operator $C_{j,B,\alpha}$ of $\Gamma L'(TX, E)$ into itself by

$$(C_{j,B,\alpha}\omega)(\eta_1,\ldots,\eta_r) = \omega(\eta_1,\ldots,B(\alpha;\alpha',\eta_j),\ldots,\eta_r).$$

Proposition 3.5 (Local Expression). *Let $\omega = \omega_U$, $\eta_j = \eta_{jU}$ etc. represent the respective objects in a chart U, omitting the subscript U to simplify the notation. Then*

$$(D_{\alpha'}\omega)(\eta_1,\ldots,\eta_r) = \omega'(\eta_1,\ldots,\eta_r) - B(\alpha;\alpha',\omega(\eta_1,\ldots,\eta_r))\delta_{E,TX}$$
$$+ \sum_{j=1}^{r} \omega(\eta_1,\ldots,B(\alpha;\alpha',\eta_j),\ldots,\eta_r)$$

or also

$$D_{\alpha'}\omega = \omega' - B(\alpha;\alpha',\omega)\delta_{E,TX} + \sum_{j=1}^{r} C_{j,B,\alpha}\omega,$$

where $\delta_{E,TX} = 1$ if $E = TX$ and 0 if $E = \mathbf{R}$.

This comes from the definition at the end of §1, and the fact that the ordinary derivative

$$\left(\omega_U(\eta_{1U},\ldots,\eta_{rU})\right)'$$

in the chart is obtained by the Leibniz rule (suppressing the index U)

$$\left(\omega(\eta_1,\ldots,\eta_r)\right)' = \omega'(\eta_1,\ldots,\eta_r) + \sum \omega(\eta_1,\ldots,\eta_j',\ldots,\eta_r).$$

Corollary 3.6. *Let $E = TX$ or $\mathbf{R}$ as above. Let $\Omega: X \to L'(TX, E)$ be a section (so a tensor field), and let $\omega(t) = \Omega(\alpha(t))$, $t \in J$. Let $t_0 \in J$. Let ξ be a vector field such that $\alpha'(t_0) = \xi(\alpha(t_0))$. Then*

$$(D_{\alpha'}\omega)(t_0) = (D_\xi\Omega)(\alpha(t_0)).$$

Proof. Immediate from the chain rule and the local representation formula.

A lift $\gamma: J \to L'(TX, E)$ is called α-**parallel** if $D_{\alpha'}\gamma = 0$. The local expression in a chart U shows that the condition $D_{\alpha'}\gamma = 0$ is locally equivalent to the condition

$$\gamma' = B(\alpha;\alpha',\gamma) - \sum_{j=1}^{r} C_{j,B,\alpha}\gamma.$$

Of course, we have suppressed the subscript U from the notation. Thus the condition of being α-parallel defines locally an ordinary linear differential equation, and we obtain from the standard existence and uniqueness theorems:

Theorem 3.7. *Let* $t_0 \in J$ *and* $\omega_0 \in \Gamma L'(T_{\alpha(t_0)}X, E_{\alpha(t_0)})$. *There exists a unique curve* $\gamma: J \to L'(TX, E)$ *which is* α-*parallel and such that* $\gamma(t_0) = \omega_0$. *Denote this curve by* γ_{ω_0}. *The map*

$$\omega_0 \mapsto \gamma_{\omega_0}$$

establishes a linear isomorphism between the Banach space $L'(T_{\alpha(t_0)}X, E_{\alpha(t_0)})$ *and the space of lifts* $\mathrm{Lift}(\alpha, L'(TX, E))$.

We have now reached a point where we have the parallelism analogous to the simplest case of the tangent bundle as in Theorem 3.4.

Theorem 3.8. *Let the notation be as in Theorem 3.7. For* $t \in J$ *define the map*

$$P_{\alpha, t_0, t} = P_{\alpha, t}: L'(T_{\alpha(t_0)}X, E_{\alpha(t_0)}) \to L'(T_{\alpha(t)}X, E_{\alpha(t)})$$

by

$$P_{\alpha, t}(\omega_0) = \gamma(t, \omega_0),$$

where $t \mapsto \gamma(t, \omega_0)$ *is the unique* α-*parallel lift of* α *with* $\gamma(0, \omega_0) = \omega_0$. *Then* $P_{\alpha, t}$ *is a linear isomorphism.*

Proof. This follows at once from the linearity of the differential equation satisfied by γ, and the uniqueness theorem for its solutions with given initial conditions.

Example. The metric g itself is a symmetric bilinear **R**-valued tensor to which the above results can be applied.

VIII, §4. THE METRIC DERIVATIVE

Let (X, g) be a pseudo Riemannian manifold. Let

$$\langle v, w \rangle_g = \langle v, w \rangle_{g(x)}$$

denote the scalar product on the tangent bundle, with $v, w \in T_x$ for some x. If ξ, η are vector fields, then $\langle \xi, \eta \rangle_g$ is a function on X, whose value at a point x is

$$\langle \xi(x), \eta(x) \rangle_g = \langle \xi(x), \eta(x) \rangle_{g(x)}.$$

If ζ is a vector field, we denote

$$\zeta \langle \xi, \eta \rangle_g = D_\zeta \langle \xi, \eta \rangle_g = \mathcal{L}_\zeta \langle \xi, \eta \rangle_g.$$

Theorem 4.1. *Let (X, g) be a pseudo Riemannian manifold. There exists a unique covariant derivative D such that for all vector fields ξ, η, ζ we have*

MD 1. $D_\xi \langle \eta, \zeta \rangle_g = \langle D_\xi \eta, \zeta \rangle_g + \langle \eta, D_\xi \zeta \rangle_g.$

This covariant derivative is called the **pseudo Riemannian derivative,** *or* **metric derivative,** *or* **Levi-Civita derivative.**

Proof. For the uniqueness, we shall express $\langle D_\xi \eta, \zeta \rangle_g$ entirely in terms of operations which do not involve the derivative D. To do this, we write down the first defining property of a connection for a cyclic permutation of the three variables:

$$\xi \langle \eta, \zeta \rangle_g = \langle D_\xi \eta, \zeta \rangle_g + \langle \eta, D_\xi \zeta \rangle_g,$$
$$\eta \langle \zeta, \xi \rangle_g = \langle D_\eta \zeta, \xi \rangle_g + \langle \zeta, D_\eta \xi \rangle_g,$$
$$\zeta \langle \xi, \eta \rangle_g = \langle D_\zeta \xi, \eta \rangle_g + \langle \xi, D_\zeta \eta \rangle_g.$$

We add the first two relations and subtract the third. Using the second defining property of a covariant derivative, the following property drops out:

MD 2. $2\langle D_\xi \eta, \zeta \rangle_g = \xi \langle \eta, \zeta \rangle_g + \eta \langle \zeta, \xi \rangle_g - \zeta \langle \xi, \eta \rangle_g$
$$+ \langle [\xi, \eta], \zeta \rangle_g - \langle [\xi, \zeta], \eta \rangle_g - \langle [\eta, \zeta], \xi \rangle_g.$$

This proves the uniqueness.

As to existence, define $\langle D_\xi \eta, \zeta \rangle_g$ to be $\frac{1}{2}$ of the right side of **MD 2**. If we view ξ, η as fixed, and ζ as variable, then this right side can be checked in a chart to give a continuous linear functional on vector fields. By Proposition 6.1 of Chapter V, such a functional can be represented by a vector, and this vector defines $D_\xi \eta$ at each point of the manifold. Thus $D_\xi \eta$ is itself a vector field. Using the basic property of the bracket product with a function φ:

$$[\xi, \varphi\eta] = \varphi[\xi, \eta] + (\xi\varphi)\eta \quad \text{and} \quad [\varphi\xi, \eta] = \varphi[\xi, \eta] - (\eta\varphi)\xi$$

it is routinely verified that $\langle D_\xi \eta, \zeta \rangle_g$ is $\mathfrak{F}$-linear in its first variable ξ, and also $\mathfrak{F}$-linear in the third variable ζ. One also verifies routinely that **COVD 2** is also satisfied, whence existence follows and the theorem is proved.

Recall that we defined $D_\xi \omega$ for any multilinear tensor ω. In particular, let $\omega = g$ be the metric. Then the defining property of the metric connection can now be phrased by stating that for all vector fields ξ,

$$D_\xi g = 0.$$

For each vector field η let $\bigvee_g \eta$ or $\eta^\vee$ be the 1-form corresponding to η under the metric, i.e. for all vector fields ζ, $(\bigvee_g \eta)(\zeta) = \langle \eta, \zeta \rangle_g$.

Corollary. *For the metric derivative D and all vector fields ξ, we have the commutation rule*

$$D_\xi \circ \bigvee_g = \bigvee_g \circ D_\xi \qquad or \quad D_\xi(\eta^\vee) = (D_\xi \eta)^\vee.$$

Proof. One line:

$$\left(\bigvee_g (D_\xi \eta)\right)(\zeta) = \langle D_\xi \eta, \zeta \rangle_g = D_\xi \langle \eta, \zeta \rangle_g - \langle \eta, D_\xi \zeta \rangle_g = D_\xi(\bigvee_g \eta)(\zeta).$$

Local representation of the metric derivative

From **MD 2**, we derive a local formula in a chart U. *In the next formula, we write ξ, η, $\zeta : U \to \mathbf{E}$ for the representatives of vector fields in the chart, instead of the correct ξ_U, η_U, ζ_U.* Omitting the index U simplifies the notation when U is fixed throughout the discussion. Here

$$g : U \to L(\mathbf{E}, \mathbf{E})$$

denotes the operator defining the metric relative to the given non-singular form on E, so that

$$\langle \xi, \eta \rangle_g = \langle g\xi, \eta \rangle = \langle \xi, g\eta \rangle.$$

Observe that in **COVD 2** and **MD 2**, we took the scalar product in the tangent space, but in the next formula, the scalar product $\langle \, , \, \rangle$ without an index is the one given by our original non-singular symmetric bilinear form on **E**.

MD 3. *Locally in a chart U, the metric derivative is determined by the formula:*

$$2\langle D_\xi \eta, g\zeta \rangle$$
$$= 2\langle g\zeta, \eta' \cdot \xi \rangle + \langle \eta, g' \cdot \xi \cdot \zeta \rangle + \langle \xi, g' \cdot \eta \cdot \zeta \rangle - \langle \xi, g' \cdot \zeta \cdot \eta \rangle.$$

Proof. We apply **MD 2**. We express a g-scalar product in terms of the standard scalar product, and we use the local representations of the

Lie derivative and the bracket from Chapter IV, Proposition 1.1 and Proposition 1.3. For instance, we have the local representation

$$\xi\langle \eta, \zeta\rangle_g = \langle \eta, g\zeta\rangle'\xi$$
$$= \langle \eta'\cdot\xi, g\zeta\rangle + \langle \eta, g'\cdot\xi\cdot\zeta\rangle + \langle \eta, g\zeta'\cdot\xi\rangle$$

by using the rule for the derivative of a product. This formula is meant to be evaluated at each point x. Note that $g'(x)\colon \mathbf{E}\times\mathbf{E}\to\mathbf{E}$ is a bilinear map, which is such that, for instance:

$$g'(x)\cdot\xi(x)\cdot\zeta(x) = g'(x)\big(\xi(x), \zeta(x)\big).$$

One can work formally without putting the (x) in the notation. Similarly,

$$\langle[\xi,\eta],\zeta\rangle_g = \langle \eta'\cdot\xi - \xi'\cdot\eta, g\zeta\rangle$$
$$= \langle g\zeta, \eta'\cdot\xi - \xi'\cdot\eta\rangle$$
$$= \langle g\zeta, \eta'\cdot\xi\rangle - \langle g\zeta, \xi'\cdot\eta\rangle.$$

Thus we can transform each term appearing on the right of **MD 2**. Then all the terms involving g (rather than g') will cancel except two of them which are equal, and add to yield $2\langle g\zeta, \eta'\cdot\xi\rangle$. The remaining terms are those which are shown on the right side of **MD 3**. This concludes the proof.

Remark. Let us denote by D^g the covariant derivative associated with the metric g. Let $c\in\mathbf{R}^+$. Then cg is also a metric, called a **scaling** of g, and it follows immediately from **MD 3** that

$$D^{cg} = D^g,$$

i.e. the covariant derivative is invariant under a scaling of the metric.

Observe that the definition of the metric derivative in Theorem 4.1 is given by a formula, namely **MD 2**, with its local representation **MD 3**. We want to see that the metric derivative is the one associated with a spray. We recall that quadratic maps and symmetric bilinear maps correspond to each other via the formulas

$$Q(v) = B(v, v) \qquad \text{and} \qquad B(v, w) = \tfrac{1}{2}[Q(v + w) - Q(v) - Q(w)].$$

The next theorem summarizes the situation.

Theorem 4.2. *Let (X, g) be a pseudo Riemannian manifold. There exists a unique spray on X satisfying the following two equivalent conditions.*

MS 1. *In a chart U, the associated bilinear map B_U satisfies the following formula for all v, w, $z \in \mathbf{E}$:*

$$-2\langle B_U(x; v, w), g(x)z \rangle =$$
$$\langle g'(x) \cdot v \cdot z, w \rangle + \langle g'(x) \cdot w \cdot z, v \rangle - \langle g'(x) \cdot z \cdot w, v \rangle.$$

Thus if we let $f_{U,2}(x, v) = B_U(x; v, v)$ and $f_U(x, v) = (v, f_{U,2}(x, v))$, then f_U represents the spray on $TU = U \times \mathbf{E}$.

MS 2. *The covariant derivative associated to the spray is the metric derivative satisfying Theorem 4.1.*

This spray is the same as the canonical spray of Chapter VII, Theorem 6.1.

Proof. First observe that B_U as defined by the formula is symmetric in (v, w). The symmetry is built in the sum of the first two terms, and to see that the third term is symmetric, one differentiates with respect to x the formula

$$\langle g(x)z, v \rangle = \langle g(x)v, z \rangle,$$

which merely expresses the symmetry of $g(x)$ itself. Thus we may form the quadratic map $f_{U,2}(x, v) = B_U(x; v, v)$ from the symmetric bilinear map $B_U(x; v, w)$. It follows that f_U as defined represents a spray F_U over TU. At this point, one may argue in two ways to globalize.

Comparing **MD 3** with **MS 1** we see that the covariant derivative on U determined by the spray F_U is precisely the metric derivative. Theorem 2.1 shows that if two sprays determine the same covariant derivative on U then they are equal. If U, V are two charts, then f_U and f_V are the local representatives of sprays F_U and F_V on U and V respectively, which must therefore coincide on $U \cap V$. Hence the family $\{F_U\}$ defines a spray F on X. Once again, Theorem 2.1 and **MD 3** show that covariant derivative determined by F is the metric derivative.

Furthermore, if we substitute $v = w$ (and $z = w_1$) in the chart formula of **MS 1**, thus giving the quadratic expression $f_{U,2}(x, v)$, then one sees that this expression coincides with the chart expression of Proposition 7.2 of Chapter VII, and hence that the spray obtained in a natural way from the metric derivative is equal to the canonical spray of Chapter VII, Theorem 7.1.

Another possibility is to admit Theorems 7.1 and 7.2 of Chapter VII, which already proved the existence of a spray whose quadratic map $f_{U,2}$ is obtained from the symmetric bilinear map B_U as defined in **MS 1**. This gives immediately the existence of a unique spray on X having the representation of **MS 1** in a chart U, and this spray is the canonical spray. That **MS 2** is equivalent to **MS 1** then follows from **MD 3**. This concludes the proof.

The spray of Theorem 4.2 will be called the **metric spray**. Since it is equal to the canonical spray, we really don't need two names for it.

Remark. To connect with other texts, note that in terms of local coordinates, the metric spray is given by a map f_2 satisfying the second order differential equation

$$\frac{d^2 x_i}{dt^2} = f_2(x, v) \qquad \text{and} \qquad v_i = \frac{dx_i}{dt}.$$

As a function of the variable v, the map f is quadratic, and minus its coefficients are functions of x, called the **Christoffel symbols**, Γ^i_{jk}. Thus by definition, the above differential equation is of type

$$\frac{d^2 x_i}{dt^2} = -\sum_{j,k} \Gamma^i_{jk}(x) \frac{dx_k}{dt} \frac{dx_j}{dt}.$$

In terms of the standard basis for $\mathbf{R}^n$, the metric is represented by a matrix

$$\bigl(g_{ij}(x)\bigr),$$

and we let (g^{ij}) be the inverse matrix. Then the formula of Theorem 4.2 can be written in terms of the local coordinates in terms of the Christoffel symbols, namely

$$\Gamma^j_{ik} = \frac{1}{2} \sum_v g^{jv} \left(\frac{\partial g_{vk}}{\partial x_i} + \frac{\partial g_{iv}}{\partial x_k} - \frac{\partial g_{ik}}{\partial x_v} \right)$$

If I gave priority to fit classical notation, I would have written $-\Gamma_U$ instead of B_U for the bilinear map associated with the spray. However, using the letter B suggests bilinearity, whereas using the letter Γ would suggest the above mess. Besides, using B is more natural for the bilinear map associated to the quadratic map of the second order differential equation, and eliminates a minus sign from that equation.

Theorem 4.3. *Let $\alpha: J \to X$ be a C^2 curve in a Riemannian manifold (X, g). For the metric derivative, and curves γ, $\zeta \in \mathrm{Lift}(\alpha, TX)$, we have the formula*

$$\langle \gamma, \zeta \rangle'_g = \langle D_{\alpha'} \gamma, \zeta \rangle_g + \langle \gamma, D_{\alpha'} \zeta \rangle_g.$$

Furthermore, parallel translation is a metric isomorphism. In particular, let $t_0 \in J$. If γ_v, γ_w are the unique α-parallel lifts of α with $\gamma_v(t_0) = v$ and $\gamma_w(t_0) = w$, then for all t,

$$\langle \gamma_v(t), \gamma_w(t) \rangle_g = \langle v, w \rangle_g.$$

Proof. The formula is proved in the same way that the computation proving Theorem 3.1 was parallel to the computation proving Theorem 2.1 (giving the behavior under changes of charts). From the formula, if $D_{\alpha'}\gamma = D_{\alpha'}\zeta = 0$, it follows that $\langle\gamma, \zeta\rangle_g$ is constant, whence the second assertion follows.

Corollary 4.4. *Let φ be a C^2 function on X. Let α be a geodesic for the metric spray. Then*

$$(\varphi \circ \alpha)'' = \langle(D_{\alpha'} \operatorname{grad} \varphi) \circ \alpha, \alpha'\rangle_g.$$

Proof. Taking the first derivative of $\varphi \circ \alpha$ yields

$$(\varphi \circ \alpha)'(t) = (d\varphi)(\alpha(t))\alpha'(t) = \langle(\operatorname{grad}\varphi)(\alpha(t)), \alpha'(t)\rangle_g.$$

Now take the next derivative using Theorem 4.3 and the fact that $D_{\alpha'}\alpha' = 0$. The desired formula drops out.

VIII, §5. MORE LOCAL RESULTS ON THE EXPONENTIAL MAP

In this section, we give further results on the exponential map obtained from a spray. We follow the same notation as in Chapter IV, §4, and at first we just deal with a spray. We do not need to know whether it comes from a metric or not.

Throughout the section, we let X be a manifold with a spray F.

Instead of looking at the exponential map restricted to the tangent space at a given point, we may consider this map in the neighborhood of a point in the whole tangent bundle. Let $\pi: TX \to X$ be the projection as always. Let $x_0 \in X$, with zero element $0_{x_0} \in T_{x_0}X$. There exists an open neighborhood V of 0_{x_0} in TX on which we can define the map

$$G: V \to X \times X \qquad \text{such that} \qquad G(v) = (\pi v, \exp_{\pi v}(v)).$$

It is sometimes useful to express this map in a different notation. Specifically, if we denote a point in the tangent bundle by a pair (x, v) if $v \in T_x X$, then

$$G(x, v) = (x, \exp_x(v)).$$

Using a pair (x, v) is certainly the way we would write a point in the tangent bundle as represented in a chart $U \times E$, with $x \in U$ and $v \in E$.

Proposition 5.1. *The map G is a local isomorphism at* $(x_0, 0)$.

Proof. The Jacobian matrix of G in a chart is given immediately from Chapter IV, Theorem 4.1 by

$$\begin{pmatrix} \text{id} & \text{id} \\ 0 & \text{id} \end{pmatrix}$$

which is invertible. The inverse mapping theorem concludes the proof.

For the next local results, it is convenient to express certain uniformities in a chart, where we can measure distances uniformly in the model Banach space **E**, with a given norm. It is irrelevant to know whether this norm has any smoothness properties or not. It will be used just to describe neighborhoods of a vector 0 in the tangent bundle. I found [Mi 63] useful.

Let $x_0 \in X$. For $\varepsilon > 0$, we let $\mathbf{E}(\varepsilon)$ denote the open ball of elements $v \in \mathbf{E}$ with $|v| < \varepsilon$. Arbitrarily small open neighborhoods of $(x_0, 0)$ in a chart for TX are of the form

$$U_0 \times \mathbf{E}(\varepsilon),$$

where U_0 is an open neighborhood of x_0 in X, and ε is arbitrarily small.

Corollary 5.2. *Given* $x_0 \in X$. *Let V be an open neighborhood of* $(x_0, 0)$ *in TX such that G induces an isomorphism of V with its image, and in a chart, for some* $\varepsilon > 0$,

$$V = U_0 \times \mathbf{E}(\varepsilon).$$

Let W be a neighborhood of x_0 *in X such that* $G(V) \supset W \times W$. *Then*:

(1) *Any two points x, y* $\in W$ *are joined by a unique geodesic in X lying in* U_0, *and this geodesic depends* C^∞ *on the pair* (x, y). *In other words, if* $t \mapsto \exp_x(tv)$ $(0 \le t \le 1)$ *is the geodesic joining x and y, with* $y = \exp_x(v)$, *then the correspondence*

$$(x, v) \leftrightarrow (x, y)$$

   is C^∞.

(2) *For each x* $\in W$ *the exponential* $\exp_x$ *maps the open set in* $T_x X$ *represented by* $(x, \mathbf{E}(\varepsilon))$ *isomorphically onto an open set* $U(x)$ *containing W.*

Proof. The properties are merely an application of the definitions and Proposition 5.1.

The pair (V, W) will be said to constitute a **normal neighborhood** of x_0 in X. Dealing with the pair rather than a single neighborhood is slightly inelegant, but to eliminate one of the neighborhoods requires a little more work, which most of the time is not necessary. It has to do with "convexity" properties, and a theorem of Whitehead [Wh 32]. We shall do the work at the end of this section for the Riemannian case.

In the Riemannian case, given $x \in X$, by a **normal chart** at x we mean an open ball $B_g(x, c)$ such that the exponential map

$$\exp_x: \mathbf{B}_g(0_x, c) \to B_g(x, c)$$

is an isomorphism.

We shall need a lemma which gives us the analogue of the commutation rule of partial derivatives in the context of covariant derivatives. Let J_1, J_2 be open intervals, and let

$$\sigma: J_1 \times J_2 \to X$$

be a C^2 map. For each fixed $t \in J_2$ we obtain a curve $\sigma_t: J_1 \to X$ such that $\sigma_t(r) = \sigma(r, t)$. We can then take the ordinary partial derivative

$$\partial_1 \sigma(r, t) = \sigma_t'(r) = \frac{\partial \sigma}{\partial r}.$$

Similarly, we can define $\partial_2(r, t) = \partial\sigma/\partial t$. Observe that for each t, the curves $r \mapsto \partial_1 \sigma(r, t)$ and $r \mapsto \partial_2 \sigma(r, t)$ are lifts of $r \mapsto \sigma(r, t)$ in TX.

More generally, let Q be a lift of σ in TX. Then one may apply the covariant derivative with respect to functions of the first variable r, with the various notation

$$(D_{\partial_1 \sigma, t} Q_t)(r) = (D_1 Q)(r, t) = \frac{DQ}{\partial r}.$$

Similarly, we have $D_2 Q(r, t)$.

Lemma 5.3. *We have the rules on lifts of σ to TX:*

(a) $D_1 \partial_2 = D_2 \partial_1$; *and*

(b) $\partial_2 \langle \partial_1 \sigma, \partial_1 \sigma \rangle_g = 2 \langle D_1 \partial_2 \sigma, \partial_1 \sigma \rangle_g.$

Proof. Let σ_U represent σ in a chart. Then from Theorem 3.1,

$$D_1 \partial_2 \sigma_U = \partial_1 \partial_2 \sigma_U - B_U(\sigma_U; \partial_1 \sigma_U, \partial_2 \sigma_U).$$

Since B_U is symmetric in the last two arguments, this proves (a). As to (b), we use the metric derivative to yield

$$\partial_2 \langle \partial_1 \sigma, \partial_1 \sigma \rangle_g = 2 \langle D_2 \partial_1 \sigma, \partial_1 \sigma \rangle_g,$$

and we use (a) to permute the partials variables on the right, to conclude the proof of (b), and therefore the proof of the lemma.

Let now (X, g) be a pseudo Riemannian manifold. For each $x \in X$ we have the scalar product $\langle v, w \rangle_g = \langle v, w \rangle_{g(x)}$ for $v, w \in T_x X$. Let $c > 0$. The equation

$$\langle v, v \rangle_g = c^2$$

defines a submanifold in $T_x X$, which may be empty. If the metric is Riemannian, the equation defines what we call a **sphere**. In the case when the metric is pseudo Riemannian, say indefinite in the finite dimensional case, then one thinks of the equation as defining something like a hyperboloid in the vector space $T_x X$. We can still define the level "**hypersurface**" $S_g(c)$ to be the set of solutions of the above equation. Even in infinite dimension, we can say that the codimension of this hypersurface is 1. Note that

$$S_g(rc) = rS_g(c) \qquad \text{for} \quad r > 0.$$

In a neighborhood of the origin 0_x in $T_x X$, the exponential map is defined, and gives an isomorphism which may be restricted to $S_g(c)$ intersected with this neighborhood. The image of this intersection is then a submanifold of a neighborhood of x in X. We look at the geodesics starting at x.

Theorem 5.4. *Let $t \mapsto u(t)$ be a curve in $S_g(1)$. Let $0 \leq r \leq b$ where b is such that the points $ru(t)$ are in the domain of the exponential $\exp_x$. Define*

$$\sigma(r, t) = \exp_x(ru(t)) \qquad \text{for} \quad 0 \leq r \leq b.$$

Then

$$\langle \partial_1 \sigma, \partial_1 \sigma \rangle_g = \langle u, u \rangle_g = 1.$$

Proof. This is immediate since parallel translation is an isometry by Theorem 4.3.

Corollary 5.5. *Assume (X, g) Riemannian. Let $v \in T_x X$. Suppose $\|v\|_g = r$, with $r > 0$. Also suppose the segment $\{tv\}$ $(0 \leq t \leq 1)$ is contained in the domain of the exponential. Let $\alpha(t) = \exp_x(tv)$. Then $L(\alpha) = r$.*

Proof. Special case of the length formula in Theorem 5.4, followed by an integration to get the length.

Remark. The corollary is also valid in the pseudo Riemannian case, if one assume that $v^2 = r^2 > 0$, so the notion of length makes sense for the curve $t \mapsto \exp_x(tv)$.

The next theorem expresses the fact that locally near x, the geodesics are orthogonal to the images of the level sets $S_g(c)$ under the exponential map.

Theorem 5.6. *Let (X, g) be pseudo Riemannian. Let $x_0 \in X$ and let W be a small open neighborhood of x_0, selected as in Corollary 5.2, with ε sufficiently small. Let $x \in W$. Then the geodesics through x are orthogonal to the image of $S_g(c)$ under $\exp_x$, for c sufficiently small positive.*

Proof. For ε sufficiently small positive, the exponential map is defined on $S_g(r)$ for $0 < r \leq \varepsilon$, and as we have seen, the level sets $S_g(r)$ are submanifolds of X. Then our assertion amounts to proving that for every curve $u \colon J \to S_g(1)$ and $0 < r < c$, if we define

$$\sigma(r, t) = \exp(ru(t)),$$

then the two curves

$$t \mapsto \exp_x(r_0 u(t)) \qquad \text{and} \qquad r \mapsto \exp_x(ru(t_0))$$

are orthogonal for any given value (r_0, t_0), which amounts to proving that

$$\langle \partial_1 \sigma, \partial_2 \sigma \rangle_g = 0.$$

Let D be the metric derivative. Then

$$\partial_1 \langle \partial_1 \sigma, \partial_2 \sigma \rangle_g = \langle D_1 \partial_1 \sigma, \partial_2 \sigma \rangle_g + \langle \partial_1 \sigma, D_1 \partial_2 \sigma \rangle_g$$
$$= 0,$$

because first, for a geodesic α, we know that the metric derivative has the property that $D_{\alpha'}\alpha' = 0$, so

$$D_1 \partial_1 \sigma = 0;$$

and second, we use Lemma 5.3 (b) and Theorem 5.4 so the derivative of the constant 1 is equal to 0. Therefore the function

$$r \mapsto \langle \partial_1 \sigma(r, t), \partial_2 \sigma(r, t) \rangle_g$$

is constant as a function of r. But for $r = 0$, we have

$$\sigma(0, t) = \exp_x(0) = x, \quad \text{independent of } t.$$

Hence

$$\partial_2 \sigma(0, t) = 0.$$

It follows finally that

$$\langle \partial_1 \sigma, \partial_2 \sigma \rangle_g = 0$$

thus concluding the proof of the theorem.

In the Riemannian case, the theorem is known as **Gauss' lemma**. Helgason [He 61] showed in the analytic case that it is valid in the pseudo Riemannian case as well. I followed the proof given in [Mi 63], which I found applicable to the present context without coordinates, and without assuming analyticity.

Convexity

We conclude this section with the more systematic study of convexity, which was bypassed in Corollary 5.2. We shall treat the Riemannian case, which is slightly simpler. So we assume that (X, g) is Riemannian.

We need to know:

Given $x \in X$, there exists $c > 0$ such that if $0 < r < c$, then the geodesic α such that $\alpha(t) = \exp_x(tv)$, with $0 \leq t \leq 1$, and $\|v\|_g = r$, is the shortest piecewise C^1 path between x and $\exp_x(v)$.

This will be proved in Theorems 6.2 and 6.4 of the next section. In particular, $\text{dist}_g(x, \exp_x(v)) = r$ for r sufficiently small. As usual, we let:

$$\mathbf{B}_g(0_x, r) = \text{open ball in } T_x X \text{ centered at } 0_x, \text{ of radius } r;$$

$$B_g(x, r) = \text{open ball in } X \text{ centered at } x, \text{ of radius } r;$$

$$\mathbf{S}_g(0_x, r) = \text{sphere of radius } r \text{ in } T_x X, \text{ centered at } 0_x; \text{ and}$$

$$S_g(x, r) = \text{sphere of radius } r \text{ in } X, \text{ centered at } x.$$

Here we shall deal only with r sufficiently small.

We define an open set U of X to be **convex** if given x, $y \in U$ there exists a unique geodesic in U joining x to y, and such that the length of the geodesic is $\text{dist}_g(x, y)$. We shall prove Whitehead's theorem [Wh 32] in the form:

Theorem 5.7. *Let (X, g) be a Riemannian manifold. Given $x \in X$, there exists $c > 0$ such that for all r with $0 < r < c$ the open neighborhood $B_g(x, r) = \exp_x \mathbf{B}_g(0_x, r)$ is convex.*

Proof. We need a lemma.

Lemma 5.8. *Given $x \in X$, there exists $c > 0$ such that if $r < c$, and if α is a geodesic in X, tangent to $S_g(x, r)$ at $y = \alpha(t_0)$, then $\alpha(t)$ lies outside $S_g(x, r)$ for $t \neq t_0$ in some neighborhood of t_0.*

Proof. We pick c such that the exponential map $\exp_x$ is a differential isomorphism on $\mathbf{B}_g(0_x, r)$ for all $r < c$ and preserves distances on rays from 0_x to $v \in T_x X$ with $\|v\|_g = r$. Without loss of generality, we can suppose $t_0 = 0$, so $\alpha(0) = y$. We shall view y as variable, so we index α by y. Also we have to look at the other initial condition $\alpha'(0) = u \in T_y Y$, so we write $\alpha_{y,u}$ for the geodesic. Now let

$$\eta_{y,u}(t) = \exp_x^{-1} \alpha_{y,u}(t) \qquad \text{and} \qquad f_{y,u}(t) = \eta_{y,u}(t)^2.$$

Then $\eta_{y,u}$ is a curve in the fixed Hilbert space $T_x X$, so

$$f'_{y,u}(t) = 2\langle \eta'_{y,u}(t), \eta_{y,u}(t)\rangle_{g(x)},$$
$$f''_{y,u}(t) = 2\eta'_{y,u}(t)^2 + 2\langle \eta''_{y,u}(t), \eta_{y,u}(t)\rangle_{g(x)}.$$

Let $h(y, u) = f''_{y,u}(0)$. Then $h(x, u) = 2u^2$, so h_x as a function on $T_x X$ is positive definite. Therefore there exists $c > 0$ such that for $0 < r < c$ and $\|y\|_g = r$ the function h_y is positive definite on $T_y Y$, and in particular $h(y, u) > 0$ for $u^2 \neq 0$. Under the assumption that $\alpha_{y,u}$ is tangent to $S_g(x, r)$ at y, we must have

$$f'_{y,u}(0) = 0 \qquad \text{and} \qquad f''_{y,u}(0) = h(y, u) > 0,$$

whence for sufficiently small $|t|$, we get

$$f_{y,u}(t) > f_{y,u}(0) = \left(\exp_x^{-1} \alpha_{y,u}(0)\right)^2 = \left(\exp_x^{-1}(u)\right)^2 = r^2,$$

which proves the lemma.

We can now conclude the proof of Theorem 5.7. Using Corollary 5.2, we can find $c_1 > 0$ such that putting $W = B_g(x, c_1)$ satisfies the condition of Corollary 5.2. Let $c < c_1$. We show that $r \leq c$ implies $B_g(x, r)$ is convex. Let $y, z \in B_g(x, r)$. Then by that corollary, there exists a unique geodesic α in the neighborhood V of x joining y and z. As in the lemma, let

$$f(t) = \left(\exp_x^{-1} \alpha(t)\right)^2, \qquad \text{with} \quad a \leq t \leq b.$$

It now suffices to prove that $f(t) < r^2$. Suppose $f(t) \geq r^2$ for some t, and let $t_0 \in [a, b]$ be the maximum of f on this interval, so $f(t_0) \geq r^2$. Then $t_0 \neq a, b$ so $f'(t_0) = 0$, whence α is tangent to the sphere $S_g(x, r_0)$ where $r_0 = f(t_0)^{1/2}$. The lemma now gives a contradiction, which concludes the proof of Theorem 5.7.

Remark. In the pseudo Riemannian case, with metric g, one has to use an auxiliary Riemannian metric h to apply a similar argument, which makes the proof slightly longer.

VIII, §6. RIEMANNIAN GEODESIC LENGTH AND COMPLETENESS

Throughout this section, we let (X, g) be a Riemannian manifold.

We return to the Riemannian case, where we use the positive definiteness of the metric. In Chapter VII, §6 we defined the length of a piecewise C^1 path. We want to compare the length locally with the length of straight lines in the tangent space at a point, under the exponential map. In the process, we shall see that locally, a geodesic is the shortest path between two points.

Thus let $x_0 \in X$ and let (V, W) be a normal neighborhood as in Corollary 5.2. Let $x \in W$. For each piecewise C^1 path

$$\gamma: [a, b] \to U(x) - \{x\},$$

with $U(x)$ being as in Corollary 5.2(2), we can use the fact that the exponential map is invertible and so there exists a unique curve $t \mapsto u(t)$ in $T_x M$ such that $\|u(t)\|_g = 1$ and

$$\gamma(t) = \exp_x(r(t)u(t)) \qquad \text{with} \quad 0 < r(t) < \varepsilon.$$

In a chart, the vector $r(t)u(t)$ is obtained by the inverse of the exponential map followed by a projection, so in particular, the functions $t \mapsto r(t)$ and $t \mapsto u(t)$ on $[a, b]$ are piecewise C^1. We call these functions the local **polar coordinates** for γ.

Lemma 6.1. *For a piecewise C^1 curve $\gamma: [a, b] \to U(x) - \{x\}$ as above, we have the inequality*

$$L(\gamma) \geq |r(b) - r(a)|.$$

Equality holds only if the function $t \mapsto r(t)$ is monotone and the map $t \mapsto u(t)$ is constant.

Proof. Let $\sigma(r, t) = \exp_x(ru(t))$. Then $\gamma(t) = \sigma(r(t), t)$. We have

$$\gamma'(t) = \frac{d\gamma}{dt} = \frac{\partial \sigma}{\partial r} r'(t) + \frac{\partial \sigma}{\partial t}.$$

By the Gauss Lemma Theorem 5.6, we know that $\partial\sigma/\partial r$ and $\partial\sigma/\partial t$ are orthogonal. Since $\|\partial\sigma/\partial r\|_g = 1$ by Lemma 5.4, it follows that

$$\|\gamma'(t)\|_g^2 = |r'(t)|^2 + \left\|\frac{\partial\sigma}{\partial t}\right\|_g^2 \geq |r'(t)|^2,$$

with equality holding only if $\partial\sigma/\partial t = 0$, or equivalently, $du/dt = 0$. Hence

$$L(\gamma) = \int_a^b \|\gamma'(t)\|_g \, dt \geq \int_a^b |r'(t)| \, dt \geq |r(b) - r(a)|;$$

and equality holds only if $t \mapsto r(t)$ is monotone and $t \mapsto u(t)$ is constant. This completes the proof.

Theorem 6.2. *Let (V, W) constitute a normal neighborhood of a point $x_0 \in X$. Let $\alpha: [0, 1] \to V$ be the geodesic (up to reparametrization) in V joining two points of W (namely $\alpha(0)$ and $\alpha(1)$). Let $\gamma: [0, 1] \to X$ be any other piecewise C^1 path in X joining these two points. Then*

$$L(\alpha) \leq L(\gamma).$$

If equality holds, then the polar component $t \mapsto v(t)$ for γ is constant, the function $t \mapsto r(t)$ is monotone, and a reparametrization of γ is equal to α.

Proof. Let $x, y \in W$ and let $y = \exp_x(ru)$ with $0 < r < \varepsilon$, and $\|u\|_g = 1$. Then for $\delta > 0$ and $0 < \delta < r$ the path γ contains a segment joining the shell $\mathrm{Sh}_g(x, \delta)$ with the shell $\mathrm{Sh}_g(x, r)$ and lying between the two shells. By Lemma 6.1, the length of this segment is $\geq r - \delta$. Letting δ tend to 0 shows that $L(\gamma) \geq r$. The same lemma proves the conditions on the polar functions as asserted.

Corollary 6.3. *Let $\alpha: [0, 1] \to X$ be a piecewise C^1 path, parametrized by arc length. If $L(\alpha) \leq L(\gamma)$ for all paths from $\alpha(0)$ to $\alpha(1)$ in X, then α is a geodesic.*

Proof. We can find a partition of $[0, 1]$ such that the image under α of each small interval in the partition is contained in some neighborhood W as in the theorem, and its length is small so the image of the segment is contained in a normal neighborhood. By Theorem 6.2, the path restricted to this segment must be a geodesic. Hence the entire path is a geodesic, as was to be shown.

Let $\alpha: [a, b] \to X$ be a geodesic. We say that α is a **minimal geodesic** if $L(\alpha) \leq L(\gamma)$ for every path γ joining $\alpha(a)$ and $\alpha(b)$ in X. Theorem 6.2

gives us the existence of minimal geodesics locally. We can then formulate another application. Let $x \in X$. Let dist_g be the Riemannian distance. Let:

$$\mathbf{B}_g(0_x, r), \ \mathbf{S}_g(0_x, r), \ B_g(x, r), \ S_g(x, r)$$

be the open balls and spheres of radius r, centered at 0_x in $T_x X$ and at x in X, respectively. We now know enough to show that $S_g(x, r)$ is the image of $\mathbf{S}_g(0_x, r)$ under the exponential map, and similarly for the open ball, for sufficiently small r.

Theorem 6.4. *Let (X, g) be a Riemannian manifold and let $x \in X$. There exists $c > 0$ such that for all $r < c$ the map $\exp_x$ is defined on $\mathbf{B}_g(0_x, c)$, gives a differential isomorphism*

$$\exp_x : \mathbf{B}_g(0_x, r) \to B_g(x, r) \qquad \text{for all } r \text{ with } \ 0 < r < c,$$

and also a differential isomorphism

$$\exp_x : \mathbf{S}_g(0_x, r) \to S_g(x, r) \qquad \text{for } 0 < r < c.$$

Proof. Immediate from Corollary 5.5 and Theorem 6.2.

Next we consider completeness. Since X is a metric space (in the ordinary sense), with respect to the distance dist_g, the notion of X being complete is standard: every Cauchy sequence for dist_g converges. On the other hand, we can now define another notion of completeness.

We say that (X, g) is **geodesically complete** if and only if the maximal interval of definition of every geodesic in X is all of $\mathbf{R}$. Alternatively, we could say that for each point $x \in X$, the exponential map $\exp_x$ is defined on all of T_x, because under one normalization of the parametrization of a geodesic, it is simply the curve $t \mapsto \exp_x(tv)$ for some $v \in T_x X$. To be systematic, let us consider the following conditions:

COM 1. *As a metric space under dist_g, X is complete.*

COM 2. *All geodesics in X are defined on $\mathbf{R}$.*

COM 3. *For every $x \in X$, the exponential $\exp_x$ is defined on all of $T_x X$.*

COM 4. *For some $x \in X$, the exponential $\exp_x$ is defined on all of $T_x X$.*

Proposition 6.5. *Each condition implies the next, i.e.*

$$\text{COM 1} \Rightarrow \text{COM 2} \Rightarrow \text{COM 3} \Rightarrow \text{COM 4.}$$

Proof. Assume **COM 1**. Let $\alpha \colon J \to X$ be a geodesic parametrized by arc length on some interval, and take J to be maximal in $\mathbf{R}$. By the existence and uniqueness theorem for differential equations, J is open in $\mathbf{R}$, and it will suffice to prove that J is closed, or in other words, that J contains its end points. For $t_1, t_2 \in J$ we have

$$\text{dist}(\alpha(t_1), \alpha(t_2)) \leq |t_2 - t_1|.$$

Suppose for instance that J is bounded above, and let $\{t_n\}$ be a sequence in J converging to the right end point of J. Then the sequence $\{\alpha(t_n)\}$ is Cauchy by the above inequality, so $\{\alpha(t_n)\}$ converges to a point x_0 by **COM 1**. Then for all n sufficiently large, $\alpha(t_n)$ lies in a small normal neighborhood of x_0, and there is some $\varepsilon > 0$, independent of n, such that the geodesic can be extended to an interval of length at least ε beyond t_n, thus contradicting the maximality of J, and proving **COM 2**. The subsequent implications are trivial, so the proposition is proved.

We are now interested when geodesic completeness implies completeness. We shall give two criteria for this. One of them is that the manifold has finite dimension, and the other one will be important for its application to conditions on curvature in Chapter IX. The finite dimensional case depends on the next result.

Theorem 6.6 (Hopf–Rinow). *Assume that (X, g) is connected geodesically complete, and finite dimensional. Then any two points in X can be joined by a minimal geodesic.*

Proof. I follow here the variation of the proof given in [Mi 63]. Let p, y be two points with $p \neq y$. Let W be a normal neighborhood of p containing the image of a small ball under the exponential map $\exp_p$. Let $r = \text{dist}(p, y)$, and let δ be small $< r$. Then the shell $\text{Sh}_g(p, \delta) = \text{Sh}(p, \delta)$ is contained in W. Since $\text{Sh}(p, \delta)$ is the image of the sphere of radius δ in T_pX, it follows that $\text{Sh}(p, \delta)$ is compact. Hence there exists a point x_0 on $\text{Sh}(p, \delta)$ which is at minimal g-distance from y, that is

$$\text{dist}(x_0, y) \leq \text{dist}(x, y) \qquad \text{for all} \quad x \in \text{Sh}(p, \delta).$$

We can write $x_0 = \exp_p(\delta u)$ for some $u \in T_x$ with $\|u\|_g = 1$. Let $\alpha(t) = \exp_p(tu)$. We shall prove that $\exp_p(ru) = y$. We prove this by "continuous induction" on t, as it were. More precisely, we shall prove:

(dist$_t$)   We have $\text{dist}(\alpha(t), y) = r - t$ \qquad for $\delta \leq t \leq r$.

Taking $t = r$ will prove the theorem. First we note that $(\mathbf{dist_\delta})$ is true. Indeed, every path from p to y intersects the shell $\mathrm{Sh}(p, \delta)$, so

$$
(1) \qquad \mathrm{dist}(p, y) = \min_x \left(\mathrm{dist}(p, x) + \mathrm{dist}(x, y) \right) \qquad \text{for} \quad x \in \mathrm{Sh}(p, \delta)
$$

$$
= \delta + \min_x \mathrm{dist}(x, y)
$$

$$
= \delta + \mathrm{dist}(x_0, y),
$$

so $(\mathbf{dist_\delta})$ is true. Now "inductively", assume that (dist_t) is true for all $t < r'$, with $\delta \leq r' \leq r$. Let r_1 be the least upper bound of such r'. Since the distance dist_g is continuous, it follows that $(\mathbf{dist}_{r_1})$ is true, and it suffices to prove that $r_1 = r$. Suppose $r_1 < r$. Pick δ_1 small so we get as usual a spherical shell $\mathrm{Sh}(\alpha(r_1), \delta_1)$ around $\alpha(r_1)$, contained in a normal neighborhood of $\alpha(r_1)$. As in (1), there is a point x_1 on $\mathrm{Sh}(\alpha(r_1), \delta_1)$ at minimal distance from y, and we have the relation as in (1), namely

$$
\mathrm{dist}(\alpha(r_1), y) = \delta_1 + \mathrm{dist}(x_1, y).
$$

Since $(\mathbf{dist}_{r_1})$ is true, we find

$$
(2) \qquad\qquad\qquad \mathrm{dist}(x_1, y) = r - r_1 - \delta_1.
$$

We claim that $x_1 = \alpha(r_1 + \delta_1)$. To see this, first observe that

$$
\mathrm{dist}(p, x_1) \geq \mathrm{dist}(p, y) - \mathrm{dist}(x_1, y) = r_1 + \delta_1.
$$

But the path consisting of the two minimal geodesics from $p = \alpha(0)$ to $\alpha(r_1)$ and from $\alpha(r_1)$ to x_1 has length $r_1 + \delta_1$, so this path (which is viewed as a broken geodesic) has minimal length, so it is an unbroken geodesic by Corollary 6.3. Hence the path is actually equal to α, so $\alpha(r_1 + \delta_1) = x_1$, and therefore

$$
\mathrm{dist}(x_1, y) = r - (r_1 + \delta_1),
$$

so $(\mathbf{dist}_{r_1 + \delta_1})$ is true, thus concluding the proof of the continuous induction, and also concluding the proof the Hopf–Rinow theorem.

Corollary 6.7. *In the finite dimensional case the four completeness conditions* **COM 1** *through* **COM 4** *are equivalent to a fifth:*

COM 5. *A closed* dist_g*-bounded subset of X is compact.*

Proof. Assume **COM 4**. Let S be closed and bounded in X, and let $x_0 \in S$. Let b be a bound for the diameter of S. Then by Theorem 6.6

(Hopf–Rinow), every point of S can be joined to x_0 by a geodesic of length $\leq b$, so S is contained in the image under $\exp_{x_0}$ of the closed ball of radius b in $T_{x_0}X$, so S is compact, thus proving **COM 5**.

Assume **COM 5**. Let $\{x_n\}$ be a Cauchy sequence in X. Then $\{x_n\}$ lies in a bounded set, whose closure is compact by assumption, so $\{x_n\}$ has a point of accumulation which is actually a limit in X. This proves **COM 1**, and concludes the proof of the corollary.

Remark. In his thesis [McA 65], McAlpin gave the following example which shows a divergence of behavior in the case of infinite dimensional Hilbert manifolds. Let **E** be a Hilbert space with orthonormal basis $\{e_n\}$ $(n \geq 0)$. Let $T: \mathbf{E} \to \mathbf{E}$ be the linear map such that for a vector $v = \sum x_n e_n \in \mathbf{E}$

$$T\left(\sum x_n e_n\right) = \sum a_n x_n e_n$$

where $a_0 = 1$ and $a_n = 1 + 1/2^n$ for $n \geq 1$. Then

$$\|v\| \leq \|Tv\| \leq \tfrac{3}{2}\|v\|,$$

and therefore T is invertible in Laut($\mathbf{E}$). Let S be the unit sphere in $\mathbf{E}$ and let $X = T(\mathbf{S})$, so X is a submanifold of $\mathbf{E}$, to which we give the induced metric. Let α be a path joining e_0 to $-e_0$ in $\mathbf{S}$. Then $T\alpha$ is a path joining e_0 to $-e_0$ in X, and T is length increasing, that is

$$L(\alpha) \leq L(T\alpha).$$

Hence the length of any path in X joining e_0 to $-e_0$ is $\geq \pi$, which is the minimal length of paths between the two points in $\mathbf{E}$. However, let α_n be the half great circle joining the two points in the upper (e_0, e_n)-half plane. Then

$$L(T\alpha_n) < (1 + 1/2^n)\pi \to \pi = \operatorname{dist}_X(e_0, -e_0).$$

Hence there is no minimal path joining the two points in X. Note that each $T\alpha_n$ is a geodesic in X joining the two points, because

$$\operatorname{Im}(T\alpha_n)u - \operatorname{Im}(T\alpha_n)$$

is the fixed point set of the isometry F_n defined by

$$F_n\left(\sum x_k e_k\right) = x_0 e_0 + x_n e_n - \sum_{k \neq 0, n} x_k e_k.$$

McAlpin refers to [Gros 64] for results on the distribution of degenerate points of the exponential map in similar examples.

Next we give another criterion for (X, g) to be complete. We start with a lemma.

Lemma 6.8. *Let $f: Y \to X$ be a C^1 map between Riemannian manifolds (Y, h) and (X, g). Assume that there is a constant $C > 0$ such that for all $y \in Y$ and $w \in T_y Y$ we have*

$$\|df(y)w\|_g \geq C\|w\|_h.$$

If $\gamma: [a, b] \to Y$ is a piecewise C^1 path in Y, then

$$L(f \circ \gamma) \geq CL(\gamma).$$

Proof. We have

$$L_g(f \circ \gamma) = \int_a^b \|(f \circ \gamma)'(t)\|_g \, dt = \int_a^b \|df(\gamma(t))\gamma'(t)\|_g \, dt$$

$$\geq \int_a^b C\|\gamma'(t)\|_h \, dt$$

$$= CL_h(\gamma),$$

as was to be shown.

Let $f: Y \to X$ be a C^1 map of manifolds. We say that f has the **unique path lifting property** if given a point $x \in X$, a piecewise C^1 path α in X starting from x, and a point $y \in Y$ such that $f(y) = x$, then there exists a unique piecewise C^1 path γ in Y such that $f \circ \gamma = \alpha$ and γ starts at y.

Theorem 6.9. *Let $f: Y \to X$ be a local C^1 isomorphism of a Riemannian manifold (Y, h) into a Riemannian manifold (X, g). Assume that (Y, h) is complete, and X is connected. Also assume that there is a constant $C > 0$ suh that for all $y \in Y$ and $w \in T_y Y$ we have*

$$\|df(y)w\|_g \geq C\|w\|_h.$$

Then f is surjective, f is a covering and has the unique path lifting property, and (X, g) is complete.

Proof. The proof is in three steps. *The first step is to prove that f is surjective and has the unique path lifting property.* Let $x \in X$, $x = f(y)$. Every point in X can be joined to x by a piecewise C^1 path. Let $\alpha: [a, b] \to X$ be such a path, joining $\alpha(a) = x$ with $\alpha(b)$. We shall prove that α can be lifted uniquely to a path in Y starting from y. This will prove the first step. Let S be the set of elements $t \in [a, b]$ such that the path α restricted to $[0, t]$ can be lifted uniquely to a path γ starting at y.

Without loss of generality, we may assume that $a < b$. The set S is not empty because $a \in S$, and it is open because f is a local isomorphism. So it remains to show that S is closed. Let $\{t_n\}$ be a sequence in S increasing to the least upper bound b_0 of S. Then $\{\alpha(t_n)\}$ converges to $\alpha(b_0)$, and by Lemma 6.8 the lengths of the lifted path between $\gamma(t_n)$ and $\gamma(t_m)$ tend to 0 as m, n tend to infinity, so the sequence $\{\gamma(t_n)\}$ is Cauchy in Y, converging to some element y_0 since Y is assumed complete. Then $f(y_0) = \alpha(b_0)$, so S is closed, whence $S = X$ by assumption. Therefore f is surjective, and we have also proved the existence and uniqueness of path liftings.

The next step in the proof is to reduce the theorem to the case when the map f is a local isometry. We do this as follows. Let $g^* = f^*(g)$ be the pull-back of the metric g from X to Y by f. Then for all $y \in Y$ and $w \in T_y Y$ we have

$$\|w\|_{g^*} = \|df(y)w\|_g \geq C\|w\|_h.$$

Hence on Y we find that $\mathrm{dist}_{g^*} \geq C\,\mathrm{dist}_h$. We now claim that Y is complete for the distance dist_{g^*}. To see this, first observe that if $\{y_n\}$ is g^*-Cauchy, then $\{y_n\}$ is also h-Cauchy, so $\{y_n\}$ is h-convergent to an element $y_0 \in Y$. Then $\{f(y_n)\}$ converges to $f(y_0)$. But f induces an isomorphism from some neighborhood V of y_0 to an open neighborhood of $f(y_0)$, and hence for all but a finite number of n, the points $f(y_n)$ lie in $f(V)$, so $\{y_n\}$ is also g^*-convergent to y_0 since $g^* = f^*(g)$. This proves that Y is g^*-complete. Furthermore, we have the inequality

$$\mathrm{dist}_{g^*}(y_1, y_2) \geq \mathrm{dist}_g\big(f(y_1), f(y_2)\big) \qquad \text{for all} \quad y_1, y_2 \in Y.$$

In this final step, we prove that f is a covering. Since Y is g^*-complete, this will also prove that (X, g) is complete, and will conclude the proof of the theorem. By the second step, we may assume without loss of generality that f is a local isometry, and that

$$(*) \qquad \mathrm{dist}_h(y_1, y_2) \geq \mathrm{dist}_g\big(f(y_1), f(y_2)\big) \qquad \text{for all} \quad y_1, y_2 \in Y.$$

Let $x \in X$. From Theorem 6.4 we know that

$$\exp_x \colon \mathbf{B}_g(0_x, r) \to B_g(x, r)$$

is an isomorphism for all r sufficiently small, say $r < c$ with $c > 0$. Let $y \in f^{-1}(x)$. Since f is a local isometry, the following diagram is commutative (using $(*)$):

$$
\begin{array}{ccc}
\mathbf{B}_h(0_y, r) & \xrightarrow{\ df(y)\ } & \mathbf{B}_g(0_x, r) \\[4pt]
{\scriptstyle \exp_y}\big\downarrow & & \big\downarrow{\scriptstyle \exp_x} \\[4pt]
B_h(y, r) & \xrightarrow[\ \ f\ \ ]{} & B_g(x, r)
\end{array}
$$

Note that the right vertical arrow is a differential isomorphism because we have picked r small enough, but so far we have made no such assertion for the left vertical arrow. For the proof of the theorem, it will suffice to show that $f^{-1}B_g(x, r)$ is the disjoint union of the balls $B_h(y, r)$ for $y \in f^{-1}(x)$, if r is taken small enough. We take r so small that given $x' \in B_g(x, r)$ there is a unique geodesic in $B_g(x, r)$ joining x to x' (namely $\exp_x(tv)$ for some v). Then, first, we have $f(B_h(y, r)) \subset B_g(x, r)$, so the union is contained in $f^{-1}B_g(x, r)$. Conversely, given a point $z \in f^{-1}B_g(x, r)$, we can join $f(z)$ to x by a geodesic of length $< r$ in $B_g(x, r)$, and by the path lifting property already proved in step 1, we can join z to a point y in $f^{-1}(x)$ by a geodesic of the same length, so

$$f^{-1}(B_g(x, r)) = \bigcup_y B_h(y, r),$$

where the union is taken over $y \in f^{-1}(x)$. Finally, let $y_1, y_2 \in f^{-1}(x)$ and suppose $y_1 \neq y_2$. We claim that $B_h(y_1, r)$ is disjoint from $B_h(y_2, r)$. Suppose there is some point z in the intersection. Then z can be joined to y_1 by a geodesic α_1 in $B_h(y_1, r)$, and z can also be joined to y_2 by a geodesic α_2 in $B_h(y_2, r)$, and these geodesics are distinct. Their images under f are geodesics in $B_g(x, r)$ joining x with $f(z)$. By the uniqueness of path lifting, this would mean we have two distinct geodesics in $B_g(x, r)$ joining x and z, and that these geodesics have length $< r$. This contradicts the local uniqueness statement, and proves that the balls $B_h(y_1, r)$ and $B_h(y_2, r)$ are disjoint. This concludes the proof of the theorem.

Remark. In the next chapter, *under a condition of seminegative curvature* (to be defined), we shall take $Y = T_xX$, and we shall prove that

$$f = \exp_x: T_xX \to X$$

satisfies the hypotheses of Theorem 6.9, and therefore in particular that geodesic completeness implies completeness. In this manner, we shall be able to replace the local compactness condition by a curvature condition to insure the equivalence between the two notions of completeness. The whole technique goes back to Hadamard [Ha 1898] in the case of surfaces with seminegative curvature, and Cartan [Ca 28] in the general case, still in this context of seminegative curvature. The notion of a "covering space" was not so clear during this early period. Except for a minor variation, the theorem is apparently due to Ambrose [Am 56], and occurs in the standard treatments of differential geometry as in [He 62] later replaced by [He 78], Chapter I, Lemma 13.4; [KoN 63], Chapter IV, Theorem 4.6 and Chapter VIII, §8, Theorem 8.1 and especially Lemma 1. The theorem is at the base of the Cartan–Hadamard theorem, to be proved later.

CHAPTER IX

Curvature

This chapter is a continuation of the preceding one, and is concerned with the iteration of covariant derivatives, from a formal point of view, and also from the point of view of their effect on the geometry of the manifold.

IX, §1. THE RIEMANN TENSOR

Let X be a manifold with a spray, and the covariant derivative D associated with the spray. If ξ, η, ζ are vector fields on X, we are concerned with the operator

$$D_\xi D_\eta - D_\eta D_\xi - D_{[\xi, \eta]} : \Gamma TX \to \Gamma TX,$$

which is a linear map of ΓTX into itself.

Proposition 1.1. *There exists a unique tensor field R, section of $L^3(TX, TX)$, i.e. arising from the functor $\mathbf{E} \mapsto L^3(\mathbf{E}, \mathbf{E})$ (continuous trilinear maps of $\mathbf{E}$ into itself) such that for all vector fields ξ, η, ζ we have*

$$R(\xi, \eta, \zeta) = D_\xi D_\eta \zeta - D_\eta D_\xi \zeta - D_{[\xi, \eta]} \zeta.$$

Proof. The expression on the right-hand side gives a well-defined vector field on X. To show that this association comes from a tensor field, we can compute in a chart. To do this, we use the local expression for the covariant derivative given in Theorem 2.1 of Chapter VIII. So for the rest of the argument, ξ, η, ζ stand for ξ_U, η_U, ζ_U in a chart U. Then, for example, we have

(1) $$D_\eta \zeta = \zeta' \cdot \eta - B(\eta, \zeta).$$

We determine $D_\xi(D_\eta\zeta)$ by substitution in this formula. As a first step, we have to write down the derivative

$$(D_\eta\zeta)' \cdot \xi = \zeta'' \cdot \xi \cdot \eta + \zeta' \cdot \eta' \cdot \xi - B(\eta'\xi, \zeta) - B(\eta, \zeta' \cdot \xi) - B(D_\eta\zeta, \xi).$$

Then it follows that

$$D_\xi(D_\eta\zeta) = \zeta'' \cdot \xi \cdot \eta + \zeta' \cdot \eta' \cdot \xi - B(\eta' \cdot \xi, \zeta) - B(\eta, \zeta' \cdot \xi) - B(\zeta' \cdot \eta, \xi)$$
$$+ B\big(B(\eta, \zeta), \xi\big).$$

Permuting ξ and η gives us the second term. Using the local expression for the bracket

$$[\xi, \eta] = \eta' \cdot \xi - \xi' \cdot \eta$$

as well as (1) will give us the third term. The reader will then verify that all the expressions containing a derivative cancel, leaving only bilinear expressions involving B (possible repeated), ξ, η, and ζ. This proves Proposition 1.1.

In addition, after the cancellation of the terms with derivatives, we obtain a local expression for R, namely:

Proposition 1.2. *Letting* ξ, η, ζ *represent vector fields in a chart*:

$$R(\xi, \eta, \zeta) = B\big(B(\eta, \zeta), \xi\big) - B\big(B(\xi, \zeta), \eta\big).$$

Remark. There is no universal convention as to the sign of R. I use the same sign as [KoN 63], [ChE 75], [He 78], and [BGV 92], but the opposite sign to [BGM 71], [HGL 87/93], and [Mi 63]. For further comments, see the discussion after the definition of sectional curvature.

Let $v, w, z \in T_xX$. It is customary to write

$$R(v, w, z) = R(v, w)z = R_x\big(\xi(x), \eta(x), \zeta(x)\big)$$
$$= R(\xi, \eta, \zeta)(x),$$

if ξ, η, ζ are any vector fields such that $\xi(x) = v$, $\eta(x) = w$, $\zeta(x) = z$. One writes

$$R(\xi, \eta): \Gamma TX \to \Gamma TX$$

for the linear map of $\Gamma T X$ into itself, given by

$$R(\xi, \eta) = D_\xi D_\eta - D_\eta D_\xi - D_{[\xi, \eta]}.$$

As a function of two variables, according to this definition, one may view R as a section of the bundle $L^2(TX, L(TX, TX))$, which is formed by applying the functor $\mathbf{E}^3 \mapsto L^2(\mathbf{E}, L(\mathbf{E}, \mathbf{E}))$ to the tangent bundle.

Next we list some identities.

Proposition 1.3.

$R(v, w) = -R(w, v)$ (skew-symmetry).

$R(v, w, z) + R(w, z, v) + R(z, v, w) = 0$ (cyclicity, **Bianchi's identity**).

Proof. The first relation is obvious from the definition. The second one is immediate from the local representation of Proposition 1.2.

For the next two properties, we assume that the spray is the one associated with a metric, so the covariant derivative is the metric derivative. We let $\langle \, , \, \rangle_g$ be the scalar product associated with the metric. Then we define a function of four variables

$$R(v, w, z, u) = \langle R(v, w)z, u \rangle_g \qquad \text{for} \quad v, w, z, u \in T_x X.$$

Then R is a tensor of type L^4, that is a section of $L^4(TX) = L^4(TX, \mathbf{R})$. We shall call R the **Riemann 4-tensor** (canonical with respect to g). We call $-R$ the **curvature tensor**. The properties of Proposition 1.3 may be formulated for this 4-tensor, and we shall see in a moment that it also satisfies two other important properties. Thus it is useful to make a general definition. A tensor R of type L^4 is called a tensor of **Riemann type** if it satisfies the following four properties:

RIEM 1. $R(v, w, z, u) = -R(w, v, z, u)$

RIEM 2. $R(v, w, z, u) = -R(v, w, u, z)$

RIEM 3. $R(v, w, z, u) + R(w, z, v, u) + R(z, v, w, u) = 0$

RIEM 4. $R(v, w, z, u) = R(z, u, v, w)$.

The first two conditions express the property of being alternating in the first two variables, and also in the last two variables. The third condition is called the **Bianchi identity**, and expresses the property that the cyclic symmetrization of the tensor is 0. The fourth property states that the tensor is symmetric in the pairs of variables (v, w) and (z, u). In particular, we note right away that from **RIEM 4**, we obtain:

$R(v, w, v, u)$ is symmetric in (w, u), that is $R(v, w, v, u) = R(v, u, v, w)$.

We shall make more comments on these properties after the next proposition, which justifies the terminology.

Proposition 1.4. *On a pseudo Riemannian manifold, the Riemann tensor satisfies all the above four properties. Furthermore,* **RIEM 4** *follows from* **RIEM 1, 2, 3**.

Proof. Properties **RIEM 1** and **RIEM 3** have been proved in Proposition 1.3. Property **RIEM 2** amounts to proving that $R(v, w, z, z) = 0$ for all v, w, z; or in terms of vector fields, $R(\xi, \eta, \zeta, \zeta) = 0$. We will need to differentiate. Since all the terms with derivatives vanish in the local formula of Proposition 1.2, we may assume without loss of generality that $[\xi, \eta] = 0$. Then

$$\langle R(\xi, \eta)\zeta, \zeta\rangle_g = \langle D_\xi D_\eta \zeta - D_\eta D_\xi \zeta, \zeta\rangle_g,$$

and we must show that the right side is symmetric in ξ, η. But $[\xi, \eta] = 0$ implies that

$$\mathscr{L}_\xi \mathscr{L}_\eta \langle \zeta, \zeta\rangle_g$$

is symmetric in ξ, η. Since we are dealing with the metric covariant derivative, it follows that

$$\mathscr{L}_\eta \langle \zeta, \zeta\rangle_g = 2\langle D_\eta \zeta, \zeta\rangle_g$$

and therefore

$$\mathscr{L}_\xi \mathscr{L}_\eta \langle \zeta, \zeta\rangle_g = 2\langle D_\xi D_\eta \zeta, \zeta\rangle_g + 2\langle D_\xi \zeta, D_\eta \zeta\rangle_g,$$

from which it follows at once that $\langle D_\xi D_\eta \zeta, \zeta\rangle_g$ is symmetric in ξ, η, thus proving **RIEM 2**.

The formula **RIEM 4** is a formal consequence of the preceding three formulas. It is basically an exercise in algebra, which we carry out. In the cyclic identity **RIEM 3**, interchange u with z, v, w successively, and add the resulting three relations. One gets, using **RIEM 1** and **RIEM 3**:

(∗) $R(u, v, w, z) + R(u, w, z, v) + R(u, z, v, w) = 0.$

From cyclicity and **RIEM 1**, one gets

$$R(z, v, u, w) = R(u, v, z, w) - R(u, z, v, w) \quad \text{or}$$
$$R(u, z, v, w) = R(u, v, z, w) - R(z, v, u, w).$$

We substitute the value on the left in (∗), and use **RIEM 1** to conclude the proof of **RIEM 4**.

We shall be dealing with a contraction of the canonical 4-tensor. We defined the **canonical 2-tensor** R_2 by

$$R_2(v, w) = R(v, w, v, w).$$

Proposition 1.5. *The canonical 2-tensor determines the Riemann tensor. Or similarly, if the canonical tensor R satisfies*

$$R(v, w, v, w) = 0 \qquad \text{for all} \quad v, w,$$

then $R = 0$.

Proof. Say we prove the second assertion first. From **RIEM 4**, which implies that $R(v, w, v, z)$ is symmetric in (w, z), if $R(v, w, v, w) = 0$ for all v, w then $R(v, w, v, z) = 0$ for all v, w, z. From the alternating properties of **RIEM 1** and **RIEM 2**, it follows that $R = 0$ identically.

To show that the canonical 2-tensor determines the Riemann tensor, we note that the problem is essentially equivalent to the other statement, but one may argue directly as when one recovers a symmetric bilinear form from a quadratic form, namely

$$\left[\frac{\partial^2}{\partial t \partial s} R(v + tz, w + su, v + tz, w + su) \right.$$

$$\left. - \frac{\partial^2}{\partial t \partial s} R(v + tu, w + sz, v + tu, w + sz) \right]_{s=t=0}$$

$$= 6R(v, w, z, u).$$

This proves the proposition.

An important case arises when $R_2 \geq 0$. We define (X, g) to have **seminegative curvature** if $R_2 \geq 0$. The following discussion explains this terminology in terms of its historical development.

Curvature discussion

A large part of the theory we are developing is fundamentally a theory of commutative rings with certain types of derivation, and possibly scalar products, in which positivity or negativity plays no role. This theory contains a number of formulas with precise equality between various terms. There would be some value in redoing this chapter and the preceding one completely in such a context of commutative differential algebra. At some point, for certain applications, the positivity or negativity properties of the real numbers are used, as in the second statement of Proposition 2.6 below. For such applications, the question arises as to what is the natural sign to be used, if indeed there is a natural sign.

Historically, the theory arose in a geometric context, based on geometric intuition. To each pair of vectors (v, w) in a tangent space $T_x X$, we define the **area square** of the parallelogram spanned by these vectors to be

$$\mathrm{Ar}_g(v, w)^2 = v^2 w^2 - \langle v, w \rangle_g^2.$$

As usual, $v^2 = \langle v, v \rangle_g$. Then, when $\mathrm{Ar}_g(v, w)^2 \neq 0$, we define the **sectional curvature** to be

$$\mathrm{Sec}_g(v, w) = -\frac{R(v, w, v, w)}{\mathrm{Ar}_g(v, w)^2}.$$

In the Riemannian case, $\mathrm{Ar}_g(v, w) \neq 0$ if and only if v, w are linearly independent. If v^2 and $w^2 > 0$, then the value on the right depends only on the unit vectors in the direction of v, w respectively; and if v, w are orthogonal unit vectors, then

$$\mathrm{Sec}_g(v, w) = -R(v, w, v, w).$$

In the Riemannian case, it is immediate that the value of the sectional curvature on (v, w) depends only on the plane generated by v and w, because of the skew-symmetry of **RIEM 1** and **RIEM 2**. For the complex analogue, see [La 87], Chapter V, §3.

Let $c \in \mathbf{R}^+$ be a positive number. The multiple cg is called a **scaling** of the metric g. Since the covariant derivative D^{cg} is the same as D^g, it follows from the definitions that under scaling, the curvature changes as

$$\mathrm{Sec}_{cg} = c^{-1} \mathrm{Sec}_g.$$

Directly from the definition, we then see in the Riemannian case that:

The sectional curvature has constant value -1 *if and only if*

$$R_2(v, w) = v^2 w^2 - \langle v, w \rangle_g^2 \qquad \textit{for all } v, w \in T_x X.$$

Viewing v as fixed, the above expression is quadratic in w, and the corresponding symmetric bilinear form is

$$R(v, w, v, z) = v^2 \langle w, z \rangle_g - \langle v, w \rangle_g \langle v, z \rangle_g.$$

Thus $R(v, w)v$ is given by

$$R(v, w)v = v^2 w - \langle v, w \rangle_g v \qquad \text{for all} \quad v, w \in T_x X.$$

Similarly, the sectional curvature has constant value $+1$ *if and only if the analogous formula holds with a minus sign inserted on one side, so that for instance*

$$-R(v, w)v = v^2 w - \langle v, w \rangle_g v.$$

In the applications of this book (the rest of this chapter, the Cartan–Hadamard theorem, the variation formulas of §4, etc.) what matters is not the "curvature" as defined above, but the canonical tensor R itself. Furthermore, formulas in these applications come out much neater with R than with "curvature" for two reasons:

First, for such formulas, dividing to normalize as in the curvature quotient is unnatural, partly because the term by which one divides, algebraically, may be equal to 0 unless extra conditions are imposed.

Second, even for inequalities as distinguished from equalities, the natural condition which arises is $R(v, w, v, w) \geqq 0$ rather than curvature $\leqq 0$. If one takes R with the sign as we have defined it, then only plus signs occur in all the formulas (cf. Lemma 2.5 and the variation formula, Theorem 4.3, for instance). This universal occurrence of plus signs is obscured if one introduces minus signs artificially. I regard this universal occurrence of plus signs as structurally important.

The naturality of R in the real case is similar to the naturality of its counterpart in the complex case, where formulas involving positivity come out neatly by using the analogue of R rather than its negative (as already noted by Griffiths). Cf. [La 87], the comments pp. 136–137 about holomorphic sectional curvature. The lesson is that the "curvature" in classical terminology is minus the natural object R (aside from questions of normalizing the dilation to the unit sphere).

Classically, starting with surface theory, people wanted some formulas such as Gauss-Bonnet or formulas relating "curvature" and Betti numbers, using $\pm R$, to come out so that on the sphere, one gets a value of certain integral to be 4π and not -4π. So they picked the minus sign, and gave the notion $-R$ (normalized) the name of curvature, which makes the sphere have positive curvature. The bottom line is that depending on what applications one makes, both R and $-R$ are "natural." However, from the point of view of universal algebraic manipulations, R is the clearest functorial notion.

One can define two other curvatures, at least. Actually, all we need is a tensor of curvature type. From such a tensor R, we obtain two other tensors. First observe that to each pair of vectors $v, z \in \mathbf{E}$ we can associate an endomorphism of $\mathbf{E}$, denoted by $\mathrm{Ric}(v, z)$, and defined by

$$\mathrm{Ric}_R(v, z)w = R(v, w)z.$$

Thus Ric gives a bilinear map

$$\mathrm{Ric}_R: \mathbf{E} \times \mathbf{E} \to L(\mathbf{E}, \mathbf{E}).$$

Applied to the tangent bundle, and the Riemann tensor R itself, Ric is called the **Ricci tensor**.

Furthermore, in the finite dimensional case, the trace

$$\text{tr: } L(\mathbf{E}, \mathbf{E}) \rightarrow \mathbf{R}$$

is a continuous linear map. Then the composite

$$\text{Sc}_R = \text{tr} \circ \text{Ric}_R$$

is a function of pairs of vectors, which when applied to the tangent bundle defines what is called the **scalar curvature**. In the infinite dimensional case, one has to give an additional structure, assuming that the Ricci tensor is of "trace class", or defining the sectional curvature with respect to a given "trace," i.e. a continuous functional on $L(\mathbf{E}, \mathbf{E})$ which is equal on products AB and BA. But this now leads far afield.

Suppose we are in the Riemannian case. We can then give an explicit formula for the scalar curvature. In the neighborhood of a point, we can find vector fields $\xi_1, \ldots, \xi_n$ (with $n = \dim X$) which are orthonormal, by the usual orthogonalization process. Such a sequence of vector fields is called an **orthonormal frame** at the point.

Proposition 1.6. *Let $\{\xi_1, \ldots, \xi_n\}$ be an orthonormal frame on an open set. Then for vector fields ξ, η we have*

$$\text{Sc}_R(\xi, \eta) = \sum_{i=1}^{n} R(\xi, \xi_i, \eta, \xi_i).$$

Proof. This is immediate from the definition of the trace of an endomorphism of a finite dimensional vector space.

The conditions **RIEM 1** and **RIEM 2** express the property of depending only on the wedge product of each pair of variables $v \wedge w$ and $z \wedge u$. Property **RIEM 4** is a symmetric property in these pairs of variables. Thus we may say that the four-variable tensor R defines a symmetric bilinear form on $\bigwedge^2 TX$, which we denote by

$$R^\wedge : \bigwedge^2 TX \wedge \bigwedge^2 TX \rightarrow \mathbf{R}, \quad \text{such that } R^\wedge(v \wedge w, z \wedge u) = R(v, w, z, u).$$

On the other hand, we also have the pseudo Riemannian metric, which induces a non-singular scalar product on $\bigwedge^2 TX$ by the formula

$$\langle v \wedge w, z \wedge u \rangle_g = \det \begin{vmatrix} \langle v, z \rangle_g & \langle v, u \rangle_g \\ \langle w, z \rangle_g & \langle w, u \rangle_g \end{vmatrix}.$$

These scalar products are of course evaluated at each point $x \in X$, v, w, z, $u \in T_x X$. The scalar product on $\bigwedge^2 TX$ with respect to the non-singular symmetric form g then corresponds to a symmetric operator which is called the **curvature operator**.

In the infinite dimensional case, from the self duality, each tangent space can be interpreted as the dual of its dual, and the wedge product is defined as in Chapter V, §3 so the above notions still make sense.

Readers wanting to pursue the topic of curvature are now referred to other books on differential geometry, including [BGM 71], [ChE 75], [doC 92], and [GHL 87/93].

IX, §2. JACOBI LIFTS

Let (X, g) be pseudo Riemannian. We write w^2 for $\langle w, w \rangle_g$, and $v \perp w$ for $\langle v, w \rangle_g = 0$. We let $\alpha: [a, b] \to X$ be a geodesic. Unless otherwised specified, (X, g) is not necessarily Riemannian.

A lift $\eta \in \text{Lift}(\alpha)$ to the tangent bundle will be called a **Jacobi lift**, or more classically a **Jacobi field**, if it satisfies the **Jacobi differential equation**

$$D_{\alpha'}^2 \eta = R(\alpha', \eta)\alpha'.$$

Theorem 3.1 of Chapter VIII and Proposition 1.2 in the preceding section of the present chapter show that locally, the above equation is a linear differential equation. Therefore, by the existence and uniqueness theorem for linear differential equations, we get:

Theorem 2.1. *Let (X, g) be pseudo Riemannian, let $\alpha: [a, b] \to X$ be a geodesic. Given vectors z, $w \in T_{\alpha(a)} X$, there exists a unique Jacobi lift $\eta = \eta_{z,w}$ of α to TX such that*

$$\eta(a) = z \qquad \text{and} \qquad D_{\alpha'} \eta(a) = w.$$

In particular, the set of Jacobi lifts of α is a vector space linearly isomorphic to $T_{\alpha(a)} \times T_{\alpha(a)}$ under the map $(z, w) \mapsto \eta_{z,w}$.

We denote the space of Jacobi lifts of α by $\text{Jac}(\alpha)$.
Let $v \in T_x$ and consider the unique geodesic

$$\alpha(t) = \exp_x(tv)$$

such that $\alpha(0) = 0$ and $\alpha'(0) = v$, with α defined on an open interval. Let $w \in T_x X$ and let η_w be the unique Jacobi lift of α such that

$$\eta_w(0) = 0 \qquad \text{and} \qquad D_{\alpha'} \eta_w(0) = w.$$

Example. Let $w = v$. Then

$$\eta_v(t) = t\alpha'(t).$$

Proof. One verifies at once that $\eta_v(0) = 0$, and since $D_{\alpha'}\alpha' = 0$, we also have

$$D_{\alpha'}\eta_v(t) = \alpha'(t) \qquad \text{and} \qquad D_{\alpha'}^2\eta_v = 0 = R(\alpha', \alpha')\alpha'.$$

Remark 1. Defining a Jacobi lift implicitly has the geodesic α in its definition. If, say, $a = 0$, this geodesic is uniquely determined by its initial condition $\alpha'(0) = v$, so the Jacobi lift is also determined by v. Thus one could write $\eta_w^{(v)}$ for the Jacobi lift. In the present discussions, this won't be necessary since we deal systematically with a fixed α.

Remark 2. In a chart, the derivative $\eta'(0)$ can be computed in the naive way since $\eta : J \to TX$ is defined on an interval. The naive derivative and the covariant derivative $D_{\alpha'}\eta$ differ locally in a chart by a term linear in η, which therefore vanishes at 0 since $\eta(0) = 0$. Hence the naive derivative and the covariant derivative have the same value at 0, that is

$$\eta'(0) = D_{\alpha'}\eta(0).$$

We note that, α being fixed, the association $w \mapsto \eta_w$ is linear. We now have the possibility of orthogonalization.

Proposition 2.2. *Let (X, g) be pseudo Riemannian. Let $\alpha : [a, b] \to X$ be a geodesic, and let η be a Jacobi lift of α. Then there are numbers c, d such that*

$$\langle \eta, \alpha' \rangle_g(t) = c(t - a) + d.$$

In fact, $d = \langle \eta, \alpha' \rangle_g(a)$ and $c = \langle D_{\alpha'}\eta, \alpha' \rangle_g(a)$. If $\eta(a)$ and $D_{\alpha'}\eta(a)$ are orthogonal to $\alpha'(a)$, then $\eta(t)$ is orthogonal to $\alpha'(t)$ for all t.

Proof. Using the metric derivative, and $D_{\alpha'}\alpha' = 0$ since α is a geodesic, we find that $\partial\langle \eta, \alpha' \rangle_g = \langle D_{\alpha'}\eta, \alpha' \rangle_g$, and then

$$\partial^2\langle \eta, \alpha' \rangle_g = \langle D_{\alpha'}^2\eta, \alpha' \rangle_g = R(\alpha', \eta, \alpha', \alpha') = 0.$$

Hence $\langle \eta, \alpha' \rangle_g$ is a linear function, whose coefficients are immediately determined to be those written down in the proposition.

Proposition 2.3. *As above, let $\alpha'(0) = v$. Write $w = cv + w_1$ with $\langle w_1, v \rangle_g = 0$. Then η_w has the decomposition*

$$\eta_w = c\eta_v + \eta_{w_1}, \qquad \text{also written} \quad \eta_w(t) = ct\alpha'(t) + \eta_{w_1}(t).$$

Furthermore η_{w_1} is orthogonal to α', that is $\langle \eta_{w_1}, \alpha' \rangle_g = 0$.

Proof. Immediate from Proposition 2.2.

Next we get the similar orthogonalization of $D_{\alpha'}\eta_w$.

Proposition 2.4. *Notation as in Proposition 2.3, we have an orthogonal decomposition*

$$D_{\alpha'}\eta_w = cD_{\alpha'}\eta_v + D_{\alpha'}\eta_{w_1} \qquad \text{also written} \qquad D_{\alpha'}\eta_w(t) = c\alpha'(t) + D_{\alpha'}\eta_{w_1}(t).$$

In other words, if $w_1 \perp \alpha'(0)$, then $D_{\alpha'}\eta_{w_1} \perp \alpha'$. Furthermore $(D_{\alpha'}\eta_w)^2$ is constant.

Proof. For the first assertion, we take the derivative and use Proposition 2.3 to get

$$0 = \partial\langle\eta_{w_1}, \alpha'\rangle_g = \langle D_{\alpha'}\eta_{w_1}, \alpha'\rangle_g.$$

For the second, we then obtain for $\eta = \eta_w$:

$$\partial\langle D_{\alpha'}\eta, D_{\alpha'}\eta\rangle_g = 2\langle D_{\alpha'}^2\eta, D_{\alpha'}\eta\rangle_g$$
$$= 2\langle R(\alpha', \eta)\alpha', D_{\alpha'}\eta\rangle_g.$$

If $\eta = \eta_v$ so $\eta_v(t) = t\alpha'(t)$, then the right side is 0 because R_4 is alternating in its last two variables. If $\eta = \eta_w$ with $w \perp \alpha'(0)$, then by the first assertion, $D_{\alpha'}\eta$ is orthogonal to α', so the right side is also 0. This concludes the proof of Proposition 2.4.

The next lemma will give us information on the rate of growth of a Jacobi lift, and the convexity of its square.

Lemma 2.5. *Assume (X, g) Riemannian. Let η be a Jacobi lift of α. Let $f(t) = \|\eta(t)\|$. Then at those values of $t > 0$ such that $\eta(t) \neq 0$, we have*

$$f'' = \frac{1}{\|\eta\|^3}((D_{\alpha'}\eta)^2\eta^2 - \langle D_{\alpha'}\eta, \eta\rangle_g^2) + \frac{1}{\|\eta\|}R_2(\alpha', \eta).$$

Proof. Straightforward calculus, using the covariant derivative. The first derivative f' is given by

$$f' = (\eta^2)^{-1/2}\langle\eta, D_{\alpha'}\eta\rangle_g = \frac{1}{\|\eta\|}\langle\eta, D_{\alpha'}\eta\rangle_g.$$

Then f'' is computed by using the rule for the derivative of a product. In the term containing $\langle D_{\alpha'}^2\eta, \eta\rangle_g$, we replace $D_{\alpha'}^2\eta$ by $R(\alpha', \eta)\alpha'$ (using the definition of a Jacobi lift) to conclude the proof of the lemma.

In the above lemma, we note that on the right side, the first term is ≥ 0, and the second term is ≥ 0 if $R_2 \geq 0$.

Proposition 2.6. *Let* $\alpha: [0, b] \to X$ *be a geodesic. Let* $v \in T_{\alpha(0)}X$, $w \neq 0$. *Let* $\eta_w = \eta_{0,w} = \eta$ *be the unique Jacobi lift satisfying*

$$\eta_w(0) = 0 \qquad and \qquad D_{\alpha'}\eta_w = w.$$

If (X, g) *is Riemannian and* $R_2 \geq 0$ $($*so* (X, g) *has seminegative curvature*$)$, *then for* $t \in [0, b]$ *we have*

$$\|\eta(t)\| \geq \|w\|t \qquad and \ in \ particular \qquad \|\eta(1)\| \geq \|w\| \ if \ b = 1.$$

Proof. Let $h(t) = \|\eta(t)\| - \|w\|t$ for $0 \leq t \leq b$. Then h is continuous, $h(0) = 0$, and by Lemma 2.5, $h'' = f'' \geq 0$ whenever $\eta(t) \neq 0$. One cannot have $\eta(t) = 0$ for arbitrarily small values of $t \neq 0$, otherwise $D_{\alpha'}\eta(0)$ would be 0 (because in a chart U, $\eta'_U(0) = D_{\alpha'}\eta(0)$). In fact, we shall prove that there is no value of $t \neq 0$ such that $\eta(t) = 0$. Suppose there is such a value, and let t_0 be the smallest value > 0. In the interval $(0, t_0)$ we have $h'' \geq 0$ by Lemma 2.5, so h' is increasing. But the beginning of the Taylor expansion of η in a chart is

$$\eta_U(t) = wt + O(t^2), \qquad so \qquad \lim_{t \to 0} f'(t) = \|w\|.$$

Furthermore, $h'(0)$ exists and is equal to 0, so $h' \geq 0$ on $[0, t_0)$, so h is increasing, and there cannot be a value $t_0 > 0$ with $\eta(t_0) = 0$. Then the above argument applies on the whole interval $[0, b]$ to prove the desired inequality on the whole interval. This concludes the proof of Proposition 2.6.

Remark. These results essentially stem from Cartan [Ca 28]. The above version without coordinates, which extends to the infinite dimensional case, comes from [BiC 64]. Readers may find it instructive to compare this version with the one involving coordinates given in [He 78], pp. 71–73.

Proposition 2.6 is used for the subsequent application to the Cartan–Hadamard theorem (Theorem 3.7), based on Theorem 6.9 of Chapter VIII, whose origin is in Hadamard for surfaces [Ha 1898] and Cartan in general. (Here and at several other places, I rely on Helgason's very useful bibliographical comments.)

Variations of geodesics

By a **variation** of a curve α one means a C^2 map

$$\sigma: [a, b] \times J \to X$$

where J is some interval containing 0, such that $\sigma(s, 0) = \alpha(s)$ for all s. One then writes

$$\sigma(s, t) = \alpha_t(s),$$

and one views $\{\alpha_t\}$ as a family of curves defined on $[a, b]$. If all curves α_t are geodesics for $t \in J$ then one says that σ is a **variation through geodesics**.

Lemma 2.7. *Let* $\sigma: J_1 \times J_2 \to X$ *be a* C^2 *map. Then on lifts of* σ *to the tangent bundle, we have the equality of operators*

$$D_1 D_2 - D_2 D_1 = R(\partial_1 \sigma, \partial_2 \sigma).$$

Proof. The formula can be verified in a chart. It follows directly from the definitions, especially using the local expression of Proposition 1.2.

Proposition 2.8. *Let* $\sigma: [a, b] \times J \to X$ *be a variation of a geodesic* α *through geodesics. Let*

$$\eta(s) = \partial_2 \sigma(s, 0).$$

Then η *is a Jacobi lift of* α.

Proof. Given σ, we have

$$D_1^2 \partial_2 \sigma = D_1 D_1 \partial_2 \sigma = D_1 D_2 \partial_1 \sigma \quad \text{by Lemma 5.3 of Chapter VIII}$$

$$= D_2 D_1 \partial_1 \sigma + R(\partial_1 \sigma, \partial_2 \sigma) \partial_1 \sigma \quad \text{by Lemma 2.2.}$$

But $D_1 \partial_1 \sigma(s, t) = 0$ because α_t is a geodesic for each t, whence

$$D_\alpha^2 \eta = R(\alpha', \eta)\alpha',$$

so η is a Jacobi lift of α, as was to be shown.

We will not need the next result, but put it in for completeness.

Theorem 2.9. *Let* $\alpha: [0, b] \to X$ *be a geodesic, parametrized by arc length. Let* η *be a Jacobi lift of* α. *Then there exists a variation*

$$\sigma: [0, b] \times (-\varepsilon, \varepsilon) \to X$$

through geodesics, such that

$$\eta(s) = \partial_2 \sigma(s, 0) \qquad \text{for all } s.$$

Proof. Let $\beta: (-\varepsilon, \varepsilon) \to X$ be the geodesic such that

$$\beta(0) = \alpha(0) \qquad \text{and} \qquad \beta'(0) = \eta(0).$$

Let $v = \alpha'(0)$ and $w = D_{\alpha'}\eta(0)$. Let

$$\zeta(t) = P_{\beta(t)}(v + tw) = P_{\beta(t)}(v) + tP_{\beta(t)}(w),$$

where $P_{\beta(t)}$ is parallel translation along β. Then from the definition of parallel translation,

$$\zeta(0) = v \quad \text{and} \quad D_{\beta'}\zeta(0) = w = D_{\alpha'}\eta(0).$$

Let

$$\sigma(s, t) = \exp_{\beta(t)} s\zeta(t).$$

We have $\sigma(s, 0) = \alpha(s)$, and for each t, the curve $s \mapsto \sigma(s, t)$ is a geodesic, so σ is a variation of α through geodesics. We still have to show that $\eta(s) = \partial_2\sigma(s, 0)$. Let $\xi(s) = \partial_2\sigma(s, 0)$. Then ξ is a Jacobi lift of α, so it suffices to prove that ξ has the same initial conditions as η. We have first

$$\xi(0) = \partial_2\sigma(0, 0) = \eta(0);$$

and in addition,

$$D_{\alpha'}\xi(0) = D_1\partial_2\sigma(0, 0) = D_2\partial_1\sigma(0, 0) \quad \text{by Chapter VIII, Lemma 5.3}$$
$$= \left(D_{\beta'}d\,\exp_{\beta(0)}(0)\zeta\right)(0)$$
$$= D_{\beta'}\zeta(0)$$
$$= D_{\alpha'}\eta(0),$$

thus concluding the proof.

Example. Constant curvature. Let (X, g) be Riemannian. As an example, we shall now determine more explicitly the Jacobi lifts when (X, g) has constant curvature. Since the covariant derivative is invariant under a scaling of the metric, we may as well assume that the curvature is 0 or ± 1. In the next three proposition, we let $x \in X$ and we let $v \in T_xX$ be a *unit vector*. As usual, we let $\alpha = \alpha_v$ be the geodesic

$$\alpha(t) = \exp_x(tv).$$

For $w \in T_xX$ we let $\eta_w = \eta_w^{(v)}$ be the *Jacobi lift* of α_v satisfying the usual initial conditions

$$\eta_w(0) = 0 \quad \text{and} \quad D_{\alpha'}\eta_w(0) = w.$$

Finally, we let $\gamma_w = \gamma_w^{(v)}$ be *parallel translation* of w along α_v, so

$$\gamma_w(0) = w \quad \text{and} \quad D_{\alpha'}\gamma_w = 0.$$

Proposition 2.10. *Assume that the curvature is 0, or equivalently that the Riemann tensor R is identically 0. Then for all $w \in T_x X$ we have*

$$\eta_w(t) = t\gamma_w(t).$$

Proof. The two curves $t \mapsto \eta_w(t)$ and $t \mapsto t\gamma_w(t)$ have the same initial conditions. Also they satisfy the same differential equation, namely

$$D_{\alpha'}^2 \eta_w = 0 \quad \text{and} \quad D_{\alpha'}^2(t\gamma_w(t)) = 0.$$

Hence they are equal, thereby proving the proposition.

The next two propositions deal with constant curvature ± 1. We recall that we wrote down the Riemann tensor explicitly in those cases in §1. We may therefore write down the differential equation for a Jacobi lift more explicitly in those cases, as follows.

Proposition 2.11. *Assume that (X, g) has constant curvature -1. Then the Jacobi differential equation has the form*

(1) $$D_{\alpha'}^2 \eta_w = \eta_w - \langle \eta_w, \alpha' \rangle_g \alpha'.$$

Furthermore, if we orthogonalize w with respect to v, so write

$$w = c_0 v + c_1 u \quad \text{with } c_0, c_1 \in \mathbf{R} \text{ and a unit vector } u \perp v,$$

then

(2) $$\eta_w(t) = c_0 t\alpha'(t) + (\sinh t)c_1 \gamma_u(t).$$

Proof. The orthogonalization of Jacobi lifts comes from Proposition 2.3, so we want to identify the orthogonal components of the Jacobi lift of α_v with scalar multiples of parallel translation. It suffices to do so when $w = v$ and $w = u \perp v$ separately. The example following Theorem 2.1 already gives us the v-component, so we may assume $w = u$. In this case, the reader will verify that the two curves

$$t \mapsto \eta_u(t) \quad \text{and} \quad t \mapsto (\sinh t)\gamma_u(t)$$

have the same initial conditions at 0 (for their value, and the value of their first covariant derivative). They also satisfy the same differential equation, namely

$$D_{\alpha'}^2 \eta_u = \eta_u$$

and similarly for the other curve, since $D_{\alpha'}\gamma_u = 0$. Hence the two curves are equal, as was to be shown.

Thirdly we deal with constant positive curvature.

Proposition 2.12. *Assume X has constant curvature $+1$. Let $x \in X$. Then the same formulas hold as in Proposition 2.11, except for a minus sign on one side in formula (1), and with $\sinh t$ replaced by $\sin t$ in formula (2).*

Proof. The arguments are the same. Using $\sin t$ instead of $\sinh t$ just guarantees that the differential equation

$$D_{\alpha'}^2 \eta_u = -\eta_u$$

is satisfied, with the minus sign.

This concludes our analysis of the Jacobi lifts in the cases of constant curvature.

The Jacobi differential equation has at least two main aspects. One of them will be applied to a study of the differential of the exponential map in the next section. The other will be applied to variational questions in §4.

IX, §3. APPLICATION OF JACOBI LIFTS TO $d\exp_x$

We continue to assume that (X, g) is pseudo Riemannian, unless otherwise specified.

We are interested in Jacobi lifts because they give precise information concerning the differential of the exponential map, for instance as in the following result. In the statement, if $v \in T_x$ then we identify $T_v T_x$ with T_x, as we usually do for a Banach space.

Theorem 3.1. *Let $x \in X$ and $v \in T_x$. Let α (defined on an open interval containing 0) be the geodesic such that $\alpha(0) = x$ and $\alpha'(0) = v$. Let $w \in T_x$ and let $\eta_w = \eta_{0,w}$ be the Jacobi lift of α such that*

$$\eta_w(0) = 0 \qquad and \qquad D_{\alpha'} \eta_w(0) = w.$$

Then for $r > 0$, in the interval of definition of α, we have the formula

$$d\exp_x(rv)w = \frac{1}{r}\eta_w(r).$$

In particular, w lies in the kernel of $d\exp_x(rv)$ if and only if $\eta_w(r) = 0$.

Proof. Let
$$\sigma(s, t) = \exp_x(s(v + tw)).$$

Then σ_t is a geodesic for each t, and $\sigma_0(s) = \exp_x(sv) = \alpha(s)$, so σ is a variation of α through geodesics. Let $\eta(s) = \partial_2\sigma(s, 0)$. Then η is a Jacobi lift of α by Proposition 2.8. Let $f(s, t) = s(v + tw)$. Then

$$\partial_2\sigma(s, t) = (d\exp_x)(f(s, t))(\partial f/\partial t) = (d\exp_x)(f(s, t))(sw).$$

Hence $\eta(0) = 0$. Furthermore this same expression yields the formula of the theorem, namely

$$\eta(r) = (d\exp_x)(f(r, 0))rw = (d\exp_x)(rv)rw.$$

Taking the limit as $r \to 0$ in the formula, noting that in a chart $D_{\alpha'}\eta(0) = \eta'(0)$, and using $d\exp_x(0) = \mathrm{id}$ proves that $D_{\alpha'}\eta(0) = w$ and concludes the proof of Proposition 3.1.

The Jacobi lifts also allow us to give a more global version of the orthogonality property as in the Gauss lemma of Chapter VIII, Theorem 5.6.

Proposition 3.2 (Gauss Lemma, Global). *Let (X, g) be pseudo Riemannian. Let $x \in X$ and $v \in T_xX$. Let the exponential map $r \mapsto \exp_x(rv)$ be defined on an open interval J. Then for all $w \in T_xX$ we have*

$$\langle d\exp_{x(rv)}v, d\exp_x(rv)w\rangle_g = \langle v, w\rangle_g.$$

Proof. Immediate from Proposition 3.1 and the orthogonalization of Proposition 2.3.

Variation of a geodesic at its end point

Next we shall give another way of constructing Jacobi lifts, which will not be used until Chapter X, Proposition 2.5. Readers interested in seeing at once the application of Jacobi lifts to the Cartan–Hadamard theorem, say, may omit the following construction.

Let x, $y \in X$ with $x \neq y$ be points such that y lies in the exponential image of a ball centered at 0_x in T_xX and such that the exponential map $\exp_x$ is an isomorphism on this ball.

Thus this ball provides a normal chart at x. Let α be the geodesic parametrized by arc length joining x to y, so there is a unit vector $u \in T_x$

such that

$$\alpha(s) = \exp_x(su) \qquad \text{and} \qquad y = \exp_x(ru) \qquad \text{for some} \quad r > 0.$$

Thus $\text{dist}_g(x, y) = r$. Let e be a unit vector in $T_y X$, and let β be the geodesic such that

$$\beta(0) = y \quad (\text{so } \beta \text{ starts at } y) \qquad \text{and} \qquad \beta'(0) = e.$$

We consider an interval of the variable t such that $\beta(t)$ is contained in the image of the previous ball around x. For each t we let α_t be the unique geodesic from x to $\beta(t)$, parametrized by arc length. Then $\{\alpha_t\}$ is a variation of α, namely $\alpha_0 = \alpha$, and it is a variation through geodesics, illustrated on the next figure, drawn when e is perpendicular to $\alpha'(r)$ to illustrate Proposition 3.3.

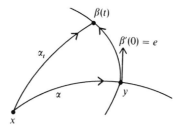

The above variation will be called the **variation of α at its end point, in the direction of** e.

Proposition 3.3. *Let* $y = \exp_x(ru)$ *be in a normal chart at* x *as above, with the unit vector* u. *Let* $\alpha(s) = \exp_x(su)$, *and let* $\{\alpha_t\}$ *be the variation of* α *at its end point* y *in the direction of the unit vector* $e \in T_y X$. *Also denote this variation by* σ, *and let* $\eta(s) = \partial_2 \sigma(s, 0)$. *Assume that* e *is orthogonal to* $\alpha'(r)$. *Then* $D_{\alpha'}\eta$ *is orthogonal to* α', *and* η *is the unique Jacobi lift of* α *such that*

$$\eta(0) = 0 \qquad \text{and} \qquad \eta(r) = e.$$

Proof. First note the uniqueness. If there is another Jacobi lift having the last stated property, then the difference vanishes at 0 and r, and by Theorem 3.1 this difference must be 0 since the exponential map is assumed to be an isomorphism from a ball to its image, which contains $y = \exp(ru)$.

Next, the variation σ is given by the formula

$$\sigma(s, t) = \alpha_t(s) = \exp_x(su(t)) \qquad \text{such that} \qquad \exp_x(s(t)u(t)) = \beta(t),$$

where $u(t)$ is a unit vector, and $s(t)u(t)$ is the vector whose exponential is $\beta(t)$. The polar coordinates $s(t)$ and $u(t)$ depend as smoothly on t as the exponential map, or its inverse. Then

$$\partial_2 \sigma(s, t) = d\exp_x(su(t))su'(t),$$

so that (since $u = u(0)$),

$$\eta(s) = d\exp_x(su)su'(0)$$
$$= \eta_{u'(0)}(s),$$

because from Theorem 3.1, we see that $D_{\alpha'}\eta(0) = u'(0)$. Since $u(t)^2 = 1$, it follows that $u'(0)$ is perpendicular to $\alpha'(0) = u$, so $D_{\alpha'}\eta$ is orthogonal to α'. Furthermore

$$\beta'(t) = d\exp(s(t)u(t))(s(t)u'(t) + s'(t)u(t)),$$

and since $s(0) = r$, we find

$$e = \beta'(0) = d\exp(ru)(ru'(0) + s'(0)u)$$
$$= d\exp(ru)ru'(0) + d\exp(ru)s'(0)u.$$

Since e is assumed orthogonal to $\alpha'(r) = d\exp(ru)u$, and $u'(0)$ is also orthogonal to u, we must have $s'(0) = 0$, whence the relation

$$e = d\exp(ru)ru'(0) \qquad \text{or} \qquad \eta_{u'(0)}(r) = e.$$

This proves the proposition.

Transpose of $d\exp_x$

In the next results we are concerned with the differential of the exponential map at arbitrary points, namely for $v \in T_x$ such that $\exp_x$ is defined on the segment $[0, v]$, we are concerned with

$$d\exp_x(v): T_x \to T_y, \qquad \text{where} \quad y = \exp_x(v),$$

especially whether this map is an isomorphism, or what is its kernel. Theorem 3.1 describes a condition for an element w to be in the kernel in terms of a zero for a suitable Jacobi lift. We shall exploit this condition to see that under some circumstances, there cannot be a non-trivial zero. We first give a lemma from Ambrose [Am 60].

Lemma 3.4. *Let (X, g) be pseudo Riemannian. Let η, ζ be Jacobi lifts of a geodesic α. Then*

$$\langle D_{\alpha'}\eta, \zeta \rangle_g - \langle \eta, D_{\alpha'}\zeta \rangle_g \quad \text{is constant.}$$

Proof. We differentiate the above expression and expect to get 0. From the defining property of the covariant derivative, the derivative of the above expression is equal to

$$\langle D_{\alpha'}^2 \eta, \zeta \rangle + \langle D_{\alpha'} \eta, D_{\alpha'} \zeta \rangle - \langle D_{\alpha'} \eta, D_{\alpha'} \zeta \rangle - \langle \eta, D_{\alpha'}^2 \zeta \rangle$$

$$= \langle D_{\alpha'}^2 \eta, \zeta \rangle - \langle D_{\alpha'}^2 \zeta, \eta \rangle.$$

$$= R(\alpha', \eta, \alpha', \zeta) - R(\alpha', \zeta, \alpha', \eta)$$

$$= 0$$

by the symmetry property of R. This proves the lemma.

The next lemma, from McAlpin's thesis [McA 65], describes the adjoint of the differential of the exponential map.

Lemma 3.5. *Let (X, g) be pseudo Riemannian. Let α (defined at least on $[0, 1]$) be the geodesic such that $\alpha(0) = x$ and $\alpha'(0) = v$. Let*

$$z \in T_{\alpha(1)}, \qquad w \in T_{\alpha(0)},$$

and let

$$v^* = -\alpha'(1) = -Pv,$$

where P is the parallel translation along α'. Then

$$\langle d\exp_{\alpha(0)}(v)w, z \rangle_{\alpha(1)} = \langle w, d\exp_{\alpha(1)}(v^*)z \rangle_{\alpha(0)}.$$

Proof. Let ζ be the Jacobi lift of α such that $\zeta(1) = 0$ and $D_{\alpha'}\zeta(1) = z$. Let η be the Jacobi lift as in Proposition 3.1. Then

$$\langle d\exp_x(v)w, z \rangle = \langle \eta(1), D_{\alpha'}\zeta(1) \rangle = \langle D_{\alpha'}\eta(1), \zeta(1) \rangle + C = C,$$

where C is the constant of Lemma 3.4. We compute C to be

$$C = -\langle D_{\alpha'}\eta(0), \zeta(0) \rangle = -\langle w, \zeta(0) \rangle.$$

Let $\mathrm{rev}(\alpha)$ be the reverse curve, so that $\mathrm{rev}(\alpha)(t) = \alpha(1 - t)$, and let ξ be the unique Jacobi lift of $\mathrm{rev}(\alpha)$ such that

$$\xi(0) = 0 \qquad \text{and} \qquad D_{\mathrm{rev}(\alpha)'}\xi(0) = z.$$

Then in fact $\xi(t) = \zeta(1 - t)$, and applying Proposition 3.1 concludes the proof.

Seminegative curvature

We apply the above results to the case of seminegative curvature. The next proposition gives us a criterion for the kernel of the differential of the exponential to be trivial, and we use Jacobi lifts in the proof.

Theorem 3.6. *Let* (X, g) *be Riemannian. Assume* (X, g) *has seminegative curvature. Then for all* $x \in X$ *and* $v \in T_x$, $v \neq 0$, *such that* $\exp_x$ *is defined on the segment* $[0, v]$ *in* T_x, *we have*

$$\|d\exp_x(v)w\|_g \geq \|w\|_g \qquad \text{for all } w \in T_x X.$$

In particular,

$$\text{Ker } d\exp_x(v) = 0.$$

Proof. Let η_w be the Jacobi lift as in Proposition 3.1, so that

$$d\exp_x(v)w = \eta_w(1).$$

The asserted inequality is then a special case of the inequality found in Proposition 2.6. This inequality implies that $\text{Ker } d\exp_x(v) = 0$, which concludes the proof.

Observe that the estimate on the differential of the exponential states that the inverse $d\exp_x(v)^{-1}$ is bounded by 1, as a continuous linear map. Of course, so far, this inverse is defined only on the image of $d\exp_x(v)$. In the finite dimensional case, invertibility is immediate. In the infinite dimensional case, it is in McAlpin's thesis [McA 65], as follows.

Theorem 3.7 (McAlpin [McA 65]). *Let* (X, g) *be a Riemannian–Hilbertian manifold with seminegative curvature, and let* $x \in X$. *Assume that* $\exp_x$ *is defined on all of* T_x *(what we called geodesically complete at* x*). Then for all* $v \in T_x$ *the map* $d\exp_x(v)$ *is a topological linear isomorphism, and in particular,* $\exp_x$ *is a local isomorphism.*

Proof. We have already proved that $d\exp_x(v)$ is injective and has a continuous inverse on its image. Lemma 3.5 shows that we can apply the same reasoning to the adjoint $(d\exp_x(v))^* = d\exp_y(v^*)$ for $y = \exp_x(v)$, so this adjoint also has kernel 0. Hence $d\exp_x(v)$ is surjective, thereby concluding the proof of the theorem.

The next theorem was proved by Hadamard for surfaces [Ha 1898], by Cartan for finite dimensional Riemannian manifolds [Ca 28], and by McAlpin in the Hilbertian case [McA 65].

Theorem 3.8 (Cartan–Hadamard). *Let (X, g) be a Riemannian manifold, connected, and such that $\exp_x$ is defined on all of T_x for some $x \in X$ (so geodesically complete. If $R_2 \geqq 0$ (i.e. X has seminegative curvature), then the exponential map $\exp_x \colon T_x X \to X$ is a covering. In particular, if X is simply connected, then $\exp_x$ is an isomorphism.*

Proof. We have already proved that $\exp_x$ is a local isomorphism. There remains to prove that $\exp_x$ is surjective, and that it is a covering. But all the work has been done, because we simply apply Theorem 6.9 of Chapter VIII with $Y = T_x$ having the given metric $h = g(x)$, for which Y is certainly complete. Theorem 3.6 guarantees that the essential estimate hypothesis is satisfied, so the proof is complete.

Corollary 3.9. *Let (X, g) be a connected Riemannian manifold with seminegative curvature. Then (X, g) is complete if and only if the exponential map $\exp_x$ is defined on all of T_x for some $x \in X$, and therefore for every $x \in X$.*

Proof. That (X, g) complete implies $\exp_x$ defined on all of T_x was proved under all circumstances in Proposition 6.5 of Chapter VIII. The converse is now immediate from Theorem 2.10 and Theorem 6.9 of Chapter VIII.

Corollary 3.10. *Let (X, g) be a complete Riemannian manifold, simply connected, and with seminegative curvature. Let $x \in X$. Then for all v, $w \in T_x X$ we have the inequality*

$$\text{dist}_g\bigl(\exp_x(v), \exp_x(w)\bigr) \geqq \|v - w\|_g.$$

Proof. By Theorem 3.7 the exponential map has an inverse

$$\varphi \colon X \to T_x X$$

and by Theorem 3.6 this inverse satisfies

$$\|d\varphi(x)\|_g \leqq 1$$

for all $z \in X$, where the norm is that of a continuous linear map from $T_z X$ to $T_{\varphi(z)} X$, with their structures of Hilbert spaces due to g. The inequality of the corollary is then immediate from the definition of the length of curves.

Corollary 3.11. *Suppose that (X, g) is Riemannian, complete, with seminegative curvature and simply connected. Then any two points can be joined by a unique geodesic whose length is the g-distance between the two points.*

Proof. Immediate from Corollary 3.10, because if x, y are the two points, then $y = \exp_x(v)$ for some $v \in T_x X$, and the geodesic α such that $\alpha(t) = \exp_x(tv)$ joins the two points, is unique by the Hadamard–Cartan theorem, and has length $\|v\|_g$.

Remark 1. The above corollary is of course mostly subsumed in the finite dimensional case by the Hopf–Rinow theorem, but it was noticed in the Hilbert case in McAlpin's thesis [McA 65]. Furthermore, McAlpin observed that one can define on the ball $\mathbf{B}(-2/c)$ with $c < 0$ of a Hilbert space $\mathbf{E}$ a bounded seminegative metric, as in the finite dimensional case, namely for $x \in \mathbf{B}(-2/c)$ and v, $w \in \mathbf{E}$ we let

$$\langle v, w \rangle_x = \frac{4\langle v, w \rangle}{4 + c^2 x^2}.$$

Then the ball has curvature $-c^2$. Note that for constant curvature K one has

$$R(v, w)z = K(\langle z, w \rangle v - \langle z, v \rangle w).$$

Similarly one shows that the sphere has constant positive curvature. Standard proofs that the only simply connected manifolds with constant curvature are all of $\mathbf{E}$, a sphere of finite radius for positive curvature, and the above example for negative curvature, work in the Hilbert case, and will be given below.

Remark 2. Proposition 3.2 can be interpreted as implying that the geodesics which come from rays starting at the origin in the tangent space are orthogonal to the g-spheres in X. Of course it may happen that the exponential map is not an injective map of T_x into X (as on the circle or 2-sphere), so the orthogonality interpretation holds only when it makes sense. In the particular case of seminegative curvature and completeness of the Cartan–Hadamard theorem, the interpretation is valid everywhere. Note that Proposition 3.2 in the case of seminegative curvature is also a special case of the "local" result on orthogonality, Theorem 5.6 of Chapter VIII, because we have a global chart coming from the Cartan-Hadamard theorem, and the previous arguments are valid for this chart.

We now work out as examples the cases of constant curvature.

Theorem 3.12. *Let X be Riemannian complete, simply connected. Let $x_0 \in X$.*

(a) *If $R = 0$, i.e. if X has 0 curvature, then the exponential map*

$$\exp_{x_0} \colon T_{x_0} X \to X$$

is an isometry.

(b) *Suppose the curvature is constant, equal to -1. Let Y also be Riemannian complete, simply connected, and let $y_0 \in Y$. Let $L: T_{x_0}X \to T_{y_0}Y$ be a linear isometry, and let $f: X \to Y$ be defined by*

$$f = \exp_{y_0} \circ L \circ \exp_{x_0}^{-1}$$

so f is a differential isomorphism according to Theorem 3.8. Then f is an isometry. In other words, up to an isometry, there is only one complete Riemannian manifold with given constant negative curvature modeled on a given Hilbert space (finite dimensional or not).

Proof. For (a), we use Theorem 3.1 and Proposition 2.10 which shows that the exponential map amounts to parallel translation, so is an isometry. For (b), we argue in a similar way, but a bit more complicated. We have to show that for each $x \in T_{x_0}X$ the map

$$df(x): T_x X \to T_{f(x)}Y$$

is a linear isometry. Since $d\exp_{x_0}(0) = \text{id}$, it follows that $df(x_0) = L$, so $df(x_0)$ is a linear isometry. Assume $x \neq x_0$. Let $x = \exp_{x_0}(rv)$ with some unit vector $v \in T_{x_0}X$ and $r > 0$. Let $\eta^{(v)}$ denote the map which to $w \in T_{x_0}X$ associates the Jacobi lift η_w of Theorem 3.1. Then

$$df(x) = d\exp_{y_0}\big(L(rv)\big) \circ L \circ d\exp_{x_0}(rv)^{-1}$$

$$= d\exp_{y_0}\big(rL(v)\big) \circ L \circ d\exp_{x_0}(rv)^{-1}$$

$$= \frac{1}{r}\eta^{L(v)}(r) \circ L \circ \left(\frac{1}{r}\eta^{(v)}(r)\right)^{-1}.$$

The map $d\exp_{x_0}(rv): T_{x_0}X \to T_xX$ is a linear isomorphism. To show that $df(x)$ preserves norms is equivalent to showing that

$$\|df(x) \circ d\exp_{x_0}(rv)w\| = \|d\exp_{x_0}(rv)w\| \qquad \text{for all } w \in T_xX.$$

But we have

$$\|df(x) \circ d\exp_{x_0}(rv)w\| = \|d\exp_{y_0}\big(rL(v)\big)L(w)\| \qquad \text{for all } \quad w \in T_{x_0}X.$$

We may now use Proposition 2.11 and Theorem 3.1 which describe $d\exp$ in terms of its components in the v-direction and a direction orthogonal to v, and parallel translation. Since in Proposition 2.11 the respective coefficients 1 and $(\sinh r)/r$ are the same whether we take $d\exp_{x_0}(rv)$ or $d\exp_{y_0}\big(rL(v)\big)$ because L is an isometry, preserving orthogonality and changing unit vectors to unit vectors, it follows that in the notation of

Proposition 2.11,

$$\|d\exp_{x_0}(rv)w\|^2 = c_0^2 + c_1^2\left(\frac{\sinh r}{r}\right)^2 = \|d\exp_{y_0}(rL(v))L(w)\|,$$

thus proving (b), and concluding the proof of the theorem.

We also have the following variation in the case of positive curvature.

Theorem 3.13. *Let X be Riemannian, complete, simply connected, with sectional curvature $+1$. Then X is isometric to the ordinary sphere of the same dimension in Hilbert space.*

Proof. The proof is similar, except that one cannot deal with the exponential defined on the whole tangent space $T_{x_0}X$. For convenience, we let X be the unit sphere in Hilbert space of a given dimension, and we let Y be Riemannian, complete simply connected with sectional curvature $+1$. We can then define the map f on the open ball of radius π. The same argument as before, replacing $\sinh r$ by $\sin r$, shows that f is a local isometry. We then pick another point $x_1 \neq \pm x_0$. We let

$$df(x_1) = L_1 \colon T_{x_1}X \to T_{f(x_1)}Y.$$

Just as we defined $f = f_{x_0}$ from x_0, we can define $f_1 = f_{x_1}$ from x_1. Then f and f_1 coincide on the intersection of their domain, and thus define a local isometry $X \to Y$. By Theorem 6.9 of Chapter VIII, this local isometry is a covering map, and since Y is assumed simply connected it follows that f is a differential isomorphism, and hence a global isometry, thus proving the theorem.

Remark. The above theorems may be viewed as fitting a special case of a theorem of Cartan, cf. [BGM 71], Proposition E.III.2.

IX, §4. THE INDEX FORM, VARIATIONS, AND THE SECOND VARIATION FORMULA

We let (X, g) be a pseudo Riemannian manifold, with the corresponding covariant derivative D. As a matter of notation, if w is a vector in a tangent space, then we write $w^2 = \langle w, w \rangle_g$. If $w^2 \geq 0$, then we define

$$\|w\| = \langle w, w \rangle_g^{1/2}.$$

We begin by a general discussion concerning the Jacobi expression defining Jacobi lifts. Let $\alpha \colon [a, b] \to X$ be a geodesic parametrized by arc length. Let $\eta \in \text{Lift}(\alpha)$. We are interested in the expression

$$D_{\alpha'}^2 \eta - R(\alpha', \eta)\alpha'$$

and its square. By definition, η is a Jacobi lift if and only if this expression is equal to 0, and in the Riemannian case, it is equal to 0 if and only if its square is equal to 0. We shall also deal with a subspace of $\text{Lift}(\alpha)$, namely:

$\text{Lift}_0(\alpha)$ = vector space of lifts η of α such that

$$\eta(a) = 0 \qquad \text{and} \qquad \eta(b) = 0.$$

For $\eta, \gamma \in \text{Lift}(\alpha)$ we define the **index**

$$I(\eta, \gamma) = \int_a^b [\langle D_{\alpha'}\eta, D_{\alpha'}\gamma \rangle_g + R(\alpha', \eta, \alpha', \gamma)](s) \, ds.$$

Then I is a symmetric bilinear form on $\text{Lift}(\alpha)$, whose corresponding quadratic form is

$$I(\eta, \eta) = \int_a^b [(D_{\alpha'}\eta)^2 + R_2(\alpha', \eta)](s) \, ds.$$

Similarly we define

$\text{Jac}_0(\alpha)$ = subspace of Jacobi lifts of α lying in $\text{Lift}_0(\alpha)$, that is, vanishing

at the end points.

Proposition 4.1. *The index form I on $\text{Lift}(\alpha)$ also has the expression*

$$I(\eta, \gamma) = -\int_a^b [\langle D_{\alpha'}^2\eta, \gamma \rangle_g - R(\alpha', \eta, \alpha', \gamma)](s) \, ds$$

$$+ \langle D_{\alpha'}\eta, \gamma \rangle_g(b) - \langle D_{\alpha'}\eta, \gamma \rangle_g(a).$$

In particular, if η is a Jacobi lift, then

$$I(\eta, \eta) = \langle D_{\alpha'}\eta, \eta \rangle_g(b) - \langle D_{\alpha'}\eta, \eta \rangle_g(a);$$

and if in addition $\eta \in \text{Jac}_0(\alpha)$, then $I(\eta, \eta) = 0$.

Proof. From the defining property of the metric derivative, we know that

$$\partial \langle D_{\alpha'}\eta, \gamma \rangle_g = \langle D_{\alpha'}^2\eta, \gamma \rangle_g + \langle D_{\alpha'}\eta, D_{\alpha'}\gamma \rangle_g.$$

Then the first formula is clear. If in addition η is a Jacobi lift, then the expression under the integral is 0 by definition, so the second formula follows; and if $\eta \in \text{Jac}_0(\alpha)$ then the expressions belonging to the end points are equal to 0, so the proposition is proved.

Theorem 4.2. *Let $\eta \in \text{Lift}(\alpha)$. Then $I(\eta, \gamma) = 0$ for all $\gamma \in \text{Lift}_0(\alpha)$ if and only if*

$$(D_{\alpha'}^2 \eta - R(\alpha', \eta)\alpha')^2 = 0.$$

In the Riemannian case, this happens if and only if η is a Jacobi lift.

Proof. If η is a Jacobi lift, then by definition

$$D_{\alpha'}^2 \eta = R(\alpha', \eta)\alpha',$$

so $I(\eta, \gamma) = 0$ for all $\gamma \in \text{Lift}_0(\alpha)$. Conversely, assume this is the case. Let φ be a C^∞ function on $[a, b]$ such that $\varphi(a) = \varphi(b) = 0$. Let

$$\gamma_1 = D_{\alpha'}^1 \eta - R(\alpha', \eta)\alpha' \quad \text{and} \quad \gamma = \varphi\gamma_1.$$

Then $\gamma \in \text{Lift}_0(\alpha)$ and by Proposition 4.1,

$$0 = I(\eta, \gamma) = \int_a^b \varphi(s)\gamma_1(s)^2 \, ds.$$

This being true for all φ as above, it follows that $\gamma_1^2 = 0$, whence the theorem follows.

The previous discussion belongs to the general theory of the Jacobi differential equation. Previously, we developed this theory to get information about the differential of the exponential map. The differential equation has another side to it, to which we now turn. We shall be interested in two functions of paths $\alpha: [a, b] \to X$:

The length function

$$L_a^b(\alpha) = L(\alpha) = \int_a^b \|\alpha'(s)\|_g \, ds \quad \text{whenever} \quad \alpha'(s)^2 \geqq 0.$$

The energy function

$$E_a^b(\alpha) = E(\alpha) = \int_a^b \alpha'(s)^2 \, ds.$$

Note that the length does not depend on the parametrization, but the energy does. We are interested in minimizing those functions.

In calculus, one applies the second derivative test at a critical point of a function, that is a point where the first derivative is 0. The second derivative then has geometric meaning. One wants to do a similar thing on function spaces, or the space of paths. Ultimately, one can define a manifold structure on this space, but there is a simple device which at first avoids defining such a structure for some specific computations. We are specifically interested here in the example of the second derivatives

$$\frac{d^2}{dt^2}L(\alpha_t) \quad \text{and} \quad \frac{d^2}{dt^2}E(\alpha_t)$$

at $t = 0$. To compute these derivatives, we don't need to give a differential structure to the path space, we need only be able to differentiate under the integral sign in the usual way. The computation of these derivatives is called the **second variation formula**, and the end result is as follows, for the variation of a geodesic. Observe how the index form enters into the result.

Theorem 4.3. *Let* $\alpha: [a, b] \to X$ *be a geodesic parametrized by arc length, that is* $\alpha'(s)^2 = 1$ *for all* s. *Let* σ *be a variation of* α, *so that* $\alpha_t(s) = \sigma(s, t)$. *Define*

$$\eta(s) = \partial_2\sigma(s, 0) \quad \text{and} \quad v(s) = D_{\alpha'}\eta(s) - \langle D_{\alpha'}\eta(s), \alpha'(s)\rangle_g\alpha'(s),$$

so $v(s)$ *is the normal projection of* $D_{\alpha'}\eta(s)$ *with respect to the unit vector* $\alpha'(s)$. *Also define a second component along* $\alpha'(s)$, *namely*

$$\gamma_2(s) = \langle D_2\partial_2\sigma, \partial_1\sigma\rangle_g(s, 0) = \langle D_2\partial_2\sigma(s, 0), \alpha'(s)\rangle_g.$$

Let $R_2(v, w) = R(v, w, v, w)$ *be the canonical 2-tensor. Then*

$$\frac{d^2}{dt^2}E(\alpha_t)\bigg|_{t=0} = I(\eta, \eta) + \gamma_2(b) - \gamma_2(a)$$

$$= \int_a^b [(D_{\alpha'}\eta)^2 + R_2(\alpha', \eta)](s)\, ds + \gamma_2(b) - \gamma_2(a).$$

As for the length, assuming the variation satisfies $\alpha_t'(s)^2 \geq 0$ *for all* t, s:

$$\frac{d^2}{dt^2}L(\alpha_t)\bigg|_{t=0} = \int_a^b [v^2 + R_2(\alpha', \eta)](s)\, ds + \gamma_2(b) - \gamma_2(a),$$

so this is the same expression as for the energy, except that $D_{\alpha'}\eta(s)$ *is replaced by the normal projection* $v(s)$.

If the curves $t \mapsto \sigma(a, t)$ and $t \mapsto \sigma(b, t)$ are geodesics, then

$$\gamma_2(b) = \gamma_2(a) = 0,$$

so the terms involving the end points are equal to 0.

Remark. The last assertion is immediate, since for any geodesic γ, we have $D_{\gamma'}\gamma' = 0$.

Proof of Theorem 4.3. Let

$$e(s, t) = \langle \partial_1\sigma, \partial_1\sigma \rangle_g(s, t) = \alpha_t'(s)^2.$$

Note that $e(s, 0) = 1$ by hypothesis. We shall compute the second derivative of the length, which if anything is harder than that of the energy because of the square root sign. The computation for the energy follows exactly the same pattern. We shall keep in mind that from the definitions,

$$D_\alpha\eta(s) = D_1\partial_2\sigma(s, 0).$$

We begin with the first derivative, also called the **first variation**,

$$\frac{d}{dt}L(\alpha_t) = \frac{d}{dt}\int_a^b e(s, t)^{1/2}\, ds$$

(1)
$$= \int_a^b \frac{1}{2}e(s, t)^{-1/2}\partial_2 e(s, t)\, ds.$$

This formula gives us the first derivative of length, about which we shall say more later. We now forge ahead to the second derivative, which is

$$\frac{d^2}{dt^2}L(\alpha_t) = \int_a^b \tfrac{1}{2}e(s, t)^{-1/2}\partial_2^2 e(s, t)\, ds - \int_a^b \tfrac{1}{4}e(s, t)^{-3/2}(\partial_2 e(s, t))^2\, ds$$

whence

(2)
$$\frac{d^2}{dt^2}L(\alpha_t)\bigg|_{t=0} = \int_a^b \tfrac{1}{2}\partial_2^2 e(s, 0)\, ds - \int_a^b \tfrac{1}{4}(\partial_2 e(s, 0))^2\, ds$$

$$= I_1 - I_2.$$

We then compute separately the expressions under the integral signs.

For the second integral, we have:

$$\partial_2 e = \partial_2 \langle \partial_1 \sigma, \partial_1 \sigma \rangle_g$$
$$= 2\langle D_2 \partial_1 \sigma, \partial_1 \sigma \rangle_g \quad \text{because } D \text{ is the metric derivative}$$
$$= 2\langle D_1 \partial_2 \sigma, \partial_1 \sigma \rangle_g \quad \text{by Lemma 5.3 of Chapter VIII.}$$

Evaluating at 0, squaring, and integrating yields the term

$$(3) \qquad\qquad I_2 = \int_a^b \langle D_{\alpha'} \eta, \alpha' \rangle_g(s) \, ds.$$

Next we compute the first integrand. We have:

$$\partial_2^2 e = \partial_2 \langle D_1 \partial_2 \sigma, \partial_1 \sigma \rangle_g$$
$$= 2\langle D_2 D_1 \partial_2 \sigma, \partial_1 \sigma \rangle_g + 2\langle D_1 \partial_2 \sigma, D_2 \partial_1 \sigma \rangle_g.$$

In the first term on the right, we use Lemma 2.7 to write

$$D_2 D_1 = D_1 D_2 - R(\partial_1 \sigma, \partial_1 \sigma).$$

In the second term on the right, we use Lemma 5.3 of Chapter VIII to write $D_2 \partial_1 = D_1 \partial_2$. Then we find

$$\partial_2^2 e = 2\langle D_1 D_2 \sigma, \partial_1 \sigma \rangle_g - 2\langle R(\partial_1 \sigma, \partial_2 \sigma)\partial_2 \sigma, \partial_1 \sigma \rangle_g + 2(D_1 \partial_2 \sigma)^2$$
$$(4) \qquad = 2\langle D_1 D_2 \partial_2 \sigma, \partial_1 \sigma \rangle_g + 2R_2(\partial_1 \sigma, \partial_2 \sigma) + 2(D_2 \partial_2 \sigma)^2.$$

Finally, we use the metric derivative again to compute:

$$(5) \qquad \partial_1 \langle D_2 \partial_2 \sigma, \partial_1 \sigma \rangle_g = \langle D_1 D_2 \partial_2 \sigma, \partial_1 \sigma \rangle_g + \langle D_2 \partial_2 \sigma, D_1 \partial_1 \sigma \rangle_g.$$

However, $D_1 \partial_1 \sigma(s, 0) = D_1 \partial_1 \alpha(s) = D_{\alpha'} \alpha'(s) = 0$, because α is assumed to be a geodesic. Hence from (3) and (4) we find

$$(6) \qquad \partial_2^2 e(s, 0) = 2\gamma_2'(s) + 2R_2(\partial_1 \sigma, \partial_1 \sigma)(s, 0) + 2(D_1 \partial_2 \sigma(s, 0))^2.$$

Now if w, u are vectors and u is a unit vector, and $v = w - (w \cdot u)u$ is the orthogonalization of w with respect to u, then we have trivially

$$v^2 = w^2 - (w \cdot u)^2.$$

Subtracting (3) from (6) and integrating yields the asserted answer, and proves the formula for the second derivative of $L(\alpha_t)$ at $t = 0$.

Remark 1. For simplicity we limited ourselves to curves rather than piecewise C^2 maps. Milnor [Mi 63] gives a thorough discussion of paths where end-point terms will appear where the path is broken. See for instance his Theorem 12.2 and Theorem 13.1 of Chapter III.

Remark 2. Observe how R_2 comes naturally in the formula. At a minimum one wants the second derivative to be semipositive, so having all plus signs in the variation formula is desirable.

Corollary 4.4. *Let η be a Jacobi lift of α, and σ a variation of α such that $\eta(s) = \partial_2\sigma(s, 0)$. Assume that $t \mapsto \sigma(a, t)$ and $t \mapsto \sigma(b, t)$ are geodesics. Then*

$$\left.\frac{d^2}{dt^2} E(\alpha_t)\right|_{t=0} = \langle D_{\alpha'}\eta(b), \eta(b)\rangle_g - \langle D_{\alpha'}\eta(a), \eta(a)\rangle_g.$$

In particular, if $D_{\alpha'}\eta$ is perpendicular to α' then this equality also holds if E is replaced by the length L.

Proof. Immediate from Theorem 4.3 and the alternative expressions of Proposition 4.1.

Concerning the orthogonality assumption which will recur, we recall that if a Jacobi lift η of α is such that $D_{\alpha'}\eta$ is orthogonal to α' at some point, then $D_{\alpha'}\eta$ is orthogonal to α' on the whole interval of definition. See Proposition 2.4.

Proposition 4.5. *Assumptions being as in Theorem 4.3, suppose that $D_{\alpha'}\eta$ is orthogonal to α'. Then*

$$\left.\frac{d}{dt} L(\alpha_t)\right|_{t=0} = 0.$$

Proof. We consider the expression for the first derivative given at the beginning of the proof in formula (1), we note that

$$\partial_2 e(s, t) = 2\langle \partial_1\sigma(s, t), \partial_2\sigma(s, t)\rangle_g,$$

whence

$$\partial_2 e(s, 0) = 2\langle \partial_1\sigma(s, 0), \partial_2\sigma(s, 0)\rangle_g = \langle \alpha'(s), \eta(s)\rangle_g.$$

By Proposition 2.4, a Jacobi lift which is perpendicular to α' somewhre remains perpendicular to α' all along, so $\partial_2 e(s, 0) = 0$, which proves the proposition.

For some applications, one wants to compute the second derivative of a composite function $f(L(\alpha_t))$, where f is a function of a real variable, for instance when we determine the Laplacian in polar coordinates later. So we give here the relevant formula, since it is essentially a corollary of the above considerations.

Proposition 4.6. *Let f be a C^2 function of a real variable. As in Theorem 4.3, let σ be a variation of α, and let $\eta(s) = \partial_2 \sigma(s, 0)$. Assume $D_{\alpha'}\eta$ orthogonal to α'. Then*

$$\frac{d^2}{dt^2} f(L(\alpha_t)) \bigg|_{t=0} = f'(L(\alpha_0))[\langle D_{\alpha'}\eta(b), \eta(b)\rangle_g - \langle D_{\alpha'}\eta(a), \eta(a)\rangle_g].$$

Proof. Let $F(t) = f(L(\alpha_t))$. Then

$$F'(t) = f'(L(\alpha_t)) \frac{d}{dt} L(\alpha_t)$$

and

$$F''(t) = f''(L(\alpha_t)) \frac{d}{dt} L(\alpha_t) + f'(L(\alpha_t)) \left(\frac{d}{dt}\right)^2 L(\alpha_t).$$

Then at $t = 0$ the first term on the right is 0 because of Proposition 4.5. The second term at $t = 0$ is the asserted one by Corollary 4.4 and the orthogonality assumption. This concludes the proof.

Example. Proposition 3.3 provides an example of the situation in Proposition 4.6. Both will be used in Chapter X, §2.

Theorem 4.3, i.e. the second variation formula, also has some topological applications which we don't prove in this book, but which we just mention. If R_2 is negative, so the curvature is positive, then one has a theorem of Synge [Sy 36]:

Let X be a compact even dimensional orientable Riemannian manifold with strictly positive sectional curvature. Then X is simply connected.

The idea is that in each homotopy class one can find a geodesic of minimal length. By the second derivative test, the expression for the second derivative of the length is 0 for such a geodesic. One has to prove that one can choose the variation such that the "orthogonal" term containing the integral of $v(s)^2$ is 0. The boundary term will vanish if one works with a variation to which we can apply Remark 1. Finally, having strictly positive curvature will yield a negative term, which gives a contradiction. Details can be found in texts on Riemannian geometry.

The same ideas and the theorem of Synge lead to a theorem of Weinstein [We 67]:

Let X be a compact oriented Riemannian manifold of positive sectional curvature. Let f be an isometry of X preserving the orientation if $\dim X$ is even, and reversing orientation if $\dim X$ is odd. Then f has a fixed point, i.e. a point x such that $f(x) = x$.

Proofs of both the Synge and the Weinstein theorems are given in [doC 92]. The Synge theorem is given in [BGM 71] and [GHL 87[93]].

This as far as we go in the direction of the calculus of variations. These are treated more completely in Morse theory, for instance in [Mi 63], [Pa 63], [Sm 64], and in differential geometry texts such as [KoN 69], [BGM 71], [ChE 75], [doC 92], [GHL 87/93].

Incidentally, I hope to have convinced the reader further about the irrelevance whether the manifold is finite dimensional or not.

IX, §5. TAYLOR EXPANSIONS

We shall deal systematically with the Taylor expansion of various curves. We consider a curve in X, say of class C^2, not necessarily a geodesic,

$$\alpha \colon J \to X,$$

and we assume $0 \in J$, so 0 is an origin. We suppose given a spray on TX, giving rise to the covariant derivative. For $w \in T_{\alpha(0)}X$ we let

$$\gamma \colon t \mapsto \gamma(t, w)$$

be the unique α-parallel curve with initial condition $\gamma(0, w) = w$. Recall that α-parallel means $D_{\alpha'}\gamma = 0$. We denote parallel translation by

$$P_t = P_{\alpha, t} = P_{\alpha, 0, t} \colon T_{\alpha(0)}X \to T_{\alpha(t)}X$$

Then P_t is a topological linear isomorphism, as we saw in Chapter VIII, §3.

Proposition 5.1. *Let $\eta \colon J \to TX$ be a lift of α in TX. Then*

$$\eta(t) = P_t \sum_{k=0}^{m} D_{\alpha}^k \eta(0) \frac{t^k}{k!} + O(t^{m+1}) \qquad \text{for } t \to 0;$$

or alternatively,

$$\eta(t) = \sum_{k=0}^{m} \gamma\big(t, D_{\alpha}^k \eta(0)\big) \frac{t^k}{k!} + O(t^{m+1}) \qquad \text{for } t \to 0.$$

Proof. The second expression is merely a reformulation of the first, taking into account the definition of parallel translation. Since $t \to 0$, the formula is local, and we may prove it in a chart, so we use η, γ to denote the vector components η_U, γ_U in a chart U, suppressing the index U. Let

$$\beta(t) = \eta(t) - \sum_{k=0}^{m} \gamma\big(t, D_{\alpha'}^k \eta(0)\big) \frac{t^k}{k!}.$$

From the existence and uniqueness of the ordinary Taylor formula, it will suffice to prove that for the ordinary derivatives of β, we have

$$\partial^k \beta(0) = \beta^{(k)}(0) = 0 \qquad \text{for} \quad k = 0, \ldots, m.$$

By definition, note that $\beta(0) = 0$. Let $w_k = D_{\alpha'}^k \beta(0)$. Since $D_{\alpha'} \gamma = 0$, we have

$$D_{\alpha'}^j \beta(t) = D_{\alpha'}^j \eta(t) - \sum_{k \geq j} \gamma(t, w_k) \frac{t^{k-j}}{(k-j)!}.$$

Therefore

$$D_{\alpha'}^j \beta(0) = D_{\alpha'}^j \eta(0) - \gamma(0, w_j)$$

$$= w_j - w_j$$

$$= 0.$$

We now need a lemma. We let $\mathbf{E}$ be the Banach space on which X is modeled.

Lemma 5.2. *Let* $\beta: J \to \mathbf{E}$ *be the vector component of a lift of* α. *If* $D_{\alpha'}^j \beta(0) = 0$ *for* $0 \leq j \leq m$ *then* $\partial^j \beta(0) = 0$ *for* $0 \leq j \leq m$.

Proof. By definition,

$$D_{\alpha'} \beta = \beta' - B(\alpha; \alpha', \beta).$$

Hence $D_{\alpha'} \beta(0) = \beta'(0)$. We can proceed by induction. Let us carry out the case of the second derivative so the reader sees what's going on. Hence suppose in addition that $D_{\alpha'}^2 \beta(0) = 0$. From the definitions, we get

$$D_{\alpha'}^2 \beta = \beta'' - [\partial_1 B(\alpha; \alpha', \beta)\alpha' + B(\alpha; \alpha'', \beta) + B(\alpha; \alpha', \beta')]$$

$$- B\big(\alpha; \alpha', \beta' - B(\alpha; \alpha', \beta)\big).$$

Since $\beta(0) = \beta'(0) = D_{\alpha'} \beta(0) = 0$ we find that

$$0 = D_{\alpha'}^2 \beta(0) = \beta''(0),$$

thus proving the assertion for $m = 2$. The inductive proof is the same in general.

We apply the above considerations to Jacobi lifts.

Proposition 5.3. *Suppose that α is a geodesic. Let $w \in T_{\alpha(0)}X$ and let η_w be the Jacobi lift of α such that $\eta_w(0) = 0$ and $D_{\alpha'}\eta_w(0) = w$. Then*

$$\eta_w(t) = P_t\left[wt + R(\alpha'(0), w, \alpha'(0))\frac{t^3}{3!}\right] + O(t^4).$$

Proof. We plug in Proposition 5.1. Since $D_{\alpha'}^2\eta_w = R(\alpha', \eta_w, \alpha')$ contains η_w linearly, the evaluation of the second term of the Taylor expansion at 0 is 0. As for the third term, we have to use the chain rule. To be sure we don't forget anything, we should write more precisely

$$R(\alpha', \eta_w, \alpha') = R(\alpha; \alpha', \eta_w, \alpha')$$

to make explicit the dependence on the extra position variable. But it turns out that it does not matter in the end, because no matter what, the chain rule gives

$$D_{\alpha'}^3\eta_w = R(\alpha', D_{\alpha'}\eta_w, \alpha') + \text{terms containing } \eta_w \text{ linearly,}$$

so $D_{\alpha'}^3\eta_w(0) = R(\alpha'(0), w, \alpha'(0))$, which proves the proposition.

From Proposition 5.3, we get information on the pull back of the metric g of a pseudo Riemannian manifold, to the tangent space at a given point.

Proposition 5.4. *Let (X, g) be a pseudo Riemannian manifold, and let $x \in X$. Fix $v, w \in T_xX$. Then*

$$\exp_x^*(tv)(g)(w, w) = w^2 + \tfrac{1}{3}R_2(v, w)t^2 + O(t^3) \qquad \text{for } t \to 0.$$

where we recall that $R_2(v, w) = R(v, w, v, w)$.

Proof. From the theory of Jacobi lifts, applied to $\alpha(t) = \exp_x(tv)$, we have the formula

$$\frac{1}{t}\eta_w(t) = d\exp_x(tv)w.$$

Therefore modulo functions which are $O(t^3)$ for $t \to 0$, we get from

Proposition 5.3

$$\exp_x^*(g)(tv)(w, w) \equiv \left\langle \frac{1}{t} \eta_w(t), \frac{1}{t} \eta_w(t) \right\rangle_{g(\alpha(t))}$$

$$\equiv \left\langle P_t \left[w + R(v, w, v) \frac{t^2}{3!} \right], P_t \left[w + R(v, w, v) \frac{t^2}{3!} \right] \right\rangle_{g(\alpha(t))}$$

$$\equiv \left\langle w + R(v, w, v) \frac{t^2}{3!}, w + R(v, w, v) \frac{t^2}{3!} \right\rangle_{g(x)}$$

$$\equiv w^2 + 2R_2(v, w) \frac{t^2}{3!},$$

which proves the proposition.

The preceding proposition gives us the Taylor expansion of the mapping

$$f(v) = \exp_x^*(g)(v) \quad \text{for} \quad v \in T_x X$$

along rays through the origin. Observe that

$$f: T_x X \to L^2_{\text{sym}}(T_x X)$$

is a map of the self-dual Banach space $T_x X$ to the space of symmetric bilinear forms on $T_x X$ (actually the open subset of non-singular forms). The map f has a Taylor expansion

$$(*) \qquad f(v) = f(0) + f_1(v) + f_2(v) + O(|v|^3) \qquad \text{for } |v| \to 0,$$

where $v \mapsto |v|$ is a Banach norm on T_x, and where f_1 and f_2 are homogeneous of degree 1 and 2 respectively. Since the homogeneous terms in the Taylor expansion are uniquely determined by f, and since we computed their restrictions to rays through the origin in Proposition 2.4, we now obtain:

Theorem 5.5. *Let (X, g) be a pseudo Riemannian manifold. Let $x \in X$. For $v \in T_x X$ let $q(v) \in L^2_{\text{sym}}(T_x X)$ be the symmetric bilinear function such that*

$$q(v)(w_1, w_2) = \frac{1}{3} R_x(v, w_1, v, w_2).$$

Let the metric g be viewed as a tensor in $L^2_{\text{sym}}(TX)$, and let f be the pull back of the metric g in a star shaped neighborhood of 0_x in $T_x X$ where the exponential map is defined. Then

$$f(v) = g(x) + q(v) + O(|v|^3) \qquad \text{for } |v| \to 0.$$

CHAPTER X

Volume Forms

For the first time we meet a strictly finite dimensional phenomenon: If X is of finite dimension n, then the n-forms $\mathscr{A}^n(X)$ play a distinguished role whose extension to the infinite dimensional case is not evident. So this chapter is devoted to these forms of maximal degree. In the next chapter, we shall study how to integrate them, so the present chapter also provides a transition from the differential theory to the integration theory.

Although for organization and reference purposes it is convenient to place together here a number of results on volume forms, only the first section giving a basic definition will be used in the next three chapters, so the other sections may be skipped by a reader wanting to get immediately into integration.

X, §1. THE RIEMANNIAN VOLUME FORM

Let V be a finite dimensional vector space over $\mathbf{R}$, of dimension n. We assume given a positive definite symmetric scalar product g, denoted by

$$(v, w) \mapsto \langle v, w \rangle_g = g(v, w) \qquad \text{for } v, w \in V.$$

The space $\bigwedge^n V$ has dimension 1. If $\{e_1, \ldots, e_n\}$ and $\{u_1, \ldots, u_n\}$ are orthonormal bases of V, then

$$e_1 \wedge \cdots \wedge e_n = \pm u_1 \wedge \cdots \wedge u_n.$$

Two such orthonormal bases are said to have the **same orientation**, or to be **orientation equivalent**, if the plus sign occurs in the above relation. A choice of an equivalence class of orthonormal bases having the same orientation is defined to be an **orientation** of V. Thus an orientation determines a basis for the one-dimensional space $\bigwedge^n V$ over $\mathbf{R}$. Such a basis will be called a **volume**. There exists a unique n-form Ω on V (alternating), also denoted by vol_g, such that for every oriented orthonormal basis $\{e_1, \ldots, e_n\}$ we have

$$\Omega(e_1, \ldots, e_n) = 1.$$

Conversely, given a non-zero n-form Ω on V, all orthonormal bases $\{e_1, \ldots, e_n\}$ such that $\Omega(e_1, \ldots, e_n) > 0$ are orientation equivalent, and on such bases Ω has a constant value.

Let (X, g) be a Riemannian manifold. By an **orientation** of (X, g) we mean a choice of a volume form Ω, and an orientation of each tangent space $T_x X$ ($x \in X$) such that for any oriented orthonormal basis $\{e_1, \ldots, e_n\}$ of $T_x X$ we have

$$\Omega_x(e_1, \ldots, e_n) = 1.$$

The form gives a coherent way of making the orientations at different points compatible. It is an exercise to show that if (X, g) has such an orientation, and X is connected, then (X, g) has exactly two orientations. In Chapter XI, we shall give a variation of this definition. By an **oriented chart**, with coordinates $x_1, \ldots, x_n$ in $\mathbf{R}^n$, we mean a chart such that with respect to these coordinates, the form has the representation

$$\Omega(x) = \varphi(x) \, dx_1 \wedge \cdots \wedge dx_n$$

with a function φ which is positive at every point of the chart. We call Ω the **Riemannian volume form**, and also denote it by vol_g, so

$$\mathrm{vol}_g(x) = \Omega(x) = \Omega_x.$$

We return to our vector space V, with positive definite metric g, and oriented.

Proposition 1.1. *Let $\Omega = \mathrm{vol}_g$. Then for all n-tuples of vectors $\{v_1, \ldots, v_n\}$ and $\{w_1, \ldots, w_n\}$ in V, we have*

$$\Omega(v_1, \ldots, v_n)\Omega(w_1, \ldots, w_n) = \det \langle v_i, w_j \rangle_g.$$

In particular,

$$\Omega(v_1, \ldots, v_n)^2 = \det \langle v_i, v_j \rangle_g.$$

Proof. The determinant on the right side of the first formula is multilinear and alternating in each n-tuple $\{v_1, \ldots, v_n\}$ and $\{w_1, \ldots, w_n\}$. Hence there exists a number $c \in \mathbf{R}$ such that

$$\det\langle v_i, w_j\rangle_g = c\Omega(v_1,\ldots,v_n)\Omega(w_1,\ldots,w_n)$$

for all such n-tuples. Evaluating on an oriented orthonormal basis shows that $c = 1$, thus proving the proposition.

Applying Proposition 1.1 to an oriented Riemannian manifold yields:

Proposition 1.2. *Let* (X, g) *be an oriented Riemannian manifold. Let* $\Omega = \mathrm{vol}_g$. *For all vector fields* $\{\xi_1,\ldots,\xi_n\}$ *and* $\{\eta_1,\ldots,\eta_n\}$ *on* X, *we have*

$$\Omega(\xi_1,\ldots,\xi_n)\Omega(\eta_1,\ldots,\eta_n) = \det\langle\xi_i, \eta_j\rangle_g.$$

In particular,

$$\Omega(\xi_1,\ldots,\xi_n)^2 = \det\langle\xi_i, \xi_j\rangle_g.$$

Furthermore, if $\xi^\vee$ *denotes the one-form dual to* ξ *(characterized by* $\xi^\vee(\eta) = \langle\xi, \eta\rangle_g$ *for all vector fields* η*), then*

$$\Omega(\xi_1,\ldots,\xi_n)\Omega = \xi_1^\vee \wedge \cdots \wedge \xi_n^\vee.$$

This last formula is merely an application of the definition of the wedge product of forms, taking into account the preceding formulas concerning the determinant.

At a point, the space of n-forms is 1-dimensional. Hence any n-form on a Riemannian manifold can be written as a product $\varphi\Omega$ where φ is a function and Ω is the Riemannian volume form.

If ξ is a vector field, then $\Omega \circ \xi$ is an $(n-1)$-form, and so there exists a function φ such that

$$d(\Omega \circ \xi) = \varphi\Omega.$$

We call φ the **divergence** of ξ with respect to Ω, or with respect to the Riemannian metric. We denote it by $\mathrm{div}_\Omega\, \xi$ or simply $\mathrm{div}\,\xi$. Thus by definition,

$$d(\Omega \circ \xi) = (\mathrm{div}\,\xi)\Omega.$$

Example. Looking back at the example of Chapter V, §8 we see that if

$$\Omega(x) = dx_1 \wedge \cdots \wedge dx_n$$

is the canonical form on $\mathbf{R}^n$ and ξ is a vector field, then its divergence is given by

$$\mathrm{div}_\Omega\, \xi = \sum_{i=1}^{n} \frac{\partial \xi_i}{\partial x_i}.$$

We shall study the divergence from a differential point of view in the next section, and from the point of view of Stokes' theorem in Chapter XIII.

On 1-forms, we define the operator

$$d^*: \mathscr{A}^1(X) \to \mathscr{A}^0(X)$$

by duality, that is if $\lambda^\vee$ denotes the vector field corresponding to λ under the Riemannian metric, then we define

$$d^*\lambda = -\operatorname{div} \lambda^\vee.$$

One defines the **Laplacian** or **Laplace operator** *on functions* by the formula

$$\Delta = d^*d = -\operatorname{div} \circ \operatorname{grad}.$$

Proposition 1.3. *For functions* φ, ψ *we have*

$$\Delta(\varphi\psi) = \varphi\Delta\psi + \psi\Delta\varphi - 2\langle d\varphi, d\psi \rangle_g.$$

Proof. The routine gives:

$$
\begin{aligned}
\Delta(\varphi\psi) = d^*d(\varphi\psi) &= d^*(\psi\, d\varphi + \varphi\, d\psi) \\
&= -\operatorname{div}(\psi\xi_{d\varphi}) - \operatorname{div}(\varphi\xi_{d\psi}) \\
&= -\psi\operatorname{div}\xi_{d\varphi} - (d\psi)\xi_{d\varphi} - \varphi\operatorname{div}\xi_{d\psi} - (d\varphi)\xi_{d\psi} \\
&= \psi\Delta\varphi + \varphi\Delta\psi - 2\langle d\varphi, d\psi \rangle_g
\end{aligned}
$$

as was to be shown.

Recall that

$$\langle d\varphi, d\psi \rangle_g = \langle \operatorname{grad} \varphi, \operatorname{grad} \psi \rangle_g,$$

so there is an alternative expression for the last term in the formula.

More formulas concerning the Laplacian will be given in the next section, using the covariant derivative and the variation formula. For applications of such formulas and theory to the heat kernel, cf. [Cha 84], especially Chapters II and III, in addition to [BGM 71].

X, §2. COVARIANT DERIVATIVES

In this section, we gather together a number of results on the covariant derivative in connection with volume forms on the oriented Riemannian manifold (X, g).

Proposition 2.1. *Let D be the metric covariant derivative. Then*

$$D_\xi \, \mathrm{vol}_g = 0$$

for all vector fields ξ.

Proof. We still write $\Omega = \mathrm{vol}_g$. Let $\xi_1, \ldots, \xi_n$ be any vector fields. We take the covariant derivative of the relation

$$(1) \qquad \Omega(\xi_1, \ldots, \xi_n)\Omega = \xi_1^\vee \wedge \cdots \wedge \xi_n^\vee$$

from Proposition 1.2. Let φ be the function $\varphi = \Omega(\xi_1, \ldots, \xi_n)$. Then by relation (1),

$$(2) \qquad \begin{aligned} D_\xi(\varphi\Omega) &= \sum \xi_1^\vee \wedge \cdots \wedge (D_\xi\xi_i)^\vee \wedge \cdots \wedge \xi_n^\vee \\ &= \sum \Omega(\xi_1, \ldots, D_\xi\xi_i, \ldots, \xi_n)\Omega. \end{aligned}$$

after using Proposition 1.2 and **COVD 3** of Chapter VIII, §1. But also

$$(3) \qquad D_\xi\varphi = (D_\xi\Omega)(\xi_1, \ldots, \xi_n) + \sum \Omega(\xi_1, \ldots, D_\xi\xi_i, \ldots, \xi_n).$$

Since $D_\xi(\varphi\Omega) = (D_\xi\varphi)\Omega + \varphi D_\xi\Omega$, from (2) and (3) we obtain

$$(4) \qquad (D_\xi\Omega)(\xi_1, \ldots, \xi_n)\Omega + \varphi D_\xi\Omega = 0.$$

We evaluate the expression on the left on $(\xi_1, \ldots, \xi_n)$, and find

$$\Omega(\xi_1, \ldots, \xi_n)(D_\xi\Omega)(\xi_1, \ldots, \xi_n) = 0.$$

This being true for all choices of vector fields $\xi_1, \ldots, \xi_n$, the proposition is proved.

Remark. The above theorem remains true suitably formulated in the non-oriented case, because the theorem is local, and locally, the absolute value of the form differs by ± 1 from the form itself.

The next theorem will give an application of Theorem 2.1.

The metric derivative D operates on vector fields and also on r-forms for all r, especially $r = 1$ and $r = n$. For any vector field ξ we let $D\xi$ be the endomorphism of ΓTX such that

$$(D\xi)\eta = D_\eta\xi.$$

At each point $x \in X$ we have the operator

$$(D\xi)_x \colon T_xX \to T_xX \qquad \text{such that} \quad (D\xi)_x(v) = (D_v\xi)(x),$$

on the finite dimensional vector space $T_x X$. This allows us to take the trace $\mathrm{tr}(D\xi)$ of this operator at each point, so to take $\mathrm{tr}(D\xi)_x$. The trace can be computed as usual by using an orthonormal basis. We recall that a finite sequence of vector fields is called an **orthonormal frame** (on some open subset of X) if they are orthonormal for the metric g, that is

$$\langle \xi_i, \xi_j \rangle_g = \begin{cases} 1 & \text{if } i = j, \\ 0 & \text{if } i \neq j. \end{cases}$$

Given a point $x \in X$, such an orthonormal frame always exists in a neighborhood of x. Similarly, we can define $D\lambda$ for a 1-form $\lambda \in \mathscr{A}^1(X)$, whereby

$$D\lambda: \Gamma TX \to \Gamma T^\vee X \qquad \text{is such that} \qquad (D\lambda)(\xi) = D_\xi \lambda.$$

Thus for each $x \in X$, $(D\lambda)_x$ may be viewed as a linear map

$$(D\lambda)_x: T_x X \to T_x^\vee X,$$

whose trace can be computed by using duality, namely

$$\mathrm{tr}(D\lambda) = \sum_i \langle D_{\xi_i} \lambda, \xi_i \rangle.$$

On the right side, we use the convenient notation $\langle \lambda, \xi \rangle = \lambda(\xi)$ for a 1-form λ and a vector field ξ. In such a case, there is no subscript g on the scalar bilinear pairing between functionals and vectors.

Theorem 2.2. *Let $\xi_1, \dots, \xi_n$ be an orthonormal frame of vector fields, and let ξ be a vector field. Then*

$$\mathrm{div}\, \xi = \sum_{i=1}^n \langle D_{\xi_i} \xi, \xi_i \rangle_g.$$

In particular, for $\lambda \in \mathscr{A}^1(X)$ we have

$$\mathrm{div}\, \lambda^\vee = \mathrm{tr}(D\lambda).$$

Proof. Let $\Omega = \mathrm{vol}_g$ be the volume form. By **COVD 6** of Chapter VIII, §1, and Proposition 2.1, we get

$$d(\Omega \circ \xi)(\xi_1, \dots, \xi_n) = \sum_{i=1}^n (-1)^{i-1} D_{\xi_i}(\Omega \circ \xi)(\xi_1, \dots, \hat{\xi}_i, \dots, \xi_n)$$

$$= \sum_{i=1}^n (-1)^{i-1} (\Omega \circ D_{\xi_i}\xi)(\xi_1, \dots, \hat{\xi}_i, \dots, \xi_n)$$

$$= \sum_{i=1}^n \Omega(\xi_1, \dots, D_{\xi_i}\xi, \dots, \xi_n)$$

and since $D_{\xi_i}\xi$ has the Fourier expression $D_{\xi_i}\xi = \sum_j \langle D_{\xi_i}\xi, \xi_j \rangle_g \xi_j$,

$$= \sum_{i=1}^{n} \langle D_{\xi_i}\xi, \xi_i \rangle_g \Omega(\xi_1, \ldots, \xi_n).$$

But also $d(\Omega \circ \xi)(\xi_1, \ldots, \xi_n) = (\text{div } \xi)\, \text{vol}_g(\xi_1, \ldots, \xi_n)$. Hence

$$\text{div } \xi = \sum_{i=1}^{n} \langle D_{\xi_i}\xi, \xi_i \rangle_g,$$

which proves the first formula. The second is a mere rephrasing, applied to the vector field $\lambda^{\vee}$.

Directly from the definition of the operator d^* in the preceding section, we now obtain:

Corollary 2.3. *On a 1-form λ, we have $d^*\lambda = -\text{tr}(D\lambda)$.*

We can then apply this to the Laplacian, to get:

Corollary 2.4. *Let $\xi_1, \ldots, \xi_n$ be an orthonormal frame as in Theorem 2.2. Let φ be a funtion. Then*

$$\Delta\varphi = -\text{tr}(D\, d\varphi) = -\sum_{i=1}^{n} \langle D_{\xi_i}\, d\varphi, \xi_i \rangle = -\sum_{i=1}^{n} \langle D_{\xi_i}(\text{grad } \varphi), \xi_i \rangle_g.$$

If $\{u_1, \ldots, u_n\}$ is an orthonormal basis of the tangent space T_xX at some point $x \in X$, and α_i is the geodesic with $\alpha_i(0) = x$ and $\alpha_i'(0) = u_i$, then

$$\Delta\varphi(x) = -\sum_{i=1}^{n} (\varphi \circ \alpha_i)''(0).$$

Proof. The first assertion comes from applying Theorem 2.2 to $\lambda = d\varphi$. The second assertion then follows by using Corollary 4.4 of Chapter VIII.

From the preceding corollary, we can obtain an expression for the Laplacian in polar coordinates. I follow [BGM 71]. *We pick a point $x \in X$ as an origin, with its tangent space T_xX. We let U_x be an open ball centered at 0_x on which $\exp_x$ induces an isomorphism to its image, and we let $y \in U_x$. We want to determine $\Delta\varphi(y)$ for a function φ which depends only on the Riemannian distance from x, say*

$$\varphi(y) = f(r(y)) \qquad \text{where} \quad r(y) = \text{dist}_g(x, y),$$

and f is a C^2 function of a real variable.

Proposition 2.5. *Let* $\alpha = \alpha_1$ *be the unique geodesic from* x *to* $y \neq x$, *parametrized by arc length, and let* $e_1 = \alpha'(r) \in T_yX$. *Let* $e_2, \ldots, e_n$ *be unit vectors in* T_yX *such that* $\{e_1, \ldots, e_n\}$ *is an orthonormal basis of* T_yX. *Let* η_i ($i = 2, \ldots, n$) *be the Jacobi lift of* α *such that*

$$\eta_i(0) = 0 \qquad and \qquad \eta_i(r) = e_i.$$

Then

$$\Delta\varphi(y) = -f''(r) - f'(r) \sum_{i=2}^{r} \langle D_{\alpha'}\eta_i(r), \eta_i(r)\rangle_g.$$

Proof. Let β_i ($i = 1, \ldots, n$) be the geodesic from y such that

$$\beta_i(0) = y \qquad and \qquad \beta_i'(0) = e_i.$$

Observe that $\beta_1(t) = \alpha_1(r + t)$ for small t, by the uniqueness of the integral curve of the corresponding differential equation. We apply Corollary 2.4 to the Laplacian at y, and the geodesics β_i ($i = 1, \ldots, n$) to get

$$\Delta\varphi(y) = -\sum_{i=1}^{n} (\varphi \circ \beta_i)''(0).$$

Since $\beta_1(t) = \alpha_1(r + t)$, we can split off the first term, to obtain

$$\Delta\varphi(y) = -f''(r) - \sum_{i=2}^{n} (\varphi \circ \beta_i)''(0).$$

Let $\alpha_{i,t}$ be the unique geodesic from x to $\beta_i(t)$ (for small t), parametrized by arc length. Thus $\alpha_{i,t}$ is what we called the variation of α at its end point, in the direction of e_i, for $i = 2, \ldots, n$. Then by Propositions 3.3 and 4.6 of Chapter IX, and the fact that

$$(\varphi \circ \beta_i)(t) = f\big(L(\alpha_{i,t})\big),$$

we obtain

$$(\varphi \circ \beta_i)''(0) = f'(r)\langle D_{\alpha'}\eta_i(r), \eta_i(r)\rangle_g,$$

which proves our proposition.

X, §3. THE JACOBIAN DETERMINANT OF THE EXPONENTIAL MAP

We continue to consider a Riemannian manifold (X, g). *We let* $x \in X$, *and we let* $\mathbf{B}_x$ *be an open ball in* T_xX *centered at* 0_x, *such that* $\exp_x$ *gives an isomorphism of* $\mathbf{B}_x$ *with its image in* X. *Thus without loss of* generality, we may assume X oriented, and we let vol_g be the volume form on X. We call $\mathbf{B}_x$ a normal chart at x. For $y \in \exp_x(\mathbf{B}_x)$. We write $y = \exp_x(v_y)$, so $v_y = \log_x(y)$, as it were.

We note that the differential

$$d\exp_x(v_y) \colon T_x \to T_y$$

is a linear isomorphism, and both T_x and T_y have the positive definite scalar products of the Riemannian metric, so we may define the absolute value of the determinant of $(d\exp_x)(v_y)$. One simply picks orthonormal bases in each one of these vector spaces, an the determinant of the matrix representing $(d\exp_x)(v_y)$ with respect to these bases. Picking oriented bases actually makes the determinant positive, so we don't need to take an absolute value. We let J denote the Jacobian determinant, so

$$\exp_x^* \text{vol}_g = J \, \text{vol}_{\text{euc}} \qquad \text{or also} \qquad \exp_x^* \text{vol}_g(v) = J(v) \, \text{vol}_{\text{euc}}(v),$$

where vol_{euc} is the euclidean volume on $T_x X$, determined by the positive definite metric $g(x)$, and v is the vector variable in $T_x X$. We shall express J in polar coordinates.

Let $\mathbf{S}(1)$ be the unit sphere in $T_x X$. Any vector $v \in T_x X$, $v \neq 0$, can be written uniquely in the form

$$v = ru,$$

where u is a unit vector in the direction of v, and $r > 0$. We call (r, u) the **polar coordinates** of v. Then the euclidean volume has the usual decomposition

$$\text{vol}_{\text{euc}}(r, u) = r^{n-1} \, dr \, d\mu(u),$$

where $d\mu(u)$ is the usual spherical measure ($d\theta$ in dimension 2). Then

$$(\exp_x^* \text{vol}_g)(ru) = J(r, u) r^{n-1} \, dr \, d\mu(u).$$

Proposition 3.1. Let u be a unit vector in $T_x X$ and let α be the geodesic parametrized by arc length such that $\alpha(0) = x$ and $\alpha'(0) = u$. Put $u = w_1$ and let $\{u, w_2, \ldots, w_n\}$ be a basis of $T_x X$ such that $w_i \perp u$ for $i = 2, \ldots, n$. Let η_i $(i = 2, \ldots, n)$ be the Jacobi lift of α such that

$$\eta_i(0) = 0 \qquad and \qquad D_{\alpha'} \eta_i(0) = w_i.$$

Then

$$r^{n-1} J(r, u) = \frac{\det(\eta_2(r), \ldots, \eta(r))}{\det(w_2, \ldots, w_n)} = \frac{\det^{1/2}\langle \eta_i(r), \eta_j(r) \rangle_g}{\det(w_2, \ldots, w_n)}.$$

The determinant on the right is taken for $i, j = 2, \ldots, n$.

Proof. Observe that we may also use η_1, which is such that $\eta_1(t) = t\alpha'(t)$. The equality between the two expressions on the right of the equality sign follows from Proposition 1.1. Let $f = \exp_x$. Then for any vectors $w_1, \ldots, w_n \in T_x X$ we have

$$
\begin{aligned}
(\exp_x^* \mathrm{vol}_g)(v)(w_1, \ldots, w_n) &= \mathrm{vol}_g\big(df(v)w_1, \ldots, df(v)w_n\big) \\
&= \det\big(df(v)w_1, \ldots, df(v)w_n\big) \\
&= J(v) \det(w_1, \ldots, w_n).
\end{aligned}
$$

We put $v = rw_1 = ru$. By Theorem 3.1 of Chapter IX we know that

$$
d\exp_x(ru)w_i = \frac{1}{r}\eta_i(r).
$$

Then for $i = 1$, $\eta_1(r)/r = \alpha'(r)$, which is a unit vector perpendicular to the others. Thus to compute the volume of the parallelotope in euclidean n-space, we may disregard this vector, and simply compute the volume of the projection on $(n-1)$-space, and thus we may compute only the $(n-1) \times (n-1)$ determinant of the vectors

$$
\det\big(\eta_2(r)/r, \ldots, \eta_n(r)/r\big) = \frac{1}{r^{n-1}} \det\big(\eta_2(r), \ldots, \eta_n(r)\big),
$$

from which the proposition falls out.

Proposition 3.1 is applied in several cases.

Corollary 3.2. *If in Proposition 3.1 all the vectors w_i are unit vectors u_i such that $\{u_1, \ldots, u_n\}$ is an orthonormal basis of $T_x X$, and $u = u_1$, then we have simply*

$$
r^{n-1} J(r, u) = \det{}^{1/2}\langle \eta_i(r), \eta_j(r)\rangle_g.
$$

From this case and the asymptotic expansion for the Jacobi lifts, we obtain:

Corollary 3.3. *Again with an orthonormal basis $\{u_1, \ldots, u_n\}$ of $T_x X$, let $u = u_1$ and*

$$
\mathrm{Ric}(u, u) = \sum_{i=2}^{n} R_2(u, u_i).
$$

Then

$$
\exp_x^* \mathrm{vol}_g(ru) = \left[1 + \mathrm{Ric}(u, u)\frac{r^2}{3!} + O(r^3)\right] \mathrm{vol}_{\mathrm{euc}}(ru) \qquad \textit{for } r \to 0.
$$

Proof. By Corollary 3.2, $J(r, u)$ is $\det^{1/2}\langle \eta_i(r)/r, \eta_j(r)/r\rangle_g$ with the determinant taken for $i, j = 1, \ldots, n$ or $i, j = 2, \ldots, n$. Using the asymptotic expansion of Chapter IX, Proposition 5.4 and the orthonormality, one gets that

$$J(r, u) = \prod_{i=2}^{n}\left(1 + 2R_2(u, u_i)\frac{r^2}{3!}\right)^{1/2} + O(r^3) \qquad \text{for } r \to 0,$$

which is immediately expanded to yield the corollary.

Example. Suppose $\dim X = 2$. Then $\text{Ric}(u, u) = R_2(u, u_2) = R_2(u_1, u_2)$. Putting $u_2 = u'$, we get

$$J(r, u) = 1 + R_2(u, u')\frac{r^2}{3!} + O(r^3) \qquad \text{for } r \to 0.$$

If we keep u fixed, and use Δ in polar coordinates, $\Delta = -\partial_r^2 - r^{-1}\partial_r$, then we see that

$$R_2(u, u') = -\tfrac{3}{2}\Delta J(0).$$

Compare with [He 78], Chapter I, Lemma 12.1 and Theorem 12.2.

For the further asymptotic expansion of the volume, see [Gray 73], as well as applications referred to in the bibliography of this paper.

On the other hand, we shall also meet a situation where $\{w_1, \ldots, w_n\}$ is not an orthonormal basis as in the next corollary. Cf. Chapter IX, Proposition 3.3.

Corollary 3.4. *Let* $\exp_x\colon \mathbf{B}_x \to X$ *be the normal chart in* X *as at the beginning of the section, and* $y = \exp_x(ru)$ *with* $ru \in \mathbf{B}_x$, *and some unit vector* u. *Let* $\alpha(s) = \exp_x(su)$ *and let* $e_1 = \alpha'(r)$. *Complete* e_1 *to an orthonormal basis* $\{e_1, \ldots, e_n\}$ *of* T_yX, *and let* η_i *be the Jacobi lift of* α *(depending on* y, *or* r *if* u_1 *is viewed as fixed), such that*

$$\eta_i(0) = 0 \qquad \text{and} \qquad \eta_i(r) = e_i \qquad \text{for } i = 2, \ldots, n.$$

Let $J'(s, u) = \partial_1 J(s, u)$. *Then*

$$J'/J(r, u) + \frac{n-1}{r} = \sum_{i=2}^{n} \langle D_{\alpha'}\eta_i(r), \eta_i(r)\rangle_g.$$

Proof. In the present case, $D_{\alpha'}\eta_i(0) = w_i$ is whatever it is, but we observe that the determinant $\det(w_2, \ldots, w_n)$ is constant, so disappears in taking the logarithmic derivative of the expression in Proposition 3.1.

We also observe that in the present case,

$$\langle \eta_i(r), \eta_j(r) \rangle_g = \delta_{ij},$$

so the matrix formed with these scalar products is the unit matrix. Taking the logarithmic derivative of one side, we obtain

$$J'/J(r, u) + (n - 1)/r.$$

Let $h_{ij} = \langle \eta_i, \eta_j \rangle_g$, and let $H = (h_{ij})$. On the other side, we obtain the logarithmic derivative

$$\frac{1}{2} \frac{(\det H)'}{\det H}.$$

Let $H_2, \ldots, H_n$ be the columns of H. By Leibniz's rule, we know that

$$(\det H)' = \sum_{i=2}^{n} \det(H_2, \ldots, H_i', \ldots, H_n).$$

Observe that

$$\langle \eta_i, \eta_j \rangle_g' = \langle D_{\alpha'} \eta_i, \eta_j \rangle_g + \langle \eta_i, D_{\alpha'} \eta_j \rangle_g.$$

and in particular,

$$\langle \eta_i, \eta_i \rangle_g' = 2 \langle D_{\alpha'} \eta_i, \eta_i \rangle_g.$$

What we want follows from a purely algebraic property of determinants, namely:

Lemma 3.5. Let $A = (A^1, \ldots, A^m)$ be a non-singular $m \times m$ matrix over a field, where $A^1, \ldots, A^m$ are the columns of A. Let $B = (B^1, \ldots, B^m)$ be any $m \times m$ matrix over the field. Then

$$\sum_i \det(A^1, \ldots, B^i, \ldots, A^m) = (\det A) \operatorname{tr}(A^{-1}B).$$

Proof. Let $X = (x_{ij})$ be the matrix such that

$$x_{1i}A^1 + \cdots + x_{mi}A^m = B^i \qquad \text{for} \quad i = 1, \ldots, m.$$

By Cramer's rule,

$$x_{ii} \det(A) = \det(A^1, \ldots, B^i, \ldots, A^m).$$

But $AX = B$ so $X = A^{-1}B$, and the lemma follows.

We apply the lemma to the case when $A = H(r)$ is the unit matrix and $B^j = H'_j(r)$ to conclude the proof.

Corollary 3.6. *Let φ be a C^2 function on a normal ball centered at the point $x \in X$. Suppose that φ depends only on the g-distance r from x, say $\varphi(y) = f(r(y))$. Let $y = \exp(ru)$, with a unit vector u. Then*

$$\Delta\varphi(y) = -f''(r) - f'(r)\left(J'/J(r, u) + \frac{n-1}{r}\right).$$

Proof. Combine Corollary 3.4 with Proposition 2.6.

For further applications of Jacobi lifts to volumes, cf. for instance [GHL 87/93], Chapter 3H.

X, §4. THE HODGE STAR ON FORMS

We already touched on the star operation on functions, and we defined $d*$ on 1-forms. We now deal systematically with the star operation on alternating forms. I shall follow **Koszul's formalism in formulas S 1 through S 8** [Ko 57], which is quite elegant. A direct very brief treatment of just what is needed to get the global duality and adjointness of d, $d*$ using Stokes' theorem, will be done in a self-contained way ad hoc in Chapter XIII, so that the reader need not go through the systematic formalism just to understand that particular application of Stokes' theorem.

Until further notice, we don't differentiate, and the theory is punctual, so:

We let T be a finite dimensional vector space over $\mathbf{R}$, of dimension n, with r-forms φ, ψ in $L_a^r(T)$, and with vectors $v \in T$. We suppose that T has a positive definite scalar product g, and is oriented so we have a volume form $\Omega_g = \Omega$. We let $v^\vee$ be the 1-form dual to v under g.

S 1. *There exists a unique isomorphism $*: L_a^r(T) \to L_a^{n-r}(T)$ such that for all $v_1, \dots, v_{n-r} \in T$ an $\varphi \in L_a^r(T)$ we have*

$$(*\varphi)(v_1, \dots, v_{n-r})\Omega = \varphi \wedge v_1^\vee \wedge \cdots \wedge v_{n-r}^\vee.$$

Proof. Given φ, the right side of the above equation is a multilinear alternating function of $v_1, \dots, v_{n-r}$ into the 1-dimensional space of n-forms, so having chosen Ω as a basis for this space, we get a real-valued form, which constitutes the coefficient of Ω on the left side. The association

$$\varphi \mapsto *\varphi$$

is obviously linear.

S 2. *We have* $*\Omega = 1$ *and* $*1 = \Omega$, *and for a function* f, $*(f\Omega) = f$.

Proof. Immediate from the definition **S 1** and Proposition 1.1.

S 3. *For* $\varphi \in L_a^r(T)$ *and* $v_1, \ldots, v_{n-r} \in T$ *we have*

$$(*\varphi)(v_1, \ldots, v_{n-r}) = *(\varphi \wedge v_1^\vee \wedge \cdots \wedge v_{n-r}^\vee).$$

Proof. Using **S 2** and **S 1**, we find:

$$*(\varphi \wedge v_1^\vee \wedge \cdots \wedge v_{n-r}^\vee) = *[(*\varphi)(v_1 \wedge \cdots \wedge v_{n-r})\Omega]$$
$$= (*\varphi)(v_1, \ldots, v_{n-r})(*\Omega)$$
$$= (*\varphi)(v_1, \ldots, v_{n-r}).$$

S 4. *For* $\varphi \in L_a^r(T)$ *and* $v \in T$ *we have*

$$*(\varphi \wedge v^\vee) = (*\varphi) \circ v.$$

Proof. Indeed,

$$\big(*(\varphi \wedge v^\vee)\big)(v_1, \ldots, v_{n-r-1}) = *(\varphi \wedge v^\vee \wedge v_1^\vee \wedge \cdots \wedge v_{n-r-1}^\vee)$$
$$= (*\varphi)(v, v_1, \ldots, v_{n-r-1})$$
$$= \big((*\varphi) \circ v\big)(v_1, \ldots, v_{n-r-1}).$$

For the next property, we need a lemma independently of the star operation.

Lemma 4.1. *For* $\varphi \in L_a^r(T)$, *and* $v, v_1, \ldots, v_{n-r+1} \in T$, *we have*

$$(\varphi \circ v) \wedge v_1^\vee \wedge \cdots \wedge v_{n-r+1}^\vee$$
$$= \sum_{i=1}^{n-r+1} (-1)^{r+i} \langle v^\vee, v_i \rangle (\varphi \wedge v_1^\vee \wedge \cdots \wedge \widehat{v_i^\vee} \wedge \cdots \wedge v_{n-1+1}^\vee).$$

Proof. The basic formalism of forms tells us that the contraction with a vector is an anti-derivation on the algebra of forms (Chapter V, §5, **CON 3**). Since $\varphi \wedge v_1^\vee \wedge \cdots \wedge v_{n-r+1}^\vee$ has degree $n+1$ and so is equal to 0, we find

$$0 = (\varphi \wedge v_1^\vee \wedge \cdots \wedge v_{n-r+1}^\vee) \circ v$$
$$= (\varphi \circ v) \wedge v_1^\vee \wedge \cdots \wedge v_{n-r+1}^\vee$$
$$+ \sum_{i=1}^{n-r+1} (-1)^{r+i-1} \varphi \wedge v_1^\vee \wedge \cdots \wedge (v_i^\vee \circ v) \wedge \cdots \wedge v_{n-r+1}^\vee.$$

We observe that

$$v_i^\vee \circ v = \langle v_i, v \rangle_g = \langle v, v_i \rangle_g = \langle v^\vee, v_i \rangle,$$

to conclude the proof of the lemma.

S 5. *For any form $\varphi \in L_a^r(T)$ and $v \in T$ we have*

$$*(\varphi \circ v) = (-1)^{n-1}(*\varphi) \wedge v^\vee.$$

Proof. First, for all $v_1, \ldots, v_{n-r} \in T$ we have

(1) $\big(*(\varphi \circ v)\big)(v_1, \ldots, v_{n-r+1}) = *\big((\varphi \circ v) \wedge v_1^\vee \wedge \cdots \wedge v_{n-r+1}^\vee\big).$

On the other hand,

$$(-1)^{n-1}(*\varphi) \wedge v^\vee = (-1)^{r+1} v^\vee \wedge *\varphi.$$

Hence

$$\big((-1)^{n-1}(*\varphi) \wedge v^\vee\big)(v_1, \ldots, v_{n-r+1})$$
$$= (-1)^{r+1}(v^\vee \wedge *\varphi)(v_1, \ldots, v_{n-r+1})$$
$$= \sum_{i=1}^{n-r+1} (-1)^{r+1} \langle v^\vee, v_i \rangle \big((*\varphi)(v_1, \ldots, \hat{v}_i, \ldots, v_{n-r+1})$$
$$= \sum_{i=1}^{n-r+1} *(-1)^{r+1} \langle v^\vee, v_i \rangle (\varphi \wedge v_1^\vee \wedge \cdots \wedge \widehat{v_i^\vee} \wedge \cdots \wedge v_{n-r+1}^\vee).$$

Using Lemma 4.1 and (1) concludes the proof.

We can do an induction on **S 5**, and also get a corollary:

S 6. *For $\varphi \in L_a^r(T)$, $** = (-1)^{r(n-1)}$ and*

$$(*\varphi) \wedge v_1^\vee \wedge \cdots \wedge v_r^\vee = (-1)^{r(n-1)} *(\varphi \circ v_1 \circ \cdots \circ v_r).$$

Proof. We have

$$(**\varphi)(v_1, \ldots, v_r) = *\big((*\varphi) \wedge v_1^\vee \wedge \cdots \wedge v_r^\vee\big),$$

[applying **S 5** repeatedly] $= (-1)^{r(n-1)} *\big(*(\varphi \circ v_1 \circ \cdots \circ v_r)\big).$

Since for any function f we have $*f = f*1$ and $**f = *f\Omega = f$, property **S 6** follows.

S 7. *Let S denote the $*$ operation. Then $S: L_a^r(T) \to L_a^{n-r}(T)$ is an isomorphism.*

This is immediate, but is stated for the record.

S 8. Let $\varphi, \psi \in L_a^r(T)$. Then

$$\varphi \wedge *\psi = \psi \wedge *\varphi.$$

Proof. The pairings of φ, ψ given by the expressions on the left and right are bilinear, so it suffices to verify the equality when

$$\varphi = v_1^\vee \wedge \cdots \wedge v_r^\vee \quad \text{and} \quad \psi = w_1^\vee \wedge \cdots \wedge w_r^\vee.$$

In this case, we obtain

$$\varphi \wedge *\psi = (-1)^{r(n-r)}(*\psi) \wedge \varphi$$

$$= (-1)^{r(n-r)}*(w_1^\vee \wedge \cdots \wedge w_r^\vee) \wedge v_1^\vee \wedge \cdots \wedge v_r^\vee$$

[by **S 6**] $\quad = (-1)^{r(n-r)}(-1)^{r(n-1)}*[(w_1^\vee \wedge \cdots \wedge w_r^\vee) \circ v_1 \circ \cdots \circ v_r]$

[by **S 2**] $\quad = \det\langle w_i, v_j \rangle_g \Omega.$

But $\det\langle w_i, v_j \rangle_g = \det\langle v_i, w_j \rangle_g$, from which **S 8** follows.

The next formula proves that the star operation is given in a simple-minded way on natural basis elements for the wedge products. We shall use this property in Chapter XIII, §4, in a self-contained way to make the results on integration independent of the general star formalism, but the next formula won't be used in the rest of this section or the next.

Proposition 4.2. *Let $\{v_1, \ldots, v_n\}$ be an orthonormal basis of T. Let ω_1, $\ldots, \omega_n$ be the dual basis of 1-forms. Let $I = (i_1, \ldots, i_r)$ with $i_1 < \cdots < i_r$ and let $J = (j_1, \ldots, j_{n-r})$ with $j_1 < \cdots < j_{n-r}$ be the complementary set such that $\{1, \ldots, n\}$ is a permutation of (I, J). Let $\varepsilon(I, J)$ be the sign of the permutation. Let $\omega_i = v_i^\vee$. Let*

$$\omega_I = \omega_{i_1} \wedge \cdots \wedge \omega_{i_r} \quad \text{and} \quad \omega_J = \omega_{j_1} \wedge \cdots \wedge \omega_{j_{n-r}}.$$

Then

$$*\omega_I = \varepsilon(I, J)\omega_J.$$

Proof. Directly from the definition of $\Omega = \Omega_g$ we have that

$$\Omega_g = \omega_1 \wedge \cdots \wedge \omega_n = v_1^\vee \wedge \cdots \wedge v_n^\vee.$$

At first, let J be an arbitrary sequence of $n - r$ indices among $(1, \ldots, n)$. Then by **S 3**,

$$(*\omega_I)(v_{j_1}, \ldots, v_{j_{n-r}}) = *(\omega_I \wedge \omega_J),$$

which is $\neq 0$ if and only if J is the complementary set, i.e. (I, J) is a permutation of $(1, \dots, n)$. In this case, the right side of the above expression is simply $\varepsilon(I, J)*\Omega = \varepsilon(I, J)$. Alternatively, one may write

$$*\omega_I = \varepsilon(I, J)\omega_J,$$

if (I, J) is a permutaiton of $(1, \dots, n)$, from which Proposition 4.2 follows.

We are now through with the punctual theory, and we pass to a Riemannian manifold (X, g), where the vector space T is replaced by the tangent bundle TX, and vectors are replaced by vector fields. We let D be the metric covariant derivative as usual. Also

$$\mathscr{A}^r(X) = \Gamma L_a^r(TX).$$

Proposition 4.3. *The star operation commutes with every D_ξ, i.e. for any vector field ξ and $\varphi \in \mathscr{A}^r(X)$, we have*

$$*D_\xi\varphi = D_\xi*\varphi.$$

Proof. For 0-forms (functions) and n-forms (functions times the volume form) the assertion is immediate by using Proposition 2.1, to the effect that $D_\xi \operatorname{vol}_g = 0$. Now let $\varphi \in \Gamma L_a^r(TX)$. Then:

$$(D_\xi *\varphi)(\xi_1, \dots, \xi_{n-r}) + \sum_{i=1}^{n-r} (*\varphi)(\xi_1, \dots, D_\xi\xi_i, \dots, \xi_{n-r})$$

$$= D_\xi\big((*\varphi)(\xi_1, \dots, \xi_{n-r})\big) \qquad \text{[because D_ξ is a derivation]}$$

$$= D_\xi *(\varphi \wedge \xi_1^\vee \wedge \cdots \wedge \xi_n^\vee) \qquad \text{[by S 3]}$$

$$= (*D_\xi)(\varphi \wedge \xi_1^\vee \wedge \cdots \wedge \xi_{n-r}^\vee) \qquad \text{[by the proposition for n-forms]}$$

$$= *(D_\xi\varphi \wedge \xi_1^\vee \wedge \cdots \wedge \xi_{n-r}^\vee) + \sum_{i=1}^{n-r} *(\varphi \wedge \xi_1^\vee \wedge \cdots \wedge D_\xi\xi_i^\vee \wedge \cdots \wedge \xi_{n-r}^\vee)$$

$$= (*D_\xi\varphi)(\xi_1, \dots, \xi_{n-r}) + \sum_{i=1}^{n-r} (*\varphi)(\xi_1, \dots, D_\xi\xi_i, \dots, \xi_{n-r}),$$

which proves the proposition.

We now define d^* in general to be

$$d^* = (-1)^{nr+n+1} *d* \qquad \text{on } \mathscr{A}^r(X).$$

In Chapter XIII, §3 we shall define a scalar product on forms with compact support for which d^* will be seen to be the adjoint of d. For the moment, we continue with an essentially differential algebraic theory.

Proposition 4.4. *For* φ, $\psi \in \mathscr{A}^r(X)$ *we have*

$$d\varphi \wedge *\psi = \varphi \wedge (*d*\psi) + d(\varphi \wedge *\psi).$$

Proof. Immediate from the definition of $d*$, **S 6**, and the basic formula for d of a wedge product (a graded derivation).

Proposition 4.5. *Let* $\xi_1, \ldots, \xi_n$ *be a frame of vector fields, and let* ξ'_1, $\ldots, \xi'_n$ *be the dual frame, that is* $\langle \xi'_i, \xi_j \rangle_g = \delta_{ij}$. *Then for any form* $\varphi \in \mathscr{A}^r(X)$ *we have*

$$d*\varphi = \sum_{i=1}^n (D_{\xi_i}\varphi) \circ \xi'_i.$$

Proof. Proposition 1.1 of Chapter VIII gives us an expression for $d(*\varphi)$ in terms of the frame. The dual frame is such that $\lambda_i^\vee = \xi'_i$. Then the formula of Proposition 4.4 is an immediate consequence of **S 5**.

Remark. *If the frame* $\xi_1, \ldots, \xi_n$ *is orthonormal, then of course* $\xi'_i = \xi_i$.

We define the **Laplacian** associated with the Riemannian manifold (X, g) to be

$$\Delta = dd* + d*d, \quad \text{operating on each } \mathscr{A}^r(X).$$

On Euclidean space $\mathbf{R}^n$ with its standard positive definite scalar product, the Laplacian on functions is the usual operator (with the minus sign)

$$\Delta = -\sum \left(\frac{\partial}{\partial x_i} \right)^2.$$

As a more general example illustrating the role of Ricci curvature, we give the one higher dimensional version of Corollary 2.4. Let $\lambda \in \mathscr{A}^1(X)$. With the Ricci curvature in mind, we define $\text{Ric}(\lambda)$ to be the scalar valued form such that, with respect to an orthonormal frame $\xi_1, \ldots, \xi_n$, and any vector field ξ we have

$$\text{Ric}(\lambda)(\xi) = \sum_i \langle (D_\xi D_{\xi_i} - D_{\xi_i} D_\xi)\lambda, \xi_i \rangle,$$

where we denote by $\langle \lambda, \xi \rangle$ the value of a 1-form λ on a vector field ξ.

Proposition 4.6. *Let* $\xi_1, \ldots, \xi_n$ *be an orthonormal frame. As an operator on 1-forms,* $\Delta \colon \mathscr{A}^1(X) \to \mathscr{A}^1(X)$ *is given by*

$$\Delta = -\sum D_{\xi_i}^2 - \text{Ric}.$$

Written in terms of the variables, this means

$$\langle \Delta\lambda, \xi \rangle = -\sum_i \langle D_{\xi_i} D_{\xi_i}\lambda, \xi \rangle - \sum_i \langle (D_\xi D_{\xi_i} - D_{\xi_i} D_\xi)\lambda, \xi_i \rangle.$$

Proof. By Proposition 4.5, we have

$$d^*\lambda = -\sum (D_{\xi_i}\lambda)(\xi_i)$$

and so by a general formula on covariant derivatives we get a value for $dd^*\lambda$, namely

$$\langle dd^*\lambda, \xi \rangle = -\sum_i \langle D_\xi D_{\xi_i}\lambda, \xi_i \rangle.$$

On the other hand, to get $d^*d\lambda$, we first note that by **COVD 6** of Chapter VIII, §1,

$$(d\lambda)(\xi, \eta) = \langle D_\xi\lambda, \eta \rangle - \langle D_\eta\lambda, \xi \rangle.$$

Again by Proposition 4.5,

$$\langle d^*d\lambda, \xi \rangle = \sum \langle D_{\xi_i} D_\xi\lambda, \xi_i \rangle - \sum \langle D_{\xi_i} D_{\xi_i}\lambda, \xi \rangle.$$

Adding the two expressions yields the formula of the proposition.

X, §5. HODGE DECOMPOSITION OF DIFFERENTIAL FORMS

In this section we carry out a bit of pure algebra, applicable to the situation of the previous section, and also applicable to other situations, especially in the complex case. See for instance [Wel 80], pp. 147–148 and [GriH 76], Chapter 0, §6. We work axiomatically. To prove the axioms **H 1** and **H 2** below requires more extensive analytical tools than we use in this book, and specifically it requires the basic theory of elliptic operators. What is needed is carried out in the above references, and the essential is done in a self-contained way in Appendix 4 of [La 75].

Since the algebraic set up which follows applies to other differential operators besides the d we have been using, I use a more neutral letter D, which in the complex theory is taken to be the so-called $\bar\partial$ operator.

None of this section will be used in the rest of the book. It is included here only for the convenience of a reader wanting to see how the theory further develops, and to isolate clearly what is purely algebraic from what demands more differential analysis.

Let A be a vector space of dimension n over $\mathbf{R}$, with a positive definite scalar product $\langle \, , \, \rangle$ and corresponding norm $\| \; \|$; or alternatively, the vector space may be over $\mathbf{C}$, with a positive definite hermitian product. Let

$$D: A \to A$$

be a linear map which has an adjoint (algebraic) D^*, that is

$$\langle Du, v \rangle = \langle u, D^*v \rangle \qquad \text{all } u, v \in A.$$

and such that $DD = 0$. We define the **Laplacian** of D to be

$$\Delta_D = \Delta = DD^* + D^*D.$$

We define $\mathbf{H}_D = \mathbf{H} = \ker \Delta$ to be the D-**harmonic space**. We assume the **Hodge Conditions**:

H 1. *The kernel* $\mathbf{H} = \ker \Delta$ *is finite dimensional.*

H 2. *We have* $\mathbf{H}^\perp = \Delta A.$

We then prove further properties as follows.

Since $\mathbf{H}$ is assumed finite dimensional, there is an orthogonal projection of A on $\mathbf{H}$, which we denote also by $\mathbf{H}$ if necessary, that is $\mathbf{H}(u)$ is the orthogonal projection of u on $\mathbf{H}$.

Theorem 5.1. *Under the above two Hodge conditions, we have*

$$\mathbf{H}^\perp = DA + D^*A,$$

and an orthogonal decomposition

$$A = \mathbf{H} \perp \Delta A = \mathbf{H} \perp DA \perp D^*A.$$

The restriction of Δ *to* $\mathbf{H}^\perp$ *is invertible, and*

$$\text{Ker } D = \mathbf{H} + DA.$$

Proof. By orthogonalization and **H 2**, given $u \in A$ we have

$$u = \mathbf{H}u + \Delta v = \mathbf{H}u + DD^*v + D^*Dv$$

with some $v \in A$. Hence A is contained in $\mathbf{H} + DA + D^*A$, so we get equality. Furthermore

$$\langle \Delta u, u \rangle = \|Du\|^2 + \|D^*u\|^2.$$

Hence $\Delta u = 0$ if and only if $Du = D^*u = 0$. (Each implication is immediate.) The adjointness relation then shows that DA, D^*A are orthogonal to $\mathbf{H}$, and $D^2 = 0$ implies that DA is orthogonal to D^*A, so we get the orthogonal decomposition

$$A = \mathbf{H} \perp DA \perp D^*A,$$

and $\Delta A = DA + D^*A$ by $\mathbf{H}\,2$. Since $\Delta\mathbf{H} = 0$ it follows that

$$\Delta : DA + D^*A \to DA + D^*A$$

is surjective, and so is an isomorphism, and thus Δ is invertible on $\mathbf{H}^\perp$. Finally $\mathbf{H} + DA$ is contained in the kernel of D, and D is injective on D^*A because

$$DD^*u = 0 \quad \Rightarrow \quad \langle DD^*u, u \rangle = 0 \quad \Rightarrow \quad \|D^*u\|^2 = 0.$$

This proves the theorem.

Remark 1. As a special case of the last formula, suppose $u \in A$ and u is perpendicular to Ker Δ. If u is D-**closed**, that is $Du = 0$, then $u = Dv$ for some $v \in A$, that is u is D-**exact**.

Remark 2. *If we denote by $H(A)$ the homology* ker $D/\mathrm{Im}\,D$ *then we get an isomorphism of the homology with the harmonic space*

$$\mathbf{H} \approx H(A).$$

We let

$$G : A \to \mathbf{H}^\perp = \Delta A$$

be equal to 0 on $\mathbf{H}$, and be the inverse of Δ on ΔA. Then by definition,

$$G\Delta = \Delta G \qquad \text{and} \qquad I = \Delta G + H.$$

Furthermore:

G and Δ commute with D and D^*.

Proof. We have

$$\Delta D = (DD^* + D^*D)D = DD^*D \qquad \text{and} \qquad D\Delta = D(DD^* + D^*D) = DD^*D$$

so D commutes with Δ. Similarly for D^*. The commutation of D and D^* with G then follows since $G = \Delta^{-1}$ on ΔA.

Graded structure

Suppose in addition that A is graded,

$$A = \bigoplus_{p=0}^{n} A^p,$$

that A^p is orthogonal to A^q for $q \neq p$, and that

$$D^p = D: A^p \to A^{p+1}$$

raises degrees by 1, so $D^*: A^p \to A^{p-1}$ lowers degrees by 1.

Under the above assumptions, we can defined the homology of D in degree p to be

$$H^p(A) = \text{Ker } D^p/\text{Im } D^{p-1},$$

where D^p is D viewed as map from A^p to A^{p+1}. Immediately from Theorem 5.1 we obtain:

Theorem 5.2. *Let* $\mathbf{H}^p = \mathbf{H} \cap H^p(A)$. *Then*

$$\mathbf{H} = \bigoplus_{p=0}^{n} \mathbf{H}^p$$

and $H^p(A) \approx \mathbf{H}^p$, *that is every class in* Ker D^p *mod* Im D^{p-1} *has a unique representative in the harmonic space* $\mathbf{H}^p$.

The star operator

We suppose given an automorphism $S: A \to A$ which is an isomorphism

$$S: A^p \to A^{n-p}.$$

We assume:

S 1. *On* A^p *we have* $S^2 = (-1)^{p(n-1)}$.

S 2. $D^* = (-1)^{np+n+1}$ *on* A^p.

Proposition 5.3. *Under these assumptions,* $D = SD^*S$ *and* $\mathbf{H}$, Δ, G *commute with* S.

Proof. We give the proof when n is even for simplicity. For $u \in A^p$, we have:

$$SD^*Su = -S^2DS^2u = -S^2D(-1)^pu$$
$$= -(-1)^p(-1)^{p+1}Du$$
$$= Du,$$

so $D = SD^*S$.

For the commutation of S with Δ, we write, using the above,

$$S\Delta = -SDSDS - SSDSD,$$
$$\Delta S = -DSDSS - SDSDS.$$

On A^p, $SS = (-1)^p$, so it is immediate that SS commutes with DSD, thus showing that S commutes with Δ.

Since S commutes with Δ, it follows that

$$S: \mathbf{H} \to \mathbf{H}$$

induces an automorphism of $\mathbf{H}$ with itself. For $u \in A$ we have:

$Su - HSu \in \mathbf{H}$ by definition of the orthogonal projection; and

$Su - SHu = S\Delta Gu = \Delta SGu$ since Δ commutes with S.

Then

$Su - SHu \perp \mathbf{H}$ since it lies in ΔA.

Subtracting shows that $HSu - SHu$ is both orthogonal to $\mathbf{H}$, and also lies in $\mathbf{H}$, so must be 0, whence H commutes with S. Since $G = \Delta^{-1}$ on $\mathbf{H}^\perp$ it follows that G also commutes with S, thus proving the proposition.

CHAPTER XI

Integration of Differential Forms

The material of this chapter is also contained in my book on real analysis [La 93], but it may be useful to the reader to have it also here in a rather self contained way, based only on standard properties of integration in Euclidean space.

Throughout this chapter, μ is Lebesgue measure on $\mathbf{R}^n$.
If A is a subset of $\mathbf{R}^n$, we write $\mathscr{L}^1(A)$ instead of $\mathscr{L}^1(A, \mu, \mathbf{C})$.
All manifolds are assumed finite dimensional.
They may have a boundary.

XI, §1. SETS OF MEASURE 0

We recall that a set has measure 0 in $\mathbf{R}^n$ if and only if, given ε, there exists a covering of the set by a sequence of rectangles $\{R_j\}$ such that $\sum \mu(R_j) < \varepsilon$. We denote by R_j the closed rectangles, and we may always assume that the interiors R_j^0 cover the set, at the cost of increasing the lengths of the sides of our rectangles very slightly (an $\varepsilon/2^n$ argument). We shall prove here some criteria for a set to have measure 0. We leave it to the reader to verify that instead of rectangles, we could have used cubes in our characterization of a set of a measure 0 (a cube being a rectangle all of whose sides have the same length).

We recall that a map f satisfies a **Lipschitz condition** on a set A if there exists a number C such that

$$|f(x) - f(y)| \leqq C|x - y|$$

for all $x, y \in A$. Any C^1 map f satisfies locally at each point a Lipschitz condition, because its derivative is bounded in a neighborhood of each point, and we can then use the mean value estimate,

$$|f(x) - f(y)| \leq |x - y| \sup |f'(z)|,$$

the sup being taken for z on the segment between x and y. We can take the neighborhood of the point to be a ball, say, so that the segment between any two points is contained in the neighborhood.

Lemma 1.1. *Let A have measure 0 in $\mathbf{R}^n$ and let $f: A \to \mathbf{R}^n$ satisfy a Lipschitz condition. Then $f(A)$ has measure 0.*

Proof. Let C be a Lipschitz constant for f. Let $\{R_j\}$ be a sequence of cubes covering A such that $\sum \mu(R_j) < \varepsilon$. Let r_j be the length of the side of R_j. Then for each j we see that $f(A \cap S_j)$ is contained in a cube R'_j whose sides have length $\leq 2Cr_j$. Hence

$$\mu(R'_j) \leq 2^n C^n r_j^n = 2^n C^n \mu(R_j).$$

Our lemma follows.

Lemma 1.2. *Let U be open in $\mathbf{R}^n$ and let $f: U \to \mathbf{R}^n$ be a C^1 map. Let Z be a set of measure 0 in U. Then $f(Z)$ has measure 0.*

Proof. For each $x \in U$ there exists a rectangle R_x contained in U such that the family $\{R_x^0\}$ of interiors covers Z. Since U is separable, there exists a denumerable subfamily covering Z, say $\{R_j\}$. It suffices to prove that $f(Z \cap R_j)$ has measure 0 for each j. But f satisfies a Lipschitz condition on R_j since R_j is compact and f' is bounded on R_j, being continuous. Our lemma follows from Lemma 1.1.

Lemma 1.3. *Let A be a subset of $\mathbf{R}^n$. Assume that $m < n$. Let*

$$f: A \to \mathbf{R}^n$$

satisfy a Lipschitz condition. Then $f(A)$ has measure 0.

Proof. We view $\mathbf{R}^m$ as embedded in $\mathbf{R}^n$ on the space of the first m coordinates. Then $\mathbf{R}^m$ has measure 0 in $\mathbf{R}^n$, so that A has also n-dimensional measure 0. Lemma 1.3 is therefore a consequence of Lemma 1.1.

Note. All three lemmas may be viewed as stating that certain parametrized sets have measure 0. Lemma 1.3 shows that parametrizing a set by strictly lower dimensional spaces always yields an image having measure 0. The other two lemmas deal with a map from one space into another of the same dimension. Observe that Lemma 1.3 would be false if f is only assumed to be continuous (Peano curves).

The next theorem will be used later only in the proof of the residue theorem, but it is worthwhile inserting it at this point.

Let $f: X \to Y$ be a morphism of class C^p, with $p \geq 1$, and assume throughout this section that X, Y are finite dimensional. A point $x \in X$ is called a **critical point** of f if f is not a submersion at x. This means that

$$T_x f: T_x X \to T_{f(x)} Y$$

is not surjective, according to our differential criterion for a submersion.

Assume that a manifold X has a countable base for its charts. Then we can say that a set has measure 0 in X if its intersection with each chart has measure 0.

Theorem 1.4 (Sard's Theorem). *Let* $f: X \to Y$ *be a* C^∞ *morphism of finite dimensional manifolds having a countable base. Let* Z *be the set of critical points of* f *in* X. *Then* $f(Z)$ *has measure 0 in* Y.

Proof. (Due to Dieudonné.) By induction on the dimension n of X. The assertion is trivial if $n = 0$. Assume $n \geq 1$. It will suffice to prove the theorem locally in the neighborhood of a point in Z. We may assume that $X = U$ is open in $\mathbf{R}^n$ and

$$f: U \to \mathbf{R}^p$$

can be expressed in terms of coordinate functions,

$$f = (f_1, \ldots, f_p).$$

We let us usual

$$D^\alpha = D_1^{\alpha_1} \cdots D_n^{\alpha_n}$$

be a differential operator, and call $|\alpha| = \alpha_1 + \cdots + \alpha_n$ its **order**. We let $Z_0 = Z$ and for $m \geq 1$ we let Z_m be the set of points $x \in Z$ such that

$$D^\alpha f_j(x) = 0$$

for all $j = 1, \ldots, p$ and all α with $1 \leq |\alpha| \leq m$. We shall prove:

(1) *For each* $m \geq 0$ *the set* $f(Z_m - Z_{m+1})$ *has measure 0.*

(2) *If* $m \geq n/p$, *then* $f(Z_m)$ *has measure 0.*

This will obviously prove Sard's theorem.

Proof of 1. Let $a \in Z_m - Z_{m+1}$. Suppose first that $m = 0$. Then for some coordinate function, say $j = 1$, and after a renumbering of the variables if necessary, we have

$$D_1 f_1(a) \neq 0.$$

The map

$$g: x \mapsto \left(f_1(x), x_2, \ldots, x_p\right)$$

obviously has an invertible derivative at $x = a$, and hence is a local isomorphism at a. Considering $f \circ g^{-1}$ instead of f, we are reduced to the case where f is given by

$$f(x) = \left(x_1, f_2(x), \ldots, f_p(x)\right) = \left(x_1, h(x)\right),$$

where h is the projection of f on the last $p - 1$ coordinates and is therefore a morphism $h: V \to \mathbf{R}^{p-1}$ defined on some open V containing a. Then

$$Df(x) = \begin{pmatrix} 1 & 0 \\ * & Dh(x) \end{pmatrix}.$$

From this it is clear that x is a critical point for f if and only if x is a critical point for h, and it follows that $h(Z \cap V)$ has measure 0 in $\mathbf{R}^{p-1}$. Since $f(Z)$ is contained in $\mathbf{R}^1 \times h(Z)$, we conclude that $f(Z)$ has measure 0 in $\mathbf{R}^p$ as desired.

Next suppose that $m \geq 1$. Then for some α with $|\alpha| = m + 1$, and say $j = 1$, we have

$$D^\alpha f_1(a) \neq 0.$$

Again after a renumbering of the indices, we may write

$$D^\alpha f_1 = D_1 g_1$$

for some function g_1, and we observe that $g_1(x) = 0$ for all $x \in Z_m$, in a neighborhood of a. The map

$$g: x \mapsto \left(g_1(x), x_2, \ldots, x_n\right)$$

is then a local isomorphism at a, say on an open set V containing a, and we see that

$$g(Z_m \cap V) \subset \{0\} \times R^{n-1}.$$

We view g as a change of charts, and considering $f \circ g^{-1}$ instead of f, together with the invariance of critical points under changes of charts, we may view f as defined on an open subset of $\mathbf{R}^{n-1}$. We can then apply induction again to conclude the proof of our first assertion.

Proof of 2. Again we work locally, and we may view f as defined on the closed n-cube of radius r centered at some point a. We denote this cube by $C_r(a)$. For $m \geq n/p$, it will suffice to prove that

$$f(Z_m \cap C_r(a))$$

has measure 0. For large N, we cut up each side of the cube into N equal segments, thus obtaining a decomposition of the cube into N^n small cubes. By Taylor's formula, if a small cube contains a critical point $x \in Z_m$, then for any point y of this small cube we have

$$|f(y) - f(x)| \leq K|x - y|^{m+1} \leq K(2r/N)^{m+1},$$

where K is a bound for the derivatives of f up to order $m + 1$, and we use the sup norm. Hence the image of Z_m contained in small cube is itself contained in a cube whose radius is given by the right-hand side, and whose volume in R^p is therefore bounded by

$$K^p(2r/N)^{p(m+1)}.$$

We have at most N^n such images to consider and we therefore see that

$$f(Z_m \cap C_r(a))$$

is contained in a union of cubes in $\mathbf{R}^p$, the sum of whose volumes is bounded by

$$K^p N^n(2r/N)^{p(m+1)} \leq K^p(2r)^{p(m+1)} N^{n-p(m+1)}.$$

Since $m \geq n/p$, we see that the right-hand side of this estimate behaves like $1/N$ as N becomes large, and hence that the union of the cubes in R^p has arbitrarily small measure, thereby proving Sard's theorem.

Sard's theorem is harder to prove in the case f is C^p with finite p [29], but $p = \infty$ already is quite useful.

XI, §2. CHANGE OF VARIABLES FORMULA

We first deal with the simplest of cases. We consider vectors $v_1, \ldots, v_n$ in $\mathbf{R}^n$ and we define the **block** B spanned by these vectors to be the set of points

$$t_1 v_1 + \cdots + t_n v_n$$

with $0 \leq t_i \leq 1$. We say that the block is **degenerate** (in $\mathbf{R}^n$) if the vectors

$v_1, \ldots, v_n$ are linearly dependent. Otherwise, we say that the block is **non-degenerate**, or is a **proper block** in $\mathbf{R}^n$.

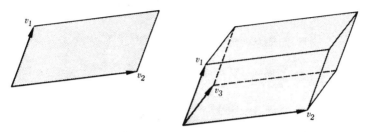

We see that a block in $\mathbf{R}^2$ is nothing but a parallelogram, and a block in $\mathbf{R}^3$ is nothing but a parallelepiped (when not degenerate).

We shall sometimes use the word volume instead of measure when applied to blocks or their images under maps, for the sake of geometry.

We denote by $\mathrm{Vol}(v_1, \ldots, v_n)$ the volume of the block B spanned by $v_1, \ldots, v_n$. We define the **oriented volume**

$$\mathrm{Vol}^0(v_1, \ldots, v_n) = \pm \mathrm{Vol}(v_1, \ldots, v_n),$$

taking the $+$ sign if $\mathrm{Det}(v_1, \ldots, v_n) > 0$ and the $-$ sign if

$$\mathrm{Det}(v_1, \ldots, v_n) < 0.$$

The determinant is viewed as the determinant of the matrix whose column vectors are $v_1, \ldots, v_n$, in that order.

We recall the following characterization of determinants. Suppose that we have a product

$$(v_1, \ldots, v_n) \mapsto v_1 \wedge v_2 \wedge \cdots \wedge v_n$$

which to each n-tuple of **vectors** associates a number, such that the product is multilinear, alternating, and such that

$$e_1 \wedge \cdots \wedge e_n = 1$$

if $e_1, \ldots, e_n$ are the unit vectors. Then this product is necessarily the determinant, that is, it is uniquely determined. "Alternating" means that if $v_i = v_j$ for some $i \neq j$, then

$$v_1 \wedge \cdots \wedge v_n = 0.$$

The uniqueness is easily proved, and we recall this short proof. We can write

$$v_i = a_{i1} e_1 + \cdots + a_{in} e_n$$

for suitable numbers a_{ij}, and then

$$
\begin{aligned}
v_1 \wedge \cdots \wedge v_n &= (a_{11}e_1 + \cdots + a_{1n}e_n) \wedge \cdots \wedge (a_{n1}e_1 + \cdots + a_{nn}e_n) \\
&= \sum_\sigma a_{1,\sigma(1)}e_{\sigma(1)} \wedge \cdots \wedge a_{n,\sigma(n)}e_{\sigma(n)} \\
&= \sum_\sigma a_{1,\sigma(1)} \cdots a_{n,\sigma(n)}e_{\sigma(1)} \wedge \cdots \wedge e_{\sigma(n)}.
\end{aligned}
$$

The sum is taken over all maps $\sigma: \{1,\ldots,n\} \to \{1,\ldots,n\}$, but because of the alternating property, whenever σ is not a permutation the term corresponding to σ is equal to 0. Hence the sum may be taken only over all permutations. Since

$$
e_{\sigma(1)} \wedge \cdots \wedge e_{\sigma(n)} = \varepsilon(\sigma)e_1 \wedge \cdots \wedge e_n
$$

where $\varepsilon(\sigma) = 1$ or -1 os a sign depending only on σ, it follows that the alternating product is completely determined by its value $e_1 \wedge \cdots \wedge e_n$, and in particular is the determinant if this value is equal to 1.

Proposition 2.1. *We have*

$$
\mathrm{Vol}^0(v_1,\ldots,v_n) = \mathrm{Det}(v_1,\ldots,v_n)
$$

and

$$
\mathrm{vol}(v_1,\ldots,v_n) = |\mathrm{Det}(v_1,\ldots,v_n)|.
$$

Proof. If $v_1,\ldots,v_n$ are linearly dependent,then the determinant is equal to 0, and the volume is also equal to 0, for instance by Lemma 1.3. So our formula holds in the case. It is clear that

$$
\mathrm{Vol}^0(e_1,\ldots,e_n) = 1.
$$

To show that Vol^0 satisfies the characteristic properties of the determinant, all we have to do now is to show that it is linear in each variable, say the first. In other words, we must prove

(*) $\mathrm{Vol}^0(cv, v_2,\ldots,v_n) = c\,\mathrm{Vol}^0(v, v_2,\ldots,v_n)$ for $c \in \mathbf{R}$,

(**) $\mathrm{Vol}^0(v + w, v_2,\ldots,v_n) = \mathrm{Vol}^0(v, v_2,\ldots,v_n) + \mathrm{Vol}^0(w, v_2,\ldots,v_n).$

As to the first assertion, suppose first that c is some positive integer k. Let B be the block spanned by v, v_2, $\ldots,v_n$. We may assume without loss of generality that v, v_2, $\ldots,v_n$ are linearly independent (otherwise, the relation is obviously true, both sides being equal to 0). We verify at once from the definition that if $B(v, v_2,\ldots,v_n)$ denotes the block spanned by v,

$v_2, \ldots, v_n$ then $B(kv, v_2, \ldots, v_n)$ is the union of the two sets

$$B((k-1)v, v_2, \ldots, v_n) \qquad \text{and} \qquad B(v, v_2, \ldots, v_n) + (k-1)v$$

which have only a set of measure 0 in common, as one verifies at once from the definitions.

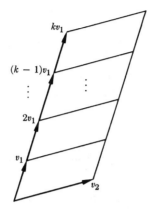

Therefore, we find that

$$\text{Vol}(kv, v_2, \ldots, v_n) = \text{Vol}((k-1)v, v_2, \ldots, v_n) + \text{Vol}(v, v_2, \ldots, v_n)$$
$$= (k-1)\,\text{Vol}(v, v_2, \ldots, v_n) + \text{Vol}(v, v_2, \ldots, v_n)$$
$$= k\,\text{Vol}(v, v_2, \ldots, v_n),$$

as was to be shown.

Now let

$$v = v_1/k$$

for a positive integer k. Then applying what we have just proved shows that

$$\text{Vol}\left(\frac{1}{k}v_1, v_2, \ldots, v_n\right) = \frac{1}{k}\text{Vol}(v_1, \ldots, v_n).$$

Writing a positive rational number in the form $m/k = m \cdot 1/k$, we conclude that the first relation holds when c is a positive rational number. If r is a positive real number, we find positive rational numbers c, c' such that $c \leqq r \leqq c'$. Since

$$B(cv, v_2, \ldots, v_n) \subset B(rv, v_2, \ldots, v_n) \subset B(c'v, v_2, \ldots, v_n),$$

we conclude that

$$c\,\text{Vol}(v, v_2, \ldots, v_n) \leqq \text{Vol}(rv, v_2, \ldots, v_n) \leqq c'\,\text{Vol}(v, v_2, \ldots, v_n).$$

Letting c, c' approach r as a limit, we conclude that for any real number $r \geq 0$ we have

$$\mathrm{Vol}(rv, v_2, \ldots, v_n) = r\, \mathrm{Vol}(v, v_2, \ldots, v_n).$$

Finally, we note that $B(-v, v_2, \ldots, v_n)$ is the translation of

$$B(v, v_2, \ldots, v_n)$$

by $-v$ so that these two blocks have the same volume. This proves the first assertion.

As for the second, we look at the geometry of the situation, which is made clear by the following picture in case $v = v_1$, $w = v_2$.

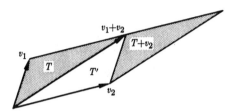

The block spanned by v_1, v_2, $\ldots$ consists of two "triangles" T, T' having only a set of measure zero in common. The block spanned by $v_1 + v_2$ and v_2 consists of T' and the translation $T + v_2$. It follows that these two blocks have the same volume. We conclude that for any number c,

$$\mathrm{Vol}^0(v_1 + cv_2, v_2, \ldots, v_n) = \mathrm{Vol}^0(v_1, v_2, \ldots, v_n).$$

Indeed, if $c = 0$ this is obvious, and if $c \neq 0$ then

$$c\, \mathrm{Vol}^0(v_1 + cv_2, v_2) = \mathrm{Vol}^0(v_1 + cv_2, cv_2)$$
$$= \mathrm{Vol}^0(v_1 + cv_2) = c\, \mathrm{Vol}^0(v_1, v_2).$$

We can then cancel c to get our conclusion.

To prove the linearity of Vol^0 with respect to its first variable, we may assume that $v_2, \ldots, v_n$ are linearly independent, otherwise both sides of (**) are equal to 0. Let v_1 be so chosen that $\{v_1, \ldots, v_n\}$ is a basis of $\mathbf{R}^n$. Then by induction, and what has been proved above,

$$\mathrm{Vol}^0(c_1 v_1 + \cdots + c_n v_n, v_2, \ldots, v_n)$$
$$= \mathrm{Vol}^0(c_1 v_1 + \cdots + c_{n-1} v_{n-1}, v_2, \ldots, v_n)$$
$$= \mathrm{Vol}^0(c_1 v_1, v_2, \ldots, v_n)$$
$$= c_1\, \mathrm{Vol}^0(v_1, \ldots, v_n).$$

From this the linearity follows at once, and the theorem is proved.

Corollary 2.2. *Let S be the unit cube spanned by the unit vectors in $\mathbf{R}^n$. Let $\lambda: \mathbf{R}^n \to \mathbf{R}^n$ be a linear map. Then*

$$\text{Vol } \lambda(S) = |\text{Det}(\lambda)|.$$

Proof. If $v_1, \ldots, v_n$ are the images of $e_1, \ldots, e_n$ under λ, then $\lambda(S)$ is the block spanned by $v_1, \ldots, v_n$. If we represent λ by the matrix $A = (a_{ij})$, then

$$v_i = a_{1i}e_1 + \cdots + a_{ni}e_n,$$

and hence $\text{Det}(v_1, \ldots, v_n) = \text{Det}(A) = \text{Det}(\lambda)$. This proves the corollary.

Corollary 2.3. *If R is any rectangle in $\mathbf{R}^n$ and $\lambda: \mathbf{R}^n \to \mathbf{R}^n$ is a linear map, then*

$$\text{Vol } \lambda(R) = |\text{Det}(\lambda)| \, \text{Vol}(R).$$

Proof. After a translation, we can assume that the rectangle is a block. If $R = \lambda_1(S)$ where S is the unit cube, then

$$\lambda(R) = \lambda \circ \lambda_1(S),$$

whence by Corollary 2.2,

$$\text{Vol } \lambda(R) = |\text{Det}(\lambda \circ \lambda_1)| = |\text{Det}(\lambda) \, \text{Det}(\lambda_1)| = |\text{Det}(\lambda)| \, \text{Vol}(R).$$

The next theorem extends Corollary 2.3 to the more general case where the linear map λ is replaced by an arbitrary C^1-invertible map. The proof then consists of replacing the linear map by its derivative and estimating the error thus introduced. For this purpose, we have the **Jacobian determinant**

$$\Delta_f(x) = \text{Det } J_f(x) = \text{Det } f'(x),$$

where $J_f(x)$ is the Jacobian matrix, and $f'(x)$ is the derivative of the map $f: U \to \mathbf{R}^n$.

Proposition 2.4. *Let R be a rectangle in $\mathbf{R}^n$, contained in some open set U. Let $f: U \to \mathbf{R}^n$ be a C^1 map, which is C^1-invertible on U. Then*

$$\mu(f(R)) = \int_R |\Delta_f| \, d\mu.$$

Proof. When f is linear, this is nothing but Corollary 2.3 of the preceding theorem. We shall prove the general case by approximating f by its derivative. Let us first assume that R is a cube for simplicity.

Given ε, let P be a partition of R, obtained by dividing each side of R into N equal segments for large N. Then R is partitioned into N^n subcubes which we denote by S_j $(j = 1, \ldots, N^n)$. We let a_j be the center of S_j.

We have

$$\mathrm{Vol}\, f(R) = \sum_j \mathrm{Vol}\, f(S_j)$$

because the images $f(S_j)$ have only sets of measure 0 in common. We investigate $f(S_j)$ for each j. The derivative f' is uniformly continuous on R. Given ε, we assume that N has been taken so large that for $x \in S_j$ we have

$$f(x) = f(a_j) + \lambda_j(x - a_j) + \varphi(x - a_j),$$

where $\lambda_j = f'(a_j)$ and

$$|\varphi(x - a_j)| \leqq |x - a_j|\varepsilon.$$

To determine $\mathrm{Vol}\, f(S_j)$ we must therefore investigate $f(S)$ where S is a cube centered at the origin, and f has the form

$$f(x) = \lambda x + \varphi(x), \qquad |\varphi(x)| \leqq |x|\varepsilon.$$

on the cube S. (We have made suitable translations which don't affect volumes.) We have

$$\lambda^{-1} \circ f(x) = x + \lambda^{-1} \circ \varphi(x),$$

so that $\lambda^{-1} \circ f$ is nearly the identity map. For some constant C, we have for $x \in S$

$$|\lambda^{-1} \circ \varphi(x)| \leqq C\varepsilon.$$

From the lemma after the proof of the inverse mapping theorem, we conclude that $\lambda^{-1} \circ f(S)$ contains a cube of radius

$$(1 - C\varepsilon)(\text{radius } S),$$

and trivial estimates show that $\lambda^{-1} \circ f(S)$ is contained in a cube of radius

$$(1 + C\varepsilon)(\text{radius } S).$$

We apply λ to these cubes, and determine their volumes. Putting indices

j on everything, we find that

$$|\text{Det } f'(a_j)| \text{ Vol}(S_j) - \varepsilon C_1 \text{Vol}(S_j)$$
$$\leq \text{Vol } f(S_j) \leq |\text{Det } f'(a_j)| \text{ Vol}(S_j) + \varepsilon C_1 \text{Vol}(S_j)$$

with some fixed constant C_1. Summing over j and estimating $|\Delta_f|$, we see that our theorem follows at once.

Remark. We assumed for simplicity that R was a cube. Actually, by changing the norm on each side, multiplying by a suitable constant, and taking the sup of the adjusting norms, we see that this involves no loss of generality. Alternatively, we can approximate a given rectangle by cubes.

Corollary 2.5. *If g is continuous on $f(R)$, then*

$$\int_{f(R)} g \, d\mu = \int_R (g \circ f)|\Delta_f| \, d\mu.$$

Proof. The functions g and $(g \circ f)|\Delta_f|$ are uniformly continuous on $f(R)$ and R respectively. Let us take a partition of R and let $\{S_j\}$ be the subrectangles of this partition. If δ is the maximum length of the sides of the subrectangles of the partition, then $f(S_j)$ is contained in a rectangle whose sides have length $\leq C\delta$ for some constant C. We have

$$\int_{f(R)} g \, d\mu = \sum_j \int_{f(S_j)} g \, d\mu.$$

The sup and inf of g of $f(S_j)$ differ only by ε if δ is taken sufficiently small. Using the theorem, applied to each S_j, and replacing g by its minimum m_j and maximum M_j on S_j, we see that the corollary follows at once.

Theorem 2.6 (Change of Variables Formula). *Let U be open in $\mathbf{R}^n$ and let $f: U \to \mathbf{R}^n$ be a C^1 map, which is C^1 invertible on U. Let g be in $\mathcal{L}^1(f(U))$. Then $(g \circ f)|\Delta_f|$ is in $\mathcal{L}^1(U)$ and we have*

$$\int_{f(U)} g \, d\mu = \int_U (g \circ f)|\Delta_f| \, d\mu.$$

Proof. Let R be a closed rectangle contained in U. We shall first prove that the restriction of $(g \circ f)|\Delta_f|$ to R is in $\mathcal{L}^1(R)$, and that the formula holds when U is replaced by R. We know that $C_c(f(U))$ is

L^1-dense in $\mathscr{L}^1(f(U))$, by [La 93], Theorem 3.1 of Chapter IX. Hence there exists a sequence $\{g_k\}$ in $C_c(f(U))$ which in L^1-convergent to g. Using [La 93], Theorem 5.2 of Chapter VI, we may assume that $\{g_k\}$ converges pointwise to g except on a set Z of measure 0 in $f(U)$. By Lemma 1.2, we know that $f^{-1}(Z)$ has measure 0.

Let $g_k^* = (g_k \circ f)|\Delta_f|$. Each function g_k^* is continuous on R. The sequence $\{g_k^*\}$ converges almost everywhere to $(g \circ f)|\Delta_f|$ restricted to R. It is in fact an L^1-Cauchy sequence in $\mathscr{L}^1(R)$. To see this, we have by the result for rectangles and continuous functions (corollary of the preceding theorem):

$$\int_R |g_k^* - g_m^*| \, d\mu = \int_{f(R)} |g_k - g_m| \, d\mu,$$

so the Cauchy nature of the sequence $\{g_k^*\}$ is clear from that of $\{g_k\}$. It follows that the restriction of $(g \circ f)|\Delta_f|$ to R is the L^1-limit of $\{g_k^*\}$, and is in $\mathscr{L}^1(R)$. It also follows that the formula of the theorem holds for R, that is

$$\int_{f(A)} g \, d\mu = \int_A (g \circ f)|\Delta_f| \, d\mu$$

when $A = R$.

The theorem is now seen to hold for any measurable subset A of R, since $f(A)$ is measurable, and since a function g in $\mathscr{L}^1(f(A))$ can be extended to a function in $\mathscr{L}^1(f(R))$ by giving it the value 0 outside $f(A)$. From this it follows that the theorem holds if A is a finite union of rectangles contained in U. We can find a sequence of rectangles $\{R_m\}$ contained in U whose union is equal to U, because U is separable. Taking the usual stepwise complementation, we can find a disjoint sequence of measurable sets

$$A_m = R_m - (R_1 \cup \cdots \cup R_{m-1})$$

whose union is U, and such that our theorem holds if $A = A_m$. Let

$$h_m = g_{f(A_m)} = g\chi_{f(A_m)} \quad \text{and} \quad h_m^* = (h_m \circ f)|\Delta_f|.$$

Then $\sum h_m$ converges to g and $\sum h_m^*$ converges to $(g \circ f)|\Delta_f|$. Our theorem follows from Corollary 5.13 of the dominated convergence theorem in [La 93].

Note. In dealing with polar coordinates or the like, one sometimes meets a map f which is invertible except on a set of measure 0, e.g. the polar coordinate map. It is now trivial to recover a result covering this type of situation.

Corollary 2.7. *Let U be open in $\mathbf{R}^n$ and let $f: U \to \mathbf{R}^n$ be a C^1 map. Let A be a measurable subset of U such that the boundary of A has measure 0, and such that f is C^1 invertible on the interior of A. Let g be in $\mathscr{L}^1(f(A))$. Then $(g \circ f)|\Delta_f|$ is in $\mathscr{L}^1(A)$ and*

$$\int_{f(A)} g \, d\mu = \int_A (g \circ f)|\Delta_f| \, d\mu.$$

Proof. Let U_0 be the interior of A. The sets $f(A)$ and $f(U_0)$ differ only by a set of measure 0, namely $f(\partial A)$. Also the sets A, U_0 differ only by a set of measure 0. Consequently we can replace the domains of integration $f(A)$ and A by $f(U_0)$ and U_0, respectively. The theorem applies to conclude the proof of the corollary.

XI, §3. ORIENTATION

Let U, V be open sets in half spaces of $\mathbf{R}^n$ and let $\varphi: U \to V$ be a C^1 isomorphism. We shall say that φ is **orientation preserving** if the Jacobian determinant $\Delta_\varphi(x)$ is > 0, all $x \in U$. If the Jacobian determinant is negative, then we say that φ is orientation **reversing**.

Let X be a C^p manifold, $p \geq 1$, and let $\{(U_i, \varphi_i)\}$ be an atlas. We say that this atlas is **oriented** if all transition maps $\varphi_j \circ \varphi_i^{-1}$ are orientation preserving. Two atlases $\{(U_i, \varphi_i)\}$ and $\{(V_\alpha, \psi_\alpha)\}$ are said to **define the same orientation**, or to be **orientation equivalent**, if their union is oriented. We can also define locally a chart (V, ψ) to be **orientation compatible** with the oriented atlas $\{(U_i, \varphi_i)\}$ if all transition maps $\varphi_i \circ \varphi^{-1}$ (defined whenever $U_i \cap V$ is not empty) are orientation preserving. An orientation equivalence class of oriented atlases is said to define an **oriented** manifold, or to be an **orientation** of the manifold. It is a simple exercise to verify that if a connected manifold has an orientation, then it has two distinct orientations.

The standard examples of the Moebius strip or projective plane show that not all manifolds admit orientations. We shall now see that the boundary of an oriented manifold with boundary can be given a natural orientation.

Let $\varphi: U \to \mathbf{R}^n$ *be an oriented chart at a boundary point of X, such that:*

(1) *if $(x_1, \ldots, x_n)$ are the local coordinates of the chart, then the boundary points correspond to those points in $\mathbf{R}^n$ satisfying $x_1 = 0$; and*

(2) *the points of U not in the boundary have coordinates satisfying $x_1 < 0$.*

Then $(x_2, \ldots, x_n)$ are the local coordinates for a chart of the boundary, namely the restriction of φ to $\partial X \cap U$, and the picture is as follows.

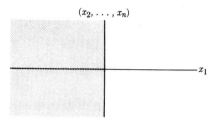

We may say that we have considered a chart φ such that the manifold lies to the left of its boundary. If the reader thinks of a domain in $\mathbf{R}^2$, having a smooth curve for its boundary, as on the following picture, the reader will see that our choice of chart corresponds to what is usually visualized as "counterclockwise" orientation.

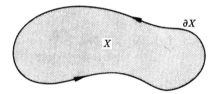

The collection of all pairs $\bigl(U \cap \partial X, \varphi|(U \cap \partial X)\bigr)$, chosen according to the criteria described above, is obviously an atlas for the boundary ∂X, and we contend that it is an oriented atlas.

We prove this easily as follows. If

$$(x_1, \ldots, x_n) = x \qquad \text{and} \qquad (y_1, \ldots, y_n) = y$$

are coordinate systems at a boundary point corresponding to choices of charts made according to our specifications, then we can write $y = f(x)$ where $f = (f_1, \ldots, f_n)$ is the transition mapping. Since we deal with oriented charts for X, we know that $\Delta_f(x) > 0$ for all x. Since f maps boundary into boundary, we have

$$f_1(0, x_2, \ldots, x_n) = 0$$

for all $x_2, \ldots, x_n$. Consequently the Jacobian matrix of f at a point $(0, x_2, \ldots, x_n)$ is equal to

$$\begin{bmatrix} D_1 f_1(0, x_2, \ldots, x_n) & 0 \cdots \cdots 0 \\ * & \\ * & \Delta_g^{(n-1)} \\ * & \end{bmatrix},$$

where $\Delta_g^{(n-1)}$ is the Jacobian matrix of the transition map g induced by f on the boundary, and given by

$$y_2 = f_2(0, x_2, \ldots, x_n),$$
$$\vdots \qquad \qquad \vdots$$
$$y_n = f_n(0, x_2, \ldots, x_n).$$

However, we have

$$D_1 f_1(0, x_2, \ldots, x_n) = \lim_{h \to 0} \frac{f_1(h, x_2, \ldots, x_n)}{h},$$

taking the limit with $h < 0$ since by prescription, points of X have coordinates with $x_1 < 0$. Furthermore, for the same reason we have

$$f_1(h, x_2, \ldots, x_n) < 0.$$

Consequently

$$D_1 f_1(0, x_2, \ldots, x_n) > 0.$$

From this it follows that $\Delta_g^{(n-1)}(x_2, \ldots, x_n) > 0$, thus proving our assertion that the atlas we have defined for ∂X is oriented.

From now on, when we deal with an oriented manifold, it is understood that its boundary is taken with orientation described above, and called the induced orientation.

XI, §4. THE MEASURE ASSOCIATED WITH A DIFFERENTIAL FORM

Let X be a manifold of class C^p with $p \geq 1$. We assume from now on that X is Hausdorff and has a countable base. Then we know that X admits C^p partitions of unity, subordinated to any given open covering.

(Actually, instead of the conditions we assumed, we could just as well have assumed the existence of C^p partitions of unity, which is the precise condition to be used in the sequel.)

We can define the **support** of a differential form as we defined the support of a function. It is the closure of the set of all $x \in X$ such that $\omega(x) \neq 0$. If ω is a form of class C^p and α is a C^q function on X, then we can form the product $\alpha\omega$, which is the form whose value at x is $\alpha(x)\omega(x)$. If α has compact support, then $\alpha\omega$ has compact support. Later, we shall study the integration of forms, and reduce this to a local problem by means of partitions of unity, in which we multiply a form by functions.

We assume that the reader is familiar with the correspondence between certain functionals on continuous functions with compact support and measures. Cf. [La 93] for this. We just recall some terminology.

We denote by $C_c(X)$ the vector space of continuous functions on X with **compact support** (i.e. vanishing outside a compact set). We write $C_c(X, \mathbf{R})$ or $C_c(X, \mathbf{C})$ if we wish to distinguish between the real or complex valued functions.

We denote by $C_K(X)$ the subspace of $C_c(X)$ consisting of those functions which vanish outside K. (Same notation $C_S(X)$ for those functions which are 0 outside any subset S of X. Most of the time, the useful subsets in this context are the compact subsets K.)

A linear map λ of $C_c(X)$ into the complex numbers (or into a normed vector space, for that matter) is said to be **bounded** if there exists some $C \geq 0$ such that we have

$$|\lambda f| \leq C \|f\|$$

for all $f \in C_c(X)$. Thus λ is bounded if and only if λ is continuous for the norm topology.

A linear map λ of $C_c(X)$ into the complex numbers is said to be **positive** if we have $\lambda f \geq 0$ whenever f is real and ≥ 0.

Lemma 4.1. *Let* $\lambda: C_c(X) \to \mathbf{C}$ *be a positive linear map. Then* λ *is bounded on* $C_K(X)$ *for any compact* K.

Proof. By the corollary of Urysohn's lemma, there exists a continuous real function $g \geq 0$ on X which is 1 on K has compact support. If $f \in C_K(X)$, let $b = \|f\|$. Say f is real. Then $bg \pm f \geq 0$, whence

$$\lambda(bg) \pm \lambda f \geq 0$$

and $|\lambda f| \leq b\lambda(g)$. Thus λg is our desired bound.

A complex valued linear map on $C_c(X)$ which is bounded on each subspace $C_K(X)$ for every compact K will be called a **C_c-functional** on $C_c(X)$, or more simply, a **functional**. A functional on $C_c(X)$ which is also continuous for the sup norm will be called a **bounded** functional. It is clear that a bounded functional is also a C_c-functional.

Lemma 4.2. *Let* $\{W_\alpha\}$ *be an open covering of* X. *For each index* α, *let* λ_α *be a functional on* $C_c(W_\alpha)$. *Assume that for each pair of indices* α, β *the functionals* λ_α *and* λ_β *are equal on* $C_c(W_\alpha \cap W_\beta)$. *Then there exists a unique functional* λ *on* X *whose restriction to each* $C_c(W_\alpha)$ *is equal to* λ_α. *If each* λ_α *is positive, then so is* λ.

Proof. Let $f \in C_c(X)$ and let K be the support of f. Let $\{h_i\}$ be a partition of unity over K subordinated to a covering of K by a finite number of the open sets W_α. Then each $h_i f$ has support in some $W_{\alpha(i)}$ and we define

$$\lambda f = \sum_i \lambda_{\alpha(i)}(h_i f).$$

We contend that this sum is independent of the choice of $\alpha(i)$, and also of the choice of partition of unity. Once this is proved, it is then obvious that λ is a functional which satisfies our requirements. We now prove this independence. First note that if $W_{\alpha'(i)}$ is another one of the open sets W_α in which the support of $h_i f$ is contained, then $h_i f$ has support in the intersection $W_{\alpha(i)} \cap W_{\alpha'(i)}$, and our assumption concerning our functionals λ_α shows that the corresponding term in the sum does not depend on the choice of index $\alpha(i)$. Next, let $\{g_k\}$ be another partition of unity over K subordinated to some covering of K by a finite number of the open sets W_α. Then for each i,

$$h_i f = \sum_k g_k h_i f,$$

whence

$$\sum_i \lambda_{\alpha(i)}(h_i f) = \sum_i \sum_k \lambda_{\alpha(i)}(g_k h_i f).$$

If the support of $g_k h_i f$ is in some W_α, then the value $\lambda_\alpha(g_k h_i f)$ is independent of the choice of index α. The expression on the right is then symmetric with respect to our two partitions of unity, whence our theorem follows.

Theorem 4.3. *Let* $\dim X = n$ *and let* ω *be an* n*-form on* X *of class* C^0, *that is continuous. Then there exists a unique positive functional* λ *on* $C_c(X)$ *having the following property. If* (U, φ) *is a chart and*

$$\omega(x) = f(x)\, dx_1 \wedge \cdots \wedge dx_n$$

is the local representation of ω *in this chart, then for any* $g \in C_c(X)$ *with support in* U, *we have*

(1) $$\lambda g = \int_{\varphi U} g_\varphi(x) |f(x)|\, dx,$$

where g_φ *represents* g *in the chart* [*i.e.* $g_\varphi(x) = g(\varphi^{-1}(x))$], *and* dx *is Lebesgue measure.*

Proof. The integral in (1) defines a positive functional on $C_c(U)$. The change of variables formula shows that if (U, φ) and (V, ψ) are two charts, and if g has support in $U \cap V$, then the value of the functional is independent of the choice of charts. Thus we get a positive functional by the general localization lemma for functionals.

The positive measure corresponding to the functional in Theorem 4.3 will be called the **measure associated with** $|\omega|$, and can be denoted by $\mu_{|\omega|}$.

Theorem 4.3 does not need any orientability assumption. With such an assumption, we have a similar theorem, obtained without taking the absolute value.

Theorem 4.4. *Let* $\dim X = n$ *and assume that* X *is oriented. Let* ω *be an n-form on* X *of class* C^0. *Then there exists a unique functional* λ *on* $C_c(X)$ *having the following property. If* (U, φ) *is an oriented chart and*

$$\omega(x) = f(x) \, dx_1 \wedge \cdots \wedge dx_n$$

is the local representation of ω *in this chart, then for any* $g \in C_c(X)$ *with support in* U, *we have*

$$\lambda g = \int_{\varphi U} g_\varphi(x) f(x) \, dx,$$

where g_φ *represents* g *in the chart, and* dx *is Lebesgue measure.*

Proof. Since the Jacobian determinant of transition maps belonging to oriented charts is positive, we see that Theorem 4.4 follows like Theorem 4.3 from the change of variables formula (in which the absolute value sign now becomes unnecessary) and the existence of partitions of unity.

If λ is the functional of Theorem 4.4, we shall call it the **functional associated with** ω. For any function $g \in C_c(X)$, we define

$$\int_X g\omega = \lambda g.$$

If in particular ω has compact support, we can also proceed directly as follows. Let $\{\alpha_i\}$ be a partition of unity over X such that each α_i has compact support. We define

$$\int_X \omega = \sum_i \int_X \alpha_i \omega,$$

all but a finite number of terms in this sum being equal to 0. As usual, it is immediately verified that this sum is in fact independent of the choice of partition of unity, and in fact, we could just as well use only a partition of unity over the support of ω. Alternatively, if α is a function in $C_c(X)$ which is equal to 1 on the support of ω, then we could also define

$$\int_X \omega = \int_X \alpha\omega.$$

It is clear that these two possible definitions are equivalent. In particular, we obtain the following variation of Theorem 4.4.

Theorem 4.5. *Let X be an oriented manifold of dimension n. Let $\mathscr{A}_c^n(X)$ be the **R**-space of differential forms with compact support. There exists a unique linear map*

$$\omega \mapsto \int_X \omega \qquad of \quad \mathscr{A}_c^n(X) \to \mathbf{R}$$

such that, if ω has support in an oriented chart U with coordinates x_1, $\ldots, x_n$ and $\omega(x) = f(x)\,dx_1 \wedge \cdots \wedge dx_n$ in this chart, then

$$\int_X \omega = \int_U f(x)\,dx_1 \cdots dx_n.$$

Let X be an oriented manifold. By **a volume form** Ω we mean a form such that in every oriented chart, the form can be written as

$$\Omega(x) = f(x)\,dx_1 \wedge \cdots \wedge dx_n$$

with $f(x) > 0$ for all x. In the next section, we shall see how to get a volume form from a Riemannian metric. Here, we shall consider the non-oriented case to get the notion of density.

Even when a manifold is not orientable, one may often reduce certain questions to the orientable case, because of the following result. We assume that readers are acquainted with basic facts about coverings.

Proposition 4.6. *Let X be a connected C^1 manifold. If X is not orientable, then there exists a covering $X' \to X$ of degree 2 such that X' is orientable.*

Sketch of Proof. Suppose first that X is simply connected. Let $x \in X$. Fix a chart (U_0, φ_0) at x such that the image of the chart is an open ball in euclidean space. Let y be any point of X, and let $\alpha: [a, b] \to X$ be a piecewise C^1 path from x to y. We select a sufficiently fine partition

$$[a, b] = [t_0, t_1, \ldots, t_n],$$

and open sets U_i containing $\alpha([t_i, t_{i+1}])$, such that U_i has an isomorphism φ_i onto an open ball in euclidean space, and such that the charts φ_i and φ_{i+1} have the same orientation. It is easy to verify that if two paths are homotopic, then the charts which we obtain at y by "continuation" as above along the two paths are orientation equivalent. This is done first for paths which are "close together," and then extended to homotopic paths, according to the standard technique which already appears in analytic continuation. Thus fixing one orientation in the neighborhood of a give point determines an orientation on all of X when X is simply connected.

Now suppose X not simply connected, and let $\tilde{X}$ be its universal covering space. Let Γ be the fundamental group. Then the subgroup of elements $\gamma \in \Gamma$ which preserve an orientation of $\tilde{X}$ is of index 2, and the covering corresponding to this subgroups has degree 2 over X and can be given an orientation by using charts which lift to oriented charts in the universal covering space. This concludes the proof.

Densities

The rest of this section will not be used, especially not for Stokes' theorem in the next chapter. However, Theorem 4.3 for the non-orientable case is important for other applications, and we make further comments about this other context.

Let s be a real number. Let $\mathbf{E}$ be a finite dimensional vector space over $\mathbf{R}$, of dimension n. We denote by E^* the set of non-zero elements of E, and by $\bigwedge^n E^*$ the set of non-zero elements of $\bigwedge^n E$. By an s-**density** on E we mean a function

$$\delta: \bigwedge^n E^* \to \mathbf{R} \qquad \text{such that} \qquad \delta(cw) = |c|^s \delta(w)$$

for all $c \neq 0$ in $\mathbf{R}$ and $w \in \bigwedge^n E^*$. Equivalently, we could say that there exists an n-form $\omega \in L_a^n(\mathbf{E}, \mathbf{R})$ such that for $v_i \in \mathbf{E}$ we have

$$\delta(v_1 \wedge \cdots \wedge v_n) = |\omega(v_1, \ldots, v_n)|^s.$$

We let $\text{den}^s(\mathbf{E})$ denote the set of densities of $\mathbf{E}$. An element of $\text{den}^s(\mathbf{E})$ amounts to picking a basis of $\bigwedge^n \mathbf{E}$, up to a factor ± 1, and assigning a number to this basis.

Let U be an open subset of $\mathbf{E}$. By a C^p **density** on U we mean a C^p morphism $\delta: U \to \text{den}^s(\mathbf{E})$. Note that $\text{den}^s(\mathbf{E})$ is an open half line, so a density on U amounts to selecting a differential form of class C^p on a neighborhood of each point of U, such that the absolute values of these forms coincide on intersections of these neighborhoods.

Let $f : U \to V$ be a C^p isomorphism. Then f induces a map on densities, by the change of variable formula on forms with the Jacobian determinant, and then taking absolute values to the s-power. Thus we may form the bundle (not vector bundle) of densities, with charts

$$U \times \mathrm{den}^s(\mathbf{E})$$

over U, and density-bundle morphisms just as we did with differential forms. For example, let $E = \mathbf{R}^n$, with coordinates $x_1, \ldots, x_n$. Then

$$|dx_1 \wedge \cdots \wedge dx_n| = dx_1 \cdots dx_n$$

defines a 1-density, and $|dx_1 \wedge \cdots \wedge dx_n|^s$ defines an s-density, denoted by $|dx|^s$.

Observe that s-densities form a cone, i.e. if δ_1, δ_2 are s-densities on a manifold X, and a_1, $a_2 \in \mathbf{R}^+$ (the set of positive real numbers), then $a_1 \delta_1 + a_2 \delta_2$ is also an s-density. In particular, continuing to assume that X admits continuous partitions of unity, we can reformulate and prove Theorem 4.3 for densities. Indeed, the differential form in Theorem 4.3 need not be globally defined, because one needs only its absolute value to define the integral. Thus with the language of densities, Theorem 4.3 reads as follows.

Theorem 4.7. *Let δ be a C^0 density on X, i.e. a continuous density. Then there is a positive functional λ on $C_c(X)$ having the following property. If U is a chart and δ is represented by the density $|f(x) \, dx_1 \wedge \cdots \wedge dx_n|$ on this chart, then for any function $\varphi \in C_c(U)$ we have*

$$\lambda(\varphi) = \int_U \varphi_U(x) |f(x)| \, dx_1 \cdots dx_n,$$

where $dx_1 \cdots dx_n$ is the usual symbol for ordinary integration on $\mathbf{R}^n$, and φ_U is the representation of φ in the chart.

Examples. We have already given the example of integration with respect to $|dx_1 \wedge \cdots \wedge dx_n|$ in euclidean space. Here is a less trivial example. Let X be a Riemannian manifold of finite dimension n, with Riemannian metric g. Locally in a chart U, we view g as a morphism

$$g : U \to L(\mathbf{E}, \mathbf{E}),$$

with $\mathbf{E}$ having a fixed positive definite scalar product. With respect to an orthonormal basis, we have a linear metric isomorphism $\mathbf{E} \approx \mathbf{R}^n$, and $g(x)$ at each point x can be represented by a matrix $(g_{ij}(x))$. If we put

$$\delta(x) = |\det g_{ij}(x)| \, dx_1 \cdots dx_n$$

then δ defines a density, called the **Riemannian density**; and

$$\delta^{1/2}(x) = |\det g_{ij}(x)|^{1/2}\, dx_1 \cdots dx_n$$

defines the **Riemannian half density**.

Remark. Locally, a manifold is always orientable. Hence a formula or result which is local, and is proved in the orientable case, also applies to densities, sometimes by inserting an absolute value sign. For example, Proposition 1.2 of Chapter X applies after inserting absolute value signs, but Proposition 2.1 of Chapter X applies as stated for the Riemannian density instead of the Riemannian volume.

CHAPTER XII

Stokes' Theorem

Throughout the chapter, all manifolds are assumed finite dimensional. They may have a boundary.

XII, §1. STOKES' THEOREM FOR A RECTANGULAR SIMPLEX

If X is a manifold and Y a submanifold, then any differential form on X induces a form on Y. We can view this as a very special case of the inverse image of a form, under the embedding (injection) map

$$\text{id}: Y \to X.$$

In particular, if Y has dimension $n - 1$, and if $(x_1, \ldots, x_n)$ is a system of coordinates for X at some point of Y such that the points of Y correspond to those coordinates satisfying $x_j = c$ for some fixed number c, and index j, and if the form on X is given in terms of these coordinates by

$$\omega(x) = f(x_1, \ldots, x_n)\, dx_1 \wedge \cdots \wedge dx_n,$$

then the restriction of ω to Y (or the form induced on Y) has the representation

$$f(x_1, \ldots, c, \ldots, x_n)\, dx_1 \wedge \cdots \wedge \widehat{dx_j} \wedge \cdots \wedge dx_n.$$

We should denote this induced form by ω_Y, although occasionally we omit the subscript Y. We shall use such an induced form especially when Y is the boundary of a manifold X.

Let

$$R = [a_1, b_1] \times \cdots \times [a_n, b_n]$$

be a rectangle in n-space, that is a product of n closed intervals. The set theoretic boundary of R consists of the union over all $i = 1, \ldots, n$ of the pieces

$$R_i^0 = [a_1, b_1] \times \cdots \times \{a_i\} \times \cdots \times \{a_n, b_n\},$$

$$R_i^1 = [a_1, b_1] \times \cdots \times \{b_i\} \times \cdots \times [a_n, b_n].$$

If

$$\omega(x_1, \ldots, x_n) = f(x_1, \ldots, x_n)\, dx_1 \wedge \cdots \wedge \widehat{dx_j} \wedge \cdots \wedge dx_n$$

is an $(n-1)$-form, and the roof over anything means that this thing is to be omitted, then we define

$$\int_{R^0} \omega = \int_{a_i}^{b_1} \cdots \widehat{\int_{a_i}^{b_1}} \cdots \int_{a_n}^{b_n} f(x_1, \ldots, a_i, \ldots, x_n)\, dx_1 \cdots \widehat{dx_j} \cdots dx_n,$$

if $i = j$, and 0 otherwise. And similarly for the integral over R_i^1. We define the integral over the oriented boundary to be

$$\int_{\partial^0 R} = \sum_{i=1}^n (-1)^i \left[\int_{R_i^0} - \int_{R_i^1} \right].$$

Stokes' Theorem for Rectangles. *Let R be a rectangle in an open set U in n-space. Let ω be an $(n-1)$-form on U. Then*

$$\int_R d\omega = \int_{\partial^0 R} \omega.$$

Proof. In two dimensions, the picture looks like this:

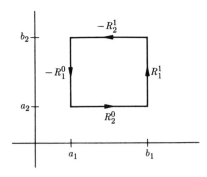

It suffices to prove the assertion when ω is a decomposable form, say

$$\omega(x) = f(x_1, \ldots, x_n)\, dx_1 \wedge \cdots \wedge \widehat{dx_j} \wedge \cdots \wedge dx_n.$$

We then evaluate the integral over the boundary of R. If $i \neq j$, then it is clear that

$$\int_{R_i^0} \omega = 0 = \int_{R_i^1} \omega,$$

so that

$$\int_{\partial^0 R} \omega =$$

$$(-1)^j \int_{a_1}^{b_1} \cdots \int_{a_j}^{\widehat{b_j}} \cdots \int_{a^n}^{b_n} [f(x_1, \ldots, a_j, \ldots, x_n) - f(x_1, \ldots, b_j, \ldots, x_n)] \, dx_1 \cdots \widehat{dx_j} \cdots dx_n.$$

On the other hand, from the definitions we find that

$$d\omega(x) = \left(\frac{\partial f}{dx_1} dx_1 + \cdots + \frac{\partial f}{\partial x_n} dx_n \right) \wedge dx_1 \wedge \cdots \wedge \widehat{dx_j} \wedge \cdots \wedge dx_n$$

$$= (-1)^{j-1} \frac{\partial f}{\partial x_j} dx_1 \wedge \cdots \wedge dx_n.$$

(The $(-1)^{j-1}$ comes from interchanging dx_j with $dx_1, \ldots, dx_{j-1}$. All other terms disappear by the alternating rule.)

Integrating $d\omega$ over R, we may use repeated integration and integrate $\partial f / \partial x_j$ with respect to x_j first. Then the fundamental theorem of calculus for one variable yields

$$\int_{a_j}^{b_j} \frac{\partial f}{\partial x_j} dx_j = f(x_1, \ldots, b_j, \ldots, x_n) - f(x_1, \ldots, a_j, \ldots, x_n).$$

We then integrate with respect to the other variables, and multiply by $(-1)^{j-1}$. This yields precisely the value found for the integral of ω over the oriented boundary $\partial^0 R$, and proves the theorem.

Remark. Stokes' theorem for a rectangle extends at once to a version in which we parametrize a subset of some space by a rectangle. Indeed, if $\sigma \colon R \to V$ is a C^1 map of a rectangle of dimension n into an open set V in $\mathbf{R}^N$, and if ω is an $(n-1)$-form in V, we may define

$$\int_\sigma d\omega = \int_R \sigma^* \, d\omega.$$

One can define

$$\int_{\partial \sigma} \omega = \int_{\partial^0 R} \sigma^* \omega,$$

and then we have a formula

$$\int_\sigma d\omega = \int_{\partial\sigma} \omega,$$

In the next section, we prove a somewhat less formal result.

XII, §2. STOKES' THEOREM ON A MANIFOLD

Theorem 2.1. *Let X be an oriented manifold of class C^2, dimension n, and let ω be an $(n-1)$-form on X, of class C^1. Assume that ω has compact support. Then*

$$\int_X d\omega = \int_{\partial X} \omega.$$

Proof. Let $\{\alpha_i\}_{i \in I}$ be a partition of unity, of class C^2. Then

$$\sum_{i \in I} \alpha_i \omega = \omega,$$

and this sum has only a finite number of non-zero terms since the support of ω is compact. Using the additivity of the operation d, and that of the integral, we find

$$\int_X d\omega = \sum_{i \in I} \int_X d(\alpha_i \omega).$$

Suppose that α_i has compact support in some open set V_i of X and that we can prove

$$\int_{V_i} d(\alpha_i \omega) = \int_{V_i \cap \partial X} \alpha_i \omega,$$

in other words we can prove Stokes' theorem locally in V_i. We can write

$$\int_{V_i \cap \partial X} \alpha_i \omega = \int_{\partial X} \alpha_i \omega,$$

and similarly

$$\int_{V_i} d(\alpha_i \omega) = \int_X d(\alpha_i \omega).$$

Using the additivity of the integral once more, we get

$$\int_X d\omega = \sum_{i \in I} \int_X d(\alpha_i \omega) = \sum_{i \in I} \int_{\partial X} \alpha_i \omega = \int_{\partial X} \omega,$$

which yields Stokes' theorem on the whole manifold. Thus our argument with partitions of unity reduces Stokes' theorem to the local case, namely it suffices to prove that for each point of X these exists an open neighborhood V such that if ω has compact support in V, then Stokes' theorem holds with X replaced by V. We now do this.

If the point is not a boundary point, we take an oriented chart (U, φ) at the point, containing an open neighborhood V of the point, satisfying the following conditions: φU is an open ball, and φV is the interior of a rectangle, whose closure is contained in φU. If ω has compact support in V, then its local representation in φU has compact support in φV. Applying Stokes' theorem for rectangles as proved in the preceding section, we find that the two integrals occurring in Stokes' formula are equal to 0 in this case (the integral over an empty boundary being equal to 0 by convention).

Now suppose that we deal with a boundary point. We take an oriented chart (U, φ) at the point, having the following properties. First, φU is described by the following inequalities in terms of local coordinates $(x_1, \ldots, x_n)$:

$$-2 < x_1 \leqq 1 \quad \text{and} \quad -2 < x_j < 2 \quad \text{for} \quad j = 2, \ldots, n.$$

Next, the given point has coordinates $(1, 0, \ldots, 0)$, and that part of U on the boundary of X, namely $U \cap \partial X$, is given in terms of these coordinates by the equation $x_1 = 1$. We then let V consist of those points whose local coordinates satisfy

$$0 < x_1 \leqq 1 \quad \text{and} \quad -1 < x_j < 1 \quad \text{for} \quad j = 2, \ldots, n.$$

If ω has compact support in V, then ω is equal to 0 on the boundary of the rectangle R equal to the closure of φV, except on the face given by $x_1 = 1$, which defines that part of the rectangle corresponding to $\partial X \cap V$. Thus the support of ω looks like the shaded portion of the following picture.

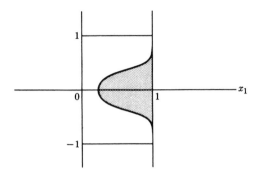

In the sum giving the integral over the boundary of a rectangle as in the previous section, only one term will give a non-zero contribution, corresponding to $i = 1$, which is

$$(-1)\left[\int_{R_1^0} \omega - \int_{R_1^1} \omega\right].$$

Furthermore, the integral over R_1^0 will also be 0, and in the contribution of the integral over R_1^1, the two minus signs will cancel, and yield the integral of ω over the part of the boundary lying in V, because our charts are so chosen that $(x_2, \ldots, x_n)$ is an oriented system of coordinates for the boundary. Thus we find

$$\int_V d\omega = \int_{V \cap \partial X} \omega,$$

which proves Stokes' theorem locally in this case, and concludes the proof of Theorem 2.7.

Corollary 2.2. *Suppose X is an oriented manifold without boundary, and ω has compact support. Then*

$$\int_X d\omega = 0.$$

For any number of reasons, some of which we consider in the next section, it is useful to formulate conditions under which Stokes' theorem holds even when the form ω does not have compact support. We shall say that ω has **almost compact support** if there exists a decreasing sequence of open sets $\{U_k\}$ in X such that the intersection

$$\bigcap_{k=1}^\infty U_k$$

is empty, and a sequence of C^1 functions $\{g_k\}$, having the following properties:

AC 1. *We have* $0 \leq g_k \leq 1$, $g_k = 1$ *outside* U_k, *and* $g_k\omega$ *has compact support.*

AC 2. *If* μ_k *is the measure associated with* $|dg_k \wedge \omega|$ *on* X, *then*

$$\lim_{k \to \infty} \mu_k(\overline{U}_k) = 0.$$

We then have the following application of Stokes' theorem.

Corollary 2.3. *Let* X *be a* C^2 *oriented manifold, of dimension n, and let* ω *be an* $(n-1)$-*form on* X, *of class* C^1. *Assume that* ω *has almost compact support, and that the measures associated with* $|d\omega|$ *on* X *and* $|\omega|$ *on* ∂X *are finite. Then*

$$\int_X d\omega = \int_{\partial X} \omega.$$

Proof. By our standard form of Stokes' theorem we have

$$\int_{\partial X} g_k\omega = \int_X d(g_k\omega) = \int_X dg_k \wedge \omega + \int_X g_k \, d\omega.$$

We estimate the left-hand side by

$$\left| \int_{\partial X} \omega - \int_{\partial X} g_k\omega \right| = \left| \int_{\partial X} (1 - g_k)\omega \right| \leq \mu_{|\omega|}(U_k \cap \partial X).$$

Since the intersection of the sets U_k is empty, it follows for a purely measure-theoretic reason that

$$\lim_{k \to \infty} \int_{\partial X} g_k\omega = \int_{\partial X} \omega.$$

Similarly,

$$\lim_{k \to \infty} \int_X g_k \, d\omega = \int_X d\omega.$$

The integral of $dg_k \wedge \omega$ over X approaches 0 as $k \to \infty$ by assumption, and the fact that $dg_k \wedge \omega$ is equal to 0 on the complement of $\overline{U}_k$ since g_k is constant on this complement. This proves our corollary.

The above proof shows that the second condition **AC 2** is a very natural one to reduce the integral of an arbitrary form to that of a form with compact support. In the next section, we relate this condition to a question of singularities when the manifold is embedded in some bigger space.

XII, §3. STOKES' THEOREM WITH SINGULARITIES

If X is a compact manifold, then of course every differential form on X has compact support. However, the version of Stokes' theorem which we have given is useful in contexts when we start with an object which is not a manifold, say as a subset of $\mathbf{R}^n$, but is such that when we remove a portion of it, what remains is a manifold. For instance, consider a cone (say the solid cone) as illustrated in the next picture.

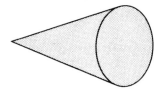

The vertex and the circle surrounding the base disc prevent the cone from being a submanifold of $\mathbf{R}^3$. However, if we delete the vertex and this circle, what remains is a submanifold with boundary embedded in $\mathbf{R}^3$. The boundary conists of the conical shell, and of the base disc (without its surrounding circle). Another example is given by polyhedra, as on the following figure.

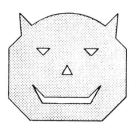

The idea is to approximate a given form by a form with compact support, to which we can apply Theorem 2.1, and then take the limit. We shall indicate one possible technique to do this.

The word "boundary" has been used in two senses: The sense of point set topology, and the sense of boundary of a manifold. Up to now, they were used in different contexts so no confusion could arise. We must now make a distinction, and therefore use the word boundary only in its manifold sense. If X is a subset of $\mathbf{R}^N$, we denote its closure by $\overline{X}$

as usual. We call the set-theoretic difference $\overline{X} - X$ the **frontier** of X in $\mathbf{R}^N$, and denote it by $\text{fr}(X)$.

Let X be a submanifold without boundary of $\mathbf{R}^N$, of dimension n. We know that this means that at each point of X there exists a chart for an open neighborhood of this point in $\mathbf{R}^N$ such that the points of X in this chart correspond to a factor in a product. A point P of $\overline{X} - X$ will be called a **regular** frontier point of X if there exists a chart at P in $\mathbf{R}^N$ with local coordinates $(x_1, \ldots, x_N)$ such that P has coordinates $(0, \ldots, 0)$; the points of X are those with coordinates

$$x_{n+1} = \cdots = x_N = 0 \qquad \text{and} \qquad x_n < 0;$$

and the points of the frontier of X which lie in the chart are those with coordinates satisfying

$$x_n = x_{n+1} = \cdots = x_N = 0.$$

The set of all regular frontier points of X will be denoted by ∂X, and will be called the **boundary** of X. We may say that $X \cup \partial X$ is a submanifold of $\mathbf{R}^N$, possibly with boundary.

A point of the frontier of X which is not regular will be called **singular**. It is clear that the set of singular points is closed in $\mathbf{R}^N$. We now formulate a version of Theorem 2.1 when ω does not necessarily have compact support in $X \cup \partial X$. Let S be a subset of $\mathbf{R}^N$. By a **fundamental sequence** of open neighborhoods of S we shall mean a sequence $\{U_k\}$ of open sets containing S such that, if W is an open set containing S, then $U_k \subset W$ for all sufficiently large k.

Let S be the set of singular frontier points of X and let ω be a form defined on an open neighborhood of $\overline{X}$, and having compact support. The intersection of $\text{supp } \omega$ with $(X \cup \partial X)$ need not be compact, so that we cannot apply Theorem 2.1 as it stands. The idea is to find a fundamental sequence of neighborhods $\{U_k\}$ of S, and a function g_k which is 0 on a neighborhood of S and 1 outside U_k so that $g_k \omega$ differs from ω only inside U_k. We can then apply Theorem 2.1 to $g_k \omega$ and we hope that taking the limit yields Stokes' theorem for ω itself. However, we have

$$\int_X d(g_k \omega) = \int_X dg_k \wedge \omega + \int_X g_k \, d\omega.$$

Thus we have an extra term on the right, which should go to 0 as $k \to \infty$ if we wish to apply this method. In view of this, we make the following definition.

Let S be a closed subset of $\mathbf{R}^N$. We shall say that S is **negligible for** X if there exists an open neighborhood U of S in $\mathbf{R}^N$, a fundamental sequence of open neighborhoods $\{U_k\}$ of S in U, with $\overline{U}_k \subset U$, and a sequence of C^1 functions $\{g_k\}$, having the following properties.

NEG 1. *We have $0 \leq g_k \leq 1$. Also, $g_k(x) = 0$ for x in some open neighborhood of S, and $g_k(x) = 1$ for $x \notin U_k$.*

NEG 2. *If ω is an $(n-1)$-form of class C^1 on U, and μ_k is the measure associated with $|dg_k \wedge \omega|$ on $U \cap X$, then μ_k is finite for large k, and*

$$\lim_{k \to \infty} \mu_k(U \cap X) = 0.$$

From our first condition, we see that $g_k\omega$ vanishes on an open neighborhood of S. Since $g_k = 1$ on the complement of $\overline{U}_k$, we have $dg_k = 0$ on this complement, and therefore our second condition implies that the measures induced on X near the singular frontier by $|dg_k \wedge \omega|$ (for $k = 1$, $2, \dots$), are concentrated on shrinking neighborhoods and tend to 0 as $k \to \infty$.

Theorem 3.1 (Stokes' Theorem with Singularities). *Let X be an oriented, C^3 submanifold without boundary of $\mathbf{R}^N$. Let $\dim X = n$. Let ω be an $(n-1)$-form of class C^1 on an open neighborhood of $\overline{X}$ in $\mathbf{R}^N$, and with compact support. Assume that:*

(i) *If S is the set of singular points in the frontier of X, then $S \cap \operatorname{supp} \omega$ is negligible for X.*

(ii) *The measures associated with $|d\omega|$ on X, and $|\omega|$ on ∂X, are finite.*

Then

$$\int_X d\omega = \int_{\partial X} \omega.$$

Proof. Let U, $\{U_k\}$, and $\{g_k\}$ satisfy conditions **NEG 1** and **NEG 2**. Then $g_k\omega$ is 0 on an open neighborhood of S, and since ω is assumed to have compact support, one verifies immediately that

$$(\operatorname{supp} g_k\omega) \cap (X \cup \partial X)$$

is compact. Thus Theorem 2.1 is applicable, and we get

$$\int_{\partial X} g_k\omega = \int_X d(g_k\omega) = \int_X dg_k \wedge \omega + \int_X g_k \, d\omega.$$

We have

$$\left| \int_{\partial X} \omega - \int_{\partial X} g_k\omega \right| \leq \left| \int_{\partial X} (1 - g_k)\omega \right|$$

$$\leq \int_{U_k \cap \partial X} 1 \, d\mu_{|\omega|} = \mu_{|\omega|}(U_k \cap \partial X).$$

Since the intersection of all sets $U_k \cap \partial X$ is empty, it follows from purely measure-theoretic reasons that the limit of the right-hand side is 0 as $k \to \infty$. Thus

$$\lim_{k \to \infty} \int_{\partial X} g_k \omega = \int_{\partial X} \omega.$$

For similar reasons, we have

$$\lim_{k \to \infty} \int_X g_k \, d\omega = \int_X d\omega.$$

Our second assumption **NEG 2** guarantees that the integral of $dg_k \wedge \omega$ over X approaches 0. This proves our theorem.

Criterion 1. *Let S, T be compact negligible sets for a submanifold X of $\mathbf{R}^N$ (assuming X without boundary). Then the union $S \cup T$ is negligible for X.*

Proof. Let U, $\{U_k\}$, $\{g_k\}$ and V, $\{V_k\}$, $\{h_k\}$ be triples associated with S and T respectively as in condition **NEG 1** and **NEG 2** (with V replacing U and h replacing g when T replaces S). Let

$$W = U \cup V, \qquad W_k = U_k \cup V_k, \qquad \text{and} \qquad f_k = g_k h_k.$$

Then the open sets $\{W_k\}$ form a fundamental sequence of open neighborhoods of $S \cup T$ in W, and **NEG 1** is trivially satisfied. As for **NEG 2**, we have

$$d(g_k h_k) \wedge \omega = h_k \, dg_k \wedge \omega + g_k \, dh_k \wedge \omega,$$

so that **NEG 2** is also trivially satisfied, thus proving our criterion.

Criterion 2. *Let X be an open set, and let S be a compact subset in $\mathbf{R}^n$. Assume that there exists a closed rectangle R of dimension $m \leq n - 2$ and a C^1 map $\sigma: R \to \mathbf{R}^n$ such that $S = \sigma(R)$. Then S is negligible for X.*

Before giving the proof, we make a couple of simple remarks. First, we could always take $m = n - 2$, since any parametrization by a rectangle of dimension $< n - 2$ can be extended to a parametrization by a rectangle of dimension $n - 2$ simply by projecting away coordinates. Second, by our first criterion, we see that a finite union of sets as described above, that is parametrized smoothly by rectangles of codimension ≥ 2, are negligible. Third, our Criterion 2, combined with the first criterion, shows that negligibility in this case is local, that is we can subdivide a rectangle into small pieces.

We now prove Criterion 2. Composing σ with a suitable linear map, we may assume that R is a unit cube. We cut up each side of the cube into k equal segments and thus get k^m small cubes. Since the derivative of σ is bounded on a compact set, the image of each small cube is contained in an n-cube in $\mathbf{R}^N$ of radius $\leq C/k$ (by the mean value theorem), whose n-dimensional volume is $\leq (2C)^n/k^n$. Thus we can cover the image by small cubes such that the sum of their n-dimensional volumes is

$$\leq (2C)^n/k^{n-m} \leq (2C)^n/k^2.$$

Lemma 3.2. *Let S be a compact subset of $\mathbf{R}^n$. Let U_k be the open set of points x such that $d(x, S) < 2/k$. There exists a C^∞ function g_k on $\mathbf{R}^N$ which is equal to 0 in some open neighborhood of S, equal to 1 outside U_k, $0 \leq g_k \leq 1$, and such that all partial derivatives of g_k are bounded by $C_1 k$, where C_1 is a constant depending only on n.*

Proof. Let φ be a C^∞ function such that $0 \leq \varphi \leq 1$, and

$$\varphi(x) = 0 \qquad \text{if} \quad 0 \leq \|x\| \leq \tfrac{1}{2},$$

$$\varphi(x) = 1 \qquad \text{if} \quad 1 \leq \|x\|.$$

We use $\|\ \|$ for the sup norm in $\mathbf{R}^n$. The graph of φ looks like this:

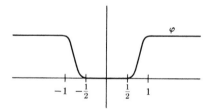

For each positive integer k, let $\varphi_k(x) = \varphi(kx)$. Then each partial derivative $D_i\varphi_k$ satisfies the bound

$$\|D_i\varphi_k\| \leq k\|D_i\varphi\|,$$

which is thus bounded by a constant times k. Let L denote the lattice of integral points in $\mathbf{R}^n$. For each $l \in L$, we consider the function

$$x \mapsto \varphi_k\left(x - \frac{l}{2k}\right).$$

This function has the same shape as φ_k but is translated to the point $l/2k$. Consider the product

$$g_k(x) = \prod \varphi_k\left(x - \frac{l}{2k}\right)$$

taken over all $l \in L$ such that $d(l/2k, S) \leq 1/k$. If x is a point of $\mathbf{R}^n$ such that $d(x, S) < 1/4k$, then we pick an l such that

$$d(x, l/2k) \leq 1/2k.$$

For this l we have $d(l/2, S) < 1/k$, so that this l occurs in the product, and

$$\varphi_k(x - l/2k) = 0.$$

Therefore g_k is equal to 0 in an open neighborhood of S. If, on the other hand, we have $d(x, S) > 2/k$ and if l occurs in the product, that is

$$d(l/2k, S) \leq 1/k,$$

then

$$d(x, l/2k) > 1/k,$$

and hence $g_k(x) = 1$. The partial derivatives of g_k are bounded in the desired manner. This is easily seen, for if x_0 is a point where g_k is not identically 1 in a neighborhood of x_0, then $\|x_0 - l_0/2k\| \leq 1/k$ for some l_0. All other factors $\varphi_k(x - 1/2k)$ will be identically 1 near x_0 unless $\|x_0 - l/2k\| \leq 1/k$. But then $\|l - l_0\| \leq 4$ whence the number of such l is bounded as a function of n (in fact by 9^n). Thus when we take the derivative, we get a sum of at most 9^n terms, each one having a derivative bounded by $C_1 k$ for some constant C_1. This proves our lemma.

We return to the proof of Criterion 2. We observe that when an $(n-1)$-form ω is expressed n terms of its coordinates,

$$\omega(x) = \sum f_j(x) \, dx_1 \wedge \cdots \wedge \widehat{dx_j} \wedge \cdots \wedge dx_n,$$

then the coefficients f_j are bounded on a compact neighborhood of S. We take U_k as in the lemma. Then for k large, each function

$$x \mapsto f_j(x) D_j g_k(x)$$

is bounded on U_k by a bound $C_2 k$, where C_2 depends on a bound for ω, and on the constant of the lemma. The Lebesgue measure of U_k is bounded by C_3/k^2, as we saw previously. Hence the measure of U_k associated with $|dg_k \wedge \omega|$ is bounded by C_4/k, and tends to 0 as $k \to \infty$. This proves our criterion.

As an example, we now state a simpler version of Stokes' theorem, applying our criteria.

Theorem 3.3. *Let X be an open subset of $\mathbf{R}^n$. Let S be the set of singular points in the closure of X, and assume that S is the finite union* *of C^1 images of m-rectangles with $m \leq n - 2$. Let ω be an $(n - 1)$-form defined on an open neighborhood of $\overline{X}$. Assume that ω has compact support, and that the measure associated with $|\omega|$ on ∂X and with $|d\omega|$ on X are finite. Then*

$$\int_X d\omega = \int_{\partial X} \omega.$$

Proof. Immediate from our two criteria and Theorem 2.

We can apply Theorem 3.3 when, for instance, X is the interior of a polyhedron, whose interior is open in $\mathbf{R}^n$. When we deal with a submanifold X of dimension n, embedded in a higher dimensional space $\mathbf{R}^N$, then one can reduce the analysis of the singular set to Criterion 2 provided that there exists a finite number of charts for X near this singular set on which the given form ω is bounded. This would for instance be the case with the surface of our cone mentioned at the beginning of the section. Criterion 2 is also the natural one when dealing with manifolds defined by algebraic inequalities. By using Hironaka's resolution of singularities, one can parametrize a compact set of algebraic singularities as in Criterion 2.

Finally, we note that the condition that ω have compact support in an open neighborhood of $\overline{X}$ is a very mild condition. If for instance X is a bounded open subset of $\mathbf{R}^n$, then $\overline{X}$ is compact. If ω is any form on some open set containing $\overline{X}$, then we can find another form η which is equal to ω on some open neighborhood of $\overline{X}$ and which has compact support. The integrals of η entering into Stokes' formula will be the same as those of ω. To find η, we simply multiply ω with a suitable C^∞ function which is 1 in a neighborhood of $\overline{X}$ and vanishes a little further away. Thus Theorem 3.3 provides a reasonably useful version of Stokes' theorem which can be applied easily to all the cases likely to arise naturally.

CHAPTER XIII

Applications of Stokes' Theorem

In this chapter we give a survey of applications of Stokes' theorem, concerning many situations. Some come just from the differential theory, such as the computation of the maximal de Rham cohomology (the space of all forms of maximal degree modulo the subspace of exact forms); some come from Riemannian geometry; and some come from complex manifolds, as in Cauchy's theorem and the Poincaré residue theorem. I hope that the selection of topics will give readers an outlook conducive for further expansion of perspectives. The sections of this chapter are logically independent of each other, so the reader can pick and choose according to taste or need.

XIII, §1. THE MAXIMAL DE RHAM COHOMOLOGY

Let X be a manifold of dimension n without boundary. Let r be an integer ≥ 0. We let $\mathscr{A}^r(X)$ be the **R**-vector space of differential forms on X of degree r. Thus $\mathscr{A}^r(X) = 0$ if $r > n$. If $\omega \in \mathscr{A}^r(X)$, we define the **support** of ω to be the closure of the set of points $x \in X$ such that $\omega(x) \neq 0$.

Examples. If $\omega(x) = f(x)\,dx_1 \wedge \cdots \wedge dx_n$ on some open subset of $\mathbf{R}^n$, then the support of ω is the closure of the set of x such that $f(x) \neq 0$.

We denote the support of a form ω by $\mathrm{supp}(\omega)$. By definition, the support is closed in X. We are interested in the space of maximal degree forms $\mathscr{A}^n(X)$. Every form $\omega \in \mathscr{A}^n(X)$ is such that $d\omega = 0$. On the other hand, $\mathscr{A}^n(X)$ contains the subspace of **exact** forms, which are defined to be those forms equal to $d\eta$ for some $\eta \in \mathscr{A}^{n-1}(X)$. The factor space is defined to be the **de Rham cohomology** $H^n(X) = H^n(X, \mathbf{R})$. The main theorem of this section can then be formulated.

Theorem 1.1. *Assume that X is compact, orientable, and connected. Then the map*

$$\omega \mapsto \int_X \omega$$

induces an isomorphism of $H^n(X)$ with $\mathbf{R}$ itself. In particular, if ω is in $\mathscr{A}^n(X)$ then there exists $\eta \in \mathscr{A}^{n-1}(X)$ such that $d\eta = \omega$ if and only if

$$\int_X \omega = 0.$$

Actually the hypothesis of compactness on X is not needed. What is needed is compactness on the support of the differential forms. Thus we are led to define $\mathscr{A}_c^r(X)$ to be the vector space of n-forms with compact support. We call a form **compactly exact** if it is equal to $d\eta$ for some $\eta \in \mathscr{A}_c^{r-1}(X)$. We let

$$H_c^n(X) = \text{factor space } \mathscr{A}_c^n(X)/d\mathscr{A}_c^{n-1}(X).$$

Then we have the more general version:

Theorem 1.2. *Let X be a manifold without boundary, of dimension n. Suppose that X is orientable and connected. Then the map*

$$\omega \mapsto \int_X \omega$$

induces an isomorphism of $H_c^n(X)$ with $\mathbf{R}$ itself.

Proof. By Stokes' theorem (Chapter XI, Corollary 2.2) the integral vanishes on exact forms (with compact support), and hence induces an $\mathbf{R}$-linear map of $H_c^n(X)$ into $\mathbf{R}$. The theorem amounts to proving the converse statement: if

$$\int_X \omega = 0,$$

then there exists some $\eta \in \mathscr{A}_c^{n-1}(X)$ such that $\omega = d\eta$. For this, we first have to prove the result locally in $\mathbf{R}^n$, which we now do.

As a matter of notation, we let

$$I^n = (0, 1)^n$$

be the open n-cube in $\mathbf{R}^n$. What we want is:

Lemma 1.3. *Let ω be an n-form on I^n, with compact support, and such that*

$$\int_{I^n} \omega = 0.$$

Then there exists a form $\eta \in \mathscr{A}_c^{n-1}(I^{n-1})$ with compact support, such that

$$\omega = d\eta.$$

We will prove Lemma 1.3 by induction, but it is necessary to load the induction to carry it out. So we need to prove a stronger version of Lemma 1.3 as follows.

Lemma 1.4. *Let ω be an $(n-1)$-form on I^{n-1} whose coefficient is a function of n variables $(x_1, \ldots, x_n)$ so*

$$\omega(x) = f(x_1, \ldots, x_n)\, dx_1 \wedge \cdots \wedge dx_{n-1}.$$

(Of course, all functions, like forms, are assumed C^∞.) Suppose that ω has compact support in I^{n-1}. Assume that

$$\int_{I^{n-1}} \omega = 0.$$

Then there exists an $(n-2)$-form η, whose coefficients are C^∞ functions of $x_1, \ldots, x_n$ such that

$$\omega(x_1, \ldots, x_{n-1}; x_n) = d_{n-1}\eta(x_1, \ldots, x_{n-1}; x_n).$$

The symbol d_{n-1} here means the usual exterior derivative taken with respect to the first $n-1$ variables.

Proof. By induction. We first prove the theorem when $n-1 = 1$. First we carry out the proof leaving out the extra variable, just to see what's going on. So let

$$\omega(x) = f(x)\, dx,$$

where f has compact support in the open interval $(0, 1)$. This means

there exists $\varepsilon > 0$ such that $f(x) = 0$ if $0 < x \leq \varepsilon$ and if $1 - \varepsilon \leq x \leq 1$.
We assume

$$\int_0^1 f(x)\, dx = 0.$$

Let

$$g(x) = \int_0^x f(t)\, dt.$$

Then $g(x) = 0$ if $0 < x \leq \varepsilon$, and also if $1 - \varepsilon \leq x \leq 1$, because for instance
if $1 - \varepsilon \leq x \leq 1$, then

$$g(x) = \int_0^1 f(t)\, dt = 0.$$

Then $f(x)\, dx = dg(x)$, and the lemma is proved in this case. Note that
we could have carried out the proof with the extra variable x_2, starting
from

$$\omega(x) = f(x_1, x_2)\, dx_1,$$

so that

$$g(x_1, x_2) = \int_0^1 f(t, x_2)\, dt.$$

We can differentiate under the integral sign to verify that g is C^∞ in the
pair of variables (x_1, x_2).

Now let $n \geq 3$ and assume the theorem proved for $n - 1$ by induction.
To simplify the notation, let us omit the extra variable x_{n+1}, and write

$$\omega(x) = f(x_1, \ldots, x_n)\, dx_1 \wedge \cdots \wedge dx_n,$$

with compact support in I^n. Then there exists $\varepsilon > 0$ such that the support of f is contained in the closed cube

$$\bar{I}^n(\varepsilon) = [\varepsilon, 1 - \varepsilon]^n.$$

The following figure illustrates this support in dimension 2.

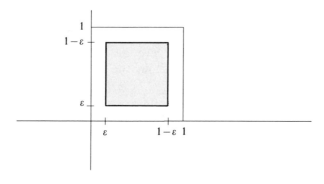

Let ψ be an $(n-1)$-form on I^{n-1}, $\psi(x) = \psi(x_1, \ldots, x_{n-1})$ such that

$$\int_{I^{n-1}} \psi = 1,$$

and ψ has compact support. Let

$$g(x_n) = \int_{I^{n-1}} f(x_1, \ldots, x_{n-1}; x_n) \, dx_1 \wedge \cdots \wedge dx_{n-1}$$

$$= \int_{\bar{I}^{n-1}(\varepsilon)} f(x_1, \ldots, x_{n-1}; x_n) \, dx_1 \wedge \cdots \wedge dx_{n-1}.$$

Note here that we do have the parameter x_n coming in at the inductive step. Let

$$\mu(x) = f(x) \, dx_1 \wedge \cdots \wedge dx_{n-1} - g(x_n)\psi(x_1, \ldots, x_{n-1}),$$

so

(*) $\mu(x) \wedge dx_n = \omega(x) - g(x_n)\psi(x) \wedge dx_n.$

Then

$$\int_{I^{n-1}} \mu = g(x_n) - g(x_n) = 0.$$

Furthermore, since f has compact support, so does g (look at the figure). By induction, there exists an $(n-1)$-form η, of the first $n-1$ variables, but depending on the parameter x_n, that is

$$\eta(x) = \eta(x_1, \ldots, x_{n-1}; x_n)$$

such that

$$\mu(x_1, \ldots, x_{n-1}; x_n) = d_{n-1}\eta(x_1, \ldots, x_{n-1}; x_n).$$

Here d_{n-1} denotes the exterior derivative with respect to the first $n-1$ variables. Then trivially,

$$\mu(x_1, \ldots, x_{n-1}; x_n) \wedge dx_n = d_{n-1}\eta(x_1, \ldots, x_{n-1}; x_n) \wedge dx_n$$

$$= d\eta(x),$$

where $d\eta$ is now the exterior derivative taken with respect to all n variables. Hence finally from equation (*) we obtain

(**) $\omega(x) = d\eta(x) + g(x_n)\psi(x_1, \ldots, x_{n-1}) \wedge dx_n.$

To conclude the proof of Lemma 1.3, it suffices to show that the second term on the right of (∗∗) is exact. We are back to a one-variable problem. Let

$$h(x_n) = \int_0^{x_n} g(t)\, dt.$$

Then $dh(x_n) = g(x_n)\, dx_n$, and h has compact support in the interval $(0, 1)$, just as in the start of the induction. Then

$$d\big(h(x_n)\psi(x_1,\dots,x_{n-1})\big) = dh(x_n) \wedge \psi(x_1,\dots,x_{n-1})$$
$$= (-1)^{n-1} g(x_n)\psi(x_1,\dots,x_{n-1}) \wedge dx_n$$

because $d\psi = 0$. Of course we could have carried along the extra parameter all the way through. This concludes the proof of Lemma 1.3.

We formulate to an immediate consequence of Lemma 1.3 directly on the manifold.

Lemma 1.5. *Let U be an open subset of X, isomorphic to I^n. Let $\psi \in \mathscr{A}_c^n(U)$ be such that*

$$\int_U \psi \neq 0.$$

Let $\omega \in \mathscr{A}_c^n(U)$. Then there exists $c \in \mathbf{R}$ and $\eta \in \mathscr{A}_c^{n-1}(U)$ such that

$$\omega - c\psi = d\eta.$$

Proof. We take $c = \int_U \omega \Big/ \int_U \psi$ and apply Lemma 1.3 to $\omega - c\psi$.

Observe that the hypothesis of connectedness has not yet entered the picture. The preceding lemmas were purely local. We now globalize.

Lemma 1.6. *Assume that X is connected and oriented. Let U, ψ be as in Lemma 1.5. Let V be the set of points $x \in X$ having the following property. There exists a neighborhood $U(x)$ of x isomorphic to I^n such that for every $\omega \in \mathscr{A}_c^n(U(x))$ there exist $c \in \mathbf{R}$ and $\eta \in \mathscr{A}_c^{n-1}(X)$ such that*

$$\omega - c\psi = d\eta.$$

Then $V = X$.

Proof. Lemma 1.5 asserts that $V \supset U$. Since X is connected, it suffices to prove that V is both open and closed. It is immediate from the definition of V that V is open, so there remains to prove its closure. Let z be in the closure of V. Let W be a neighborhood of z isomorphic to I^n. There exists a point $x \in V \cap W$. There exists a neighborhood $U(x)$ as in the definition of V such that $U(x) \subset W$. For instance, we may take

$$U(x) \approx (a_1, b_1) \times \cdots \times (a_n, b_n) \approx I^n$$

with a_i sufficiently close to 0 and b_i sufficiently close to 1, and of course $0 < a_i < b_i$ for $i = 1, \ldots, n$. Let $\psi_1 \in \mathcal{A}_c^n(U(x))$ be such that

$$\int_{U(x)} \psi_1 = 1.$$

Let $\omega \in \mathcal{A}_c^n(W)$. By the definition of V, there exist $c_1 \in \mathbf{R}$ and $\eta_1 \in \mathcal{A}_c^n(X)$ such that

$$\psi_1 - c_1 \psi = d\eta_1.$$

By Lemma 1.5, there exists $c_2 \in \mathbf{R}$ and $\eta_2 \in \mathcal{A}_c^n(X)$ such that

$$\omega - c_2 \psi_1 = d\eta_2.$$

Then

$$\omega - c_2 c_1 \psi = d(\eta_2 + c_2 \eta_1),$$

thus concluding the proof of Lemma 1.6.

We have now reached the final step in the proof of Theorem 5.2, namely we first fix a form $\psi \in \mathcal{A}_c^n(U)$ with $U \approx I^n$ and $\int_X \psi \neq 0$. Let $\omega \in \mathcal{A}_c^n(X)$. It suffices to prove that there exist $c \in \mathbf{R}$ and $\eta \in \mathcal{A}_c^{n-1}(X)$ such that

$$\omega - c\psi = d\eta.$$

Let K be the compact support of ω. Cover K be a finite number of open neighborhoods $U(x_1), \ldots, U(x_m)$ satisfying the property of Lemma 1.6. Let $\{\varphi_i\}$ be a partition of unity subordinated to this covering, so that we can write

$$\omega = \sum \varphi_i \omega.$$

Then each form $\varphi_i \omega$ has support in some $U(x_j)$. Hence by Lemma 5.6, there exist $c_i \in \mathbf{R}$ and $\eta_i \in \mathcal{A}_c^{n-1}(X)$ such that

$$\varphi_i \omega - c_i \psi = d\eta_i,$$

whence $\omega - c\psi = d\eta$, with $c = \sum c_i$ and $\eta = \sum \eta_i$. This concludes the proof of Theorems 1.1 and 1.2.

XIII, §2. MOSER'S THEOREM

We return here to the techniques of proof in Chapter V, as for Poincaré's lemma, Theorem 5.1 and Darboux's Theorem 7.3 of that chapter. However, we now have a similar theorem in the context of integration.

We first make the general comment, similar to the one we made previously, for general forms. Let $\mathbf{E}$ be a Banach space, and let ω be an r-multilinear alternating form on $\mathbf{E}$ (so $\mathbf{R}$-valued). We say that ω is non-singular if for each vector $v \in \mathbf{E}$, defining ω_v by

$$\omega_v \colon (v_1, \ldots, v_{r-1}) \mapsto (v, v_1, \ldots, v_{r-1}),$$

the map $v \mapsto \omega_v$ is a toplinear isomorphismm between $\mathbf{E}$ and $L_a^{r-1}(\mathbf{E})$. We previously considered bilinear forms, in Chapter V, §6.

We can globalize the notion to a manifold, so a form $\omega \in \mathscr{A}^r(X)$ is called **non-singular** if $\omega(x)$ is non-singular for each x. It is clear that in the finite dimensional case, a volume form is non-singular. With this globalization, we obtain:

Proposition 2.1. *Let ω be a non-singular r-form on X. Given a form $\eta \in \mathscr{A}^{r-1}(X)$, there exists a unique vector field ξ such that*

$$\omega \circ \xi = \eta.$$

We could also write the relation with the contraction notation, i.e. $C_\xi \omega = \eta$.

We now come to Moser's theorem [Mo 65].

Theorem 2.2. *Let X be a compact, connected oriented manifold of dimension n. Let $\omega, \psi \in \mathscr{A}^n(X) \left(= \mathscr{A}_c^n(X)\right)$ be volume forms such that*

$$\int_X \omega = \int_X \psi.$$

Then there exists an automorphism $f \colon X \to X$ of X such that $\omega = f^*\psi$.

Proof. Let

$$\omega_s = (1 - s)\omega + s\psi \qquad \text{for} \quad 0 \le s \le 1.$$

Then ω_s is a volume form for each s, and in particular is non-singular. By Theorem 1.1, there exists $\eta \in \mathscr{A}^{n-1}(X)$ such that $\psi - \omega = d\eta$. Note also that $\psi - \omega = d\omega_s/ds$. Since ω_s is non-singular, there exists a unique vector field ξ_s such that

$$\omega_s \circ \xi_s = -\eta.$$

Let α_s be the flow of ξ_s. Then α_s is defined on $\mathbf{R} \times X$ by Corollary 2.4 of Chapter IV. Then we get:

$$\frac{d}{ds}(\alpha_s^* \omega_s) = \frac{d}{du}(\alpha_u^* \omega_s)\bigg|_{u=s} + \alpha_s^*\left(\frac{d\omega_s}{ds}\right)$$

$$= \alpha_s^* d(\omega_s \circ \xi_s) + \alpha_s^*(\psi - \omega) \quad \text{by Proposition 5.2 of Chapter V}$$

$$= -\alpha_s^* \, d\eta + \alpha_s^* \, d\eta$$

$$= 0.$$

Therefore $\alpha_s^* \omega_s$ is constant as a function of s, so we find

$$\omega = \alpha_0^* \omega_0 = \alpha_1^* \omega_1 = f^* \psi, \qquad \text{with} \quad f = \alpha_1,$$

thereby proving the theorem.

XIII, §3. THE DIVERGENCE THEOREM

Let X be an oriented manifold of dimension n possibly with boundary, and let Ω be an n-form on X. Let ξ be a vector field on X. Then $d\Omega = 0$, and hence the basic formula for the Lie derivative (Chapter V, Proposition 5.3) shows that

$$\mathscr{L}_\xi \Omega = d(\Omega \circ \xi).$$

Consequently in this case, Stokes' theorem yields:

Theorem 3.1 (Divergence Theorem).

$$\boxed{\int_X \mathscr{L}_\xi \Omega = \int_{\partial X} \Omega \circ \xi.}$$

Remark. Even if the manifold is not orientable, it is possible to use the notion of density to formulate a Stokes theorem for densities. Cf. Loomis–Sternberg [Los 68] for the formulation, due to Rasala. However, this formulation reduces at once to a local question (using partitions of unity on densities). Since locally every manifold is orientable, and a density then amounts to a differential form, this more general formulation again reduces to the standard one on an orientable manifold.

Suppose that (X, g) is a Riemannian manifold, assumed oriented for simplicity. We let Ω or vol_g be the volume form defined in Chapter X, §1. Let ω be the canonical Riemannian volume form on ∂X for the metric induced by g on the boundary. Let $\mathbf{n}_x$ be a unit vector in the tangent space $T_x(X)$ such that u is perpendicular to $T_x(\partial X)$. Such a unit vector is determined up to sign. Denote by $\mathbf{n}_x^\vee$ its dual functional, i.e. the component on the projection along $\mathbf{n}_x$. *We select $\mathbf{n}_x$ with the sign such that*

$$\mathbf{n}_x^\vee \wedge \omega(x) = \Omega(x).$$

We then shall call $\mathbf{n}_x$ the **unit outward normal vector** to the boundary at x. In an oriented chart, it looks like this.

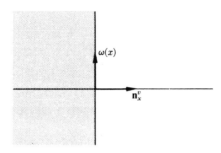

Then by formula **CON 3** of Chapter V, §5 we find

$$\Omega \circ \xi = \langle \mathbf{n}, \xi \rangle \omega - \mathbf{n}^\vee \wedge (\omega \circ \xi),$$

and the restriction of this form to ∂X is simply $\langle \mathbf{n}, \xi \rangle \omega$. Thus we get:

Theorem 3.2 (Gauss Theorem). *Let X be a Riemannian manifold. Let ω be the canonical Riemannian volume form on ∂X and let Ω be the canonical Riemannian volume form on X itself. Let $\mathbf{n}$ be the unit outward normal vector field to the boundary, and let ξ be a C^1 vector field on X, with compact support. Then*

$$\int_X (\mathrm{div}_\Omega \, \xi)\Omega = \int_{\partial X} \langle \mathbf{n}, \xi \rangle \omega.$$

The next thing is to show that the map d^* from Chapter X, §1 is the adjoint for a scalar product defined by integration. First we expand slightly the formalism of d^* for this application. Recall that for any vector field ξ, the divergence of ξ is defined by the property

(1) $$d(\mathrm{vol}_g \circ \xi) = (\mathrm{div} \, \xi) \, \mathrm{vol}_g.$$

Note the trivial derivation formula for a function φ:

(2) $$\mathrm{div}(\varphi \xi) = \varphi \, \mathrm{div} \, \xi + (d\varphi)(\xi).$$

If λ is a 1-form, i.e. in $\Gamma L^1(TX) = \mathcal{A}^1(X)$, we have the corresponding vector field $\xi_\lambda = \lambda^\vee$ uniquely determined by the condition that

$$\langle \xi_\lambda, \eta \rangle_g = \lambda(\eta) \qquad \text{for all vector fields } \eta.$$

For a 1-form λ, we define the operator

$$d^*\colon \mathcal{A}^1(X) \to \mathcal{A}^0(X) = \mathfrak{F}(X) \qquad \text{by} \qquad d^*\lambda = -\operatorname{div} \xi_\lambda,$$

so by (1),

(3) $$(d^*\lambda) \operatorname{vol}_g = d(\operatorname{vol}_g \circ \xi_\lambda).$$

We get a formula analogous to (2) for d^*, namely

(4) $$d^*(\varphi\lambda) = \varphi d^*\lambda - \langle d\varphi, \lambda \rangle.$$

Indeed, $d^*(\varphi\lambda) = -\operatorname{div} \xi_{\varphi\lambda} = -\operatorname{div}(\varphi\xi_\lambda) = -\varphi \operatorname{div} \xi_\lambda - (d\varphi)(\xi_\lambda)$ by (2), which proves the formula.

Let $\lambda, \omega \in \mathcal{A}^1(TX)$. We define the **scalar product** via duality

$$\langle \lambda, \omega \rangle_g = \langle \xi_\lambda, \xi_\omega \rangle_g.$$

Then for a function φ we have the formula

(5) $$\langle d\varphi, \lambda \rangle_g \operatorname{vol}_g = (\varphi d^*\lambda) \operatorname{vol}_g - d(\operatorname{vol}_g \circ \varphi\xi_\lambda).$$

Indeed,

$$\langle d\varphi, \lambda \rangle_g \operatorname{vol}_g = [\varphi d^*\lambda - d^*(\varphi\lambda)] \operatorname{vol}_g \qquad \text{by (4)}$$

$$= (\varphi d^*\lambda) \operatorname{vol}_g - d(\operatorname{vol}_g \circ \varphi\xi_\lambda) \quad \text{by (3)}$$

thus proving (5). Note that the congruence of the two forms $\langle d\varphi, \lambda \rangle_g \operatorname{vol}_g$ and $(\varphi d^*\lambda) \operatorname{vol}_g$ modulo exact forms is significant, and is designed for Proposition 3.3 below.

Observe that the scalar product between two forms above is a function, which when multiplied by the volume form vol_g may be integrated over X. Thus we define the **global scalar product** on 1-forms with compact support to be

$$\langle \lambda, \omega \rangle_{(X,g)} = \langle \lambda, \omega \rangle_X = \int_X \langle \lambda, \omega \rangle_g \operatorname{vol}_g.$$

Applying Stokes' theorem, we then find:

Proposition 3.3. *Let* (X, g) *be a Riemannian manifold, oriented and without boundary. Then* d^* *is the adjoint of* d *with respect to the global scalar product, i.e.*

$$\langle d\varphi, \lambda \rangle_X = \langle \varphi, d^*\lambda \rangle_X.$$

We define the **Laplacian** (operating on functions) to be the operator

$$\Delta = d^*d.$$

For the Laplacian operating on higher degree forms, we shall give the expression $d^*d + dd^*$ in the next section, but here for functions, the second term disappears.

For a manifold with boundary, we define the **normal derivative** of a function φ to be the function *on the boundary* given by

$$\partial_\mathbf{n}\varphi = \langle \mathbf{n}, \xi_{d\varphi} \rangle_g = \langle \mathbf{n}, \text{grad}_g\, \varphi \rangle_g.$$

Theorem 3.4 (Green's Formula). *Let* (X, g) *be an oriented Riemannian manifold possibly with boundary, and let* φ, ψ *be functions on* X *with compact support. Let* ω *be the canonical volume form associated with the induced metric on the boundary. Then*

$$\int_X (\varphi\Delta\psi - \psi\Delta\varphi)\, \text{vol}_g = -\int_{\partial X} (\varphi\partial_\mathbf{n}\psi - \psi\partial_\mathbf{n}\varphi)\omega.$$

Proof. From formula (4) we get

$$d^*(\varphi\, d\psi) = \varphi\Delta\psi - \langle d\varphi, d\psi \rangle_g,$$

whence

$$\varphi\Delta\psi - \psi\Delta\varphi = d^*(\varphi\, d\psi) - d^*(\psi\, d\varphi)$$
$$= -\text{div}(\varphi\, d\psi) + \text{div}(\psi\, d\varphi).$$

We apply Theorem 3.2 to conclude the proof.

Remark. Of course, if X has no boundary in Theorem 3.7, then the integral on the left side is equal to 0.

Corollary 3.5 (E. Hopf). *Let* X *be a Riemannian manifold without boundary, and let* f *be a* C^2 *function on* X *with compact support, such that* $\Delta f \geqq 0$. *Then* f *is constant. In particular, every harmonic function with compact support is constant.*

Proof. We first give the proof assuming that X is oriented. By Green's formula we get

$$\int_X \Delta f \, \mathrm{vol}_g = 0.$$

Since $\Delta f \geqq 0$, it follows that in fact $\Delta f = 0$, so we are reduced to the harmonic case. We now apply Green's formula to f^2, and get

$$0 = \int_X \Delta f^2 \, \mathrm{vol}_g = \int_X 2f\Delta f \, \mathrm{vol}_g - \int_X 2(\mathrm{grad} \, f)^2 \, \mathrm{vol}_g.$$

Hence $(\mathrm{grad} \, f)^2 = 0$ because $\Delta f = 0$, and finally $\mathrm{grad} \, f = 0$, so $df = 0$ and f is constant, thus proving the corollary in the oriented case. For the non-oriented case, by Proposition 4.6 of Chapter XI, there exists a covering of degree 2 of X which is oriented, and then one can pull back all the objects from X to this covering to conclude the proof in this case.

XIII, §4. THE ADJOINT OF d FOR HIGHER DEGREE FORMS

We extend the results of the preceding section to arbitrary forms. Given the vector space V of dimension n over **R**, with a positive definite scalar product g, we note that the exterior powers $\bigwedge^r V$ are self dual, with a positive definite scalar product such that

$$(V_1 \wedge \cdots \wedge v_r, w_1 \wedge \cdots \wedge w_r) \mapsto \det \langle v_i, w_j \rangle_g.$$

We defined the notion of orientation on V in Chapter VIII, §5, and we now assume that V is oriented.

Proposition 4.1. *Given $1 \leq r \leq n$, there exists a unique isomorphism*

$$*: \bigwedge^r V \to \bigwedge^{n-r} V$$

such that for $\varphi, \psi \in \bigwedge^r V$ we have

$$\langle \varphi, \psi \rangle_g \, \mathrm{vol}_g = \varphi \wedge *\psi.$$

Proof. The proof will give an explicit determination of the isomorphism on the usual basis for $\bigwedge^r V$. Let $I = [i_1 < i_2 < \cdots < i_r]$ be an ordered set of r indices. We let

$$e_I = e_{i_1} \wedge \cdots \wedge e_{i_r}.$$

If I' is another such ordered set with $n - r$ elements, and $I \cup I' = \{1, \ldots, n\}$ then we let ϵ_I be the sign of the permutation (I, J) of $(1, \ldots, n)$. We then define

$$*e_I = \epsilon_I e_{I'},$$

and extend this operation by linearity to all of $\bigwedge^r V$. Then directly from the definition, we see that if J is an ordered set of r indices, then

$$e_I \wedge *e_J = \langle e_I, e_J \rangle e_1 \wedge \cdots \wedge e_n$$
$$= \delta_{IJ} e_1 \wedge \cdots \wedge e_n.$$

Thus on the standard basis elements of $\bigwedge^r V$ the desired relation of the proposition is satisfied. The same relation is therefore satisfied for all elements of $\bigwedge^r V$, as desired.

We define the operator $\mathbf{w}$ on the direct sum $\bigoplus_r \bigwedge^r V$ to have the effect

$$\mathbf{w} = (-1)^{nr+r} \quad \text{on } \bigwedge^r V.$$

Proposition 4.2. *We have* $*\mathbf{w} = \mathbf{w}*$. *If* n *is even, then* $\mathbf{w} = (-1)^r$ *on* $\bigwedge^r V$. *Furthermore,* $** = \mathbf{w}$.

Proof. Direct, simple computations.

We now apply the above to a Riemannian manifold X of dimension n, and to real differential forms. We let

$$\mathscr{A}_c^r(X)$$

be the space of C^∞ differential forms of degree r, with compact support on the manifold. At each point $x \in X$, we use the space $V = T_x^\vee$ (the dual space of the tangent space). The usual operator

$$d: \mathscr{A}_c^r(X) \to \mathscr{A}_c^{r+1}(X)$$

is **R**-linear. By Stokes' theorem, if ω has compact support, then

$$\int_X d\omega = 0.$$

We shall give an application of this fact in a Riemannian context. We have the volume form vol_g (which does not necessarily have compact support) and we define a **scalar product** on $\mathscr{A}_c^r(X)$ by the formula

$$\langle \varphi, \psi \rangle_{X,g} = \langle \varphi, \psi \rangle_X = \int_X \varphi \wedge *\psi = \int_X \langle \varphi, \psi \rangle_g \, \text{vol}_g,$$

where we usually omit the index g and merely write X as in $\langle \varphi, \psi \rangle_X$.

Proposition 4.3. *The exterior derivative d has an adjoint $d*$ with respect to the scalar product $\langle \ , \ \rangle_X$, namely for $\varphi \in \mathscr{A}_c^{r-1}(X)$ and $\psi \in \mathscr{A}_c^r(X)$ we have*

$$\langle d\varphi, \psi \rangle_X = \langle \varphi, d*\psi \rangle_X.$$

Furthermore, the adjoint is given by the explicit formula

$$d* = (-1)^{nr+n+1} *d* \quad \text{on } \mathscr{A}_c^r(X).$$
$$= -*d* \qquad \text{if } n \text{ is even.}$$

Proof. By Stokes' theorem, we have:

$$\int_X d\varphi \wedge *\psi = \int_X d(\varphi \wedge *\psi) - (-1)^{r-1} \int_X \varphi \wedge d*\psi$$
$$= (-1)^r \int_X \varphi \wedge d*\psi.$$

Now

$$(-1)^r \varphi \wedge d*\psi = (-1)^r \varphi \wedge **\mathbf{w}d*\psi$$
$$= (-1)^r \varphi \wedge \mathbf{w}*(*d*)\psi$$
$$= (-1)^{nr+n+1} \varphi \wedge *(*d*)\psi,$$

which proves the proposition.

XIII, §5. CAUCHY'S THEOREM

It is possible to define a complex analytic (analytic, for short) manifold, using open sets in $\mathbf{C}^n$ and charts such that the transition mappings are analytic. Since analytic maps are C^∞, we see that we get a C^∞ manifold, but with an additional structure, and we call such a manifold **complex analytic**. It is verified at once that the analytic charts of such a manifold define an orientation. Indeed, under a complex analytic change of charts, the Jacobian changes by a complex number times its complex conjugate, so changes by a positive real number.

If $z_1, \ldots, z_n$ are the complex coordinates of $\mathbf{C}^n$, then

$$(z_1, \ldots, z_n, \bar{z}_1, \ldots, \bar{z}_n)$$

can be used a C^∞ local coordinates, viewing $\mathbf{C}^n$ as $\mathbf{R}^{2n}$. If $z_k = x_k + iy_k$, then

$$dz_k = dx_k + i\, dy_k \qquad \text{and} \qquad d\bar{z}_k = dx_k - i\, dy_k.$$

Differential forms can then be expressed in terms of wedge products of the dz_k and $d\bar{z}_k$. For instance

$$dz_k \wedge d\bar{z}_k = 2i\, dy_k \wedge dx_k.$$

The complex standard expression for a differential form is then

$$\omega(z) = \sum_{(i,j)} \varphi_{(i,j)}(z)\, dz_{i_1} \wedge \cdots \wedge dz_{i_r} \wedge d\bar{z}_{j_1} \wedge \cdots \wedge d\bar{z}_{j_s}.$$

Under an analytic change of coordinates, one sees that the numbers r and s remain unchanged, and that if $s = 0$ in one analytic chart, then $s = 0$ in any other analytic chart. Similarly for r. Thus we can speak of a form of type (r, s). A form is said to be **analytic** if $s = 0$, that is if it is of type $(r, 0)$.

We can decompose the exterior derivative d into two components. Namely, we note that if ω is of type (r, s), then $d\omega$ is a sum of forms of type $(r + 1, s)$ and $(r, s + 1)$, say

$$d\omega = (d\omega)_{(r+1,s)} + (d\omega)_{(r,s+1)}.$$

We define

$$\partial\omega = (d\omega)_{(r+1,s)} \qquad \text{and} \qquad \bar{\partial}\omega = (d\omega)_{(r,s+1)}.$$

In terms of local coordinates, it is then easy to verify that if ω is decomposable, and is expressed as

$$\omega(z) = \varphi(z)\, dz_{i_1} \wedge \cdots \wedge dz_{i_r} \wedge d\bar{z}_{j_1} \wedge \cdots \wedge d\bar{z}_{j_s} = \varphi\tilde{\omega},$$

then

$$\partial\omega = \sum \frac{\partial\varphi}{\partial z_k}\, dz_k \wedge \tilde{\omega}$$

and

$$\bar{\partial}\omega = \sum \frac{\partial\varphi}{\partial \bar{z}_k}\, d\bar{z}_k \wedge \tilde{\omega}.$$

In particular, we have

$$\frac{\partial}{\partial z_k} = \frac{1}{2}\left(\frac{\partial}{\partial x_k} - i\frac{\partial}{\partial y_k}\right) \qquad \text{and} \qquad \frac{\partial}{\partial \bar{z}_k} = \frac{1}{2}\left(\frac{\partial}{\partial x_k} + i\frac{\partial}{\partial y_k}\right).$$

(*Warning*: Note the position of the plus and minus signs in these expressions.)

Thus we have

$$d = \partial + \bar{\partial},$$

and operating with ∂ or $\bar{\partial}$ follows rules similar to the rules for operating with d.

Note that f is analytic if and only if $\bar{\partial}f = 0$. Similarly, we say that a differential form is **analytic** if in its standard expression, the functions $\varphi_{(i,j)}$ are analytic and the form is of type $(r, 0)$, that is there are no $d\bar{z}_j$ present. Equivalently, this amounts to saying that $\bar{\partial}\omega = 0$. The following extension of Cauchy's theorem to several variables is due to Martinelli.

We let $|z|$ be the euclidean norm,

$$|z| = (z_1\bar{z}_1 + \cdots + z_n\bar{z}_n)^{1/2}.$$

Theorem 5.1 (Cauchy's Theorem). *Let f be analytic on an open set in $\mathbf{C}^n$ containing the closed ball of radius R centered at a point ζ. Let*

$$\omega_k(z) = dz_1 \wedge \cdots \wedge dz_n \wedge d\bar{z}_1 \wedge \cdots \wedge \widehat{d\bar{z}_k} \wedge \cdots \wedge d\bar{z}_n$$

and

$$\omega(z) = \sum_{k=1}^{n} (-1)^k \bar{z}_k \omega_k(z).$$

Let S_R be the sphere of radius R centered at ζ. Then

$$f(\zeta) = \varepsilon(n)\frac{(n-1)!}{(2\pi i)^n} \int_{S_R} \frac{f(z)}{|z-\zeta|^{2n}}\omega(z-\zeta)$$

where $\varepsilon(n) = (-1)^{n(n+1)/2}$.

Proof. We may assume $\zeta = 0$. First note that

$$\bar{\partial}\omega(z) = \sum_{k=1}^{n} (-1)^k\, d\bar{z}_k \wedge \omega_k(z) = (-1)^{n+1}n\, dz \wedge d\bar{z},$$

where $dz = dz_1 \wedge \cdots \wedge dz_n$ and similarly for $d\bar{z}$. Next, observe that if

$$\psi(z) = \frac{f(z)}{|z|^{2n}}\omega(z),$$

then

$$d\psi = 0.$$

This is easily seen. On the one hand, $\partial\psi = 0$ because ω already has

$dz_1 \wedge \cdots \wedge dz_n$, and any further dz_i wedged with this gives 0. On the other hand, since f is analytic, we find that

$$\bar{\partial}\psi(z) = f(z)\bar{\partial}\left(\frac{\omega(z)}{|z|^{2n}}\right) = 0$$

by the rule for differentiating a product and a trivial computation.

Therefore, by Stokes' theorem, applied to the annulus between two spheres, for any r with $0 < r \le R$ we get

$$\int_{S_R} \psi - \int_{S_r} \psi = 0,$$

or in other words,

$$\int_{S_R} f(z)\frac{\omega(z)}{|z|^{2n}} = \int_{S_r} f(z)\frac{\omega(z)}{|z|^{2n}}$$

$$= \frac{1}{r^{2n}}\int_{S_r} f(z)\omega(z).$$

Using Stokes' theorem once more, and the fact that $\partial\omega = 0$, we see that this is

$$= \frac{1}{r^{2n}}\int_{B_r} \bar{\partial}(f\omega) = \frac{1}{r^{2n}}\int_{B_r} f\,\bar{\partial}\omega.$$

We can write $f(z) = f(0) + g(z)$, where $g(z)$ tends to 0 as z tends to 0. Thus in taking the limit as $r \to 0$, we may replace f by $f(0)$. Hence our last expression has the same limit as

$$f(0)\frac{1}{r^{2n}}\int_{B_r} \bar{\partial}\omega = f(0)\frac{1}{r^{2n}}\int_{B_r} (-1)^{n+1}n\,dz \wedge d\bar{z}.$$

But

$$dz \wedge d\bar{z} = (-1)^{n(n-1)/2}i^n 2^n\,dy_1 \wedge dx_1 \wedge \cdots \wedge dy_n \wedge dx_n.$$

Interchanging dy_k and dx_k to get the proper orientation gives another contribution of $(-1)^n$, together with the form giving Lebesgue measure. Hence our expression is equal to

$$f(0)(-1)^{n(n+1)/2}n(2i)^n\frac{1}{r^{2n}}V(B_r),$$

where $V(B_r)$ is the Lebesgue volume of the ball of radius r in $\mathbf{R}^{2n}$, and is classically known to be equal to $\pi^n r^{2n}/n!$. Thus finally we see that our expression is equal to

$$f(0)(-1)^{n(n+1)/2} \frac{(2\pi i)^n}{(n-1)!}.$$

This proves Cauchy's theorem.

XIII, §6. THE RESIDUE THEOREM

Let f be an analytic function in an open set U of $\mathbf{C}^n$. The set of zeros of f is called a **divisor**, which we denote by $V = V_f$. In the neighborhood of a regular point a, that is a point where $f(a) = 0$ but some complex partial derivative of f is not zero, the set V is a complex submanifold of U. In fact, if, say, $D_n f(a) \neq 0$, then the map

$$(z_1, \ldots, z_n) \mapsto (z_1, \ldots, z_{n-1}, f(z))$$

gives a local analytic chart (analytic isomorphism) in a neighborhood of a. Thus we may use f as the last coordinate, and locally V is simply obtained by the projection on the set $f = 0$. This is a special case of the complex analytic inverse function theorem.

It is always true that the function $\log|f|$ is locally in $\mathcal{L}^1$. We give the proof only in the neighborhood of a regular point a. In this case, we can change f by a chart (which is known as a change-of-variable formula), and we may therefore assume that $f(z) = z_n$. Then $\log|f| = \log|z_n|$, and the Lebesgue integral decomposes into a simple product integral, which reduces our problem to the case of one variable, that is to the fact that $\log|z|$ is locally integrable near 0 in the ordinary complex plane. Writing $z = re^{i\theta}$, our assertion is obvious since the function $r \log r$ is locally integrable near 0 on the real line.

Note. In a neighborhood of a singular point, the fastest way and formally clearest, is to invoke Hironaka's resolution of singularities, which reduces the question to the non-singular case.

For the next theorem, it is convenient to let

$$d^c = \frac{1}{4\pi i}(\partial - \bar{\partial}).$$

Note that

$$dd^c = \frac{i}{2\pi}\partial\bar{\partial}.$$

The advantage of dealing with d and d^c is that they are real operators.

The next theorem, whose proof consists of repeated applications of Stokes' theorem, is due to Poincaré. It relates integration on V and U by a suitable kernel.

Theorem 6.1 (Residue Theorem). *Let f be analytic on an open set U of $\mathbf{C}^n$ and let V be its divisor of zeros in U. Let ψ be a C^∞ form with compact support in U, of degree $2n - 2$ and type $(n - 1, n - 1)$. Then*

$$\int_V \psi = \int_U \log|f|^2 dd^c\psi.$$

(As usual, the integral on the left is the integral of the restriction of ψ to V, and by definition, it is taken over the regular points of V.)

Proof. Since ψ and $dd^c\psi$ have compact support, the theorem is local (using partitions of unity). We give the proof only in the neighborhood of a regular point. Therefore we may assume that U is selected sufficiently small so that every point of the divisor of f in U is regular, and such that, for small ε, the set of points

$$U_\varepsilon = \{z \in U, |f(z)| \geqq \varepsilon\}$$

is a submanifold with boundary in U. The boundary of U_ε is then the set of points z such that $|f(z)| = \varepsilon$. (Actually to make this set a submanifold we only need to select ε to be a regular value, which can be done for arbitrarily small ε by Sard's theorem.) For convenience we let S_ε be the boundary of U_ε, that is the set of points z such that $|f(z)| = \varepsilon$. Since $\log|f|$ is locally in $\mathscr{L}^1$, it follows that

$$\int_{U_\varepsilon} \log|f| dd^c\psi = \lim_{\varepsilon \to 0} \int_{U_\varepsilon} \log|f| dd^c\psi.$$

Using the trivial identity

$$d(\log|f| d^c\psi) = d \log|f| \wedge d^c\psi + \log|f| dd^c\psi,$$

we conclude by Stokes' theorem that this limit is equal to

$$\lim_{\varepsilon \to 0} \left[\int_{S_\varepsilon} \log|f| d^c\psi - \int_{U_\varepsilon} d \log|f| \wedge d^c\psi \right].$$

The first integral under the limit sign approaches 0. Indeed, we may assume that $f(z) = z_n = re^{i\theta}$. On S_ε we have $|f(z)| = \varepsilon$, so $\log|f| = \log \varepsilon$.

There exist forms ψ_1, ψ_2 in the first $n-1$ variables such that

$$d^c\psi = \psi_1 \wedge dz_n + \psi_2 \wedge d\bar{z}_n,$$

and the restriction of dz_n to S_ε is equal to

$$\varepsilon i e^{i\theta}\, d\theta,$$

with a similar expression for $d\bar{z}_n$. Hence our boundary integral is of type

$$\varepsilon \log \varepsilon \int_{S_\varepsilon} \omega,$$

where ω is a bounded form. From this it is clear that the limit is 0.

Now we compute the second integral. Since ψ is assumed to be of type $(n-1, n-1)$ it follows that for any function g,

$$\partial g \wedge \partial \psi = 0 \qquad \text{and} \qquad \bar{\partial} g \wedge \bar{\partial} \psi = 0.$$

Replacing d and d^c by their values in terms of ∂ and $\bar{\partial}$, it follows that

$$-\int_{U_\varepsilon} d\log|f| \wedge d^c\psi = \int_{U_\varepsilon} d^c \log|f| \wedge d\psi.$$

We have

$$d(d^c \log|f| \wedge \psi) = dd^c \log|f| \wedge \psi - d^c \log|f| \wedge d\psi.$$

Furthermore dd^c is a constant times $\partial\bar{\partial}$, and $dd^c \log|f|^2 = 0$ in any open set where $f \neq 0$, because

$$\partial\bar{\partial} \log|f|^2 = \partial\bar{\partial}(\log f + \log \bar{f}) = 0$$

since $\partial \log \bar{f} = 0$ and $\bar{\partial} \log f = 0$ by the local analyticity of $\log f$. Hence we obtain the following values for the second integral by Stokes:

$$\int_{U_\varepsilon} d^c \log|f|^2 \wedge d\psi = \int_{S_\varepsilon} d^c \log|f|^2 \wedge \psi.$$

Since

$$d^c \log|f|^2 = -\frac{i}{4\pi}(\partial - \bar{\partial})(\log f + \log\bar{f})$$

$$= -\frac{i}{4\pi}\left(\frac{dz_n}{z_n} - \frac{d\bar{z}_n}{\bar{z}_n}\right)$$

(always assuming $f(z) = z_n$), we conclude that if $z_n = re^{i\theta}$, then the restriction of $d^c \log |f|^2$ to S_ε is given by

$$\text{res}_{S_\varepsilon} d^c \log f = \frac{d\theta}{2\pi}.$$

Now write ψ in the form

$$\psi = \psi_1 + \psi_2$$

where ψ_1 contains only $dz_j, d\bar{z}_j$ for $j = 1, \ldots, n-1$ and ψ_2 contains dz_n or $d\bar{z}_n$. Then the restriction of ψ_2 to S_ε contains $d\theta$, and consequently

$$\int_{S_\varepsilon} d^c \log |f|^2 \wedge \psi = \int_{S_\varepsilon} \frac{d\theta}{2\pi} \wedge (\psi_1|S_\varepsilon).$$

The integral over S_ε decomposes into a product integral, we respect to the first $n-1$ variables, and with respect to $d\theta$. Let

$$\int^{(n-1)} \psi_1(z)|S_\varepsilon = g(z_n).$$

Then simply by the continuity of g we get

$$\lim_{\varepsilon \to 0} \frac{1}{2\pi} \int_0^{2\pi} g(\varepsilon e^{i\theta}) \, d\theta = g(0).$$

Hence

$$\lim_{\varepsilon \to 0} \int_{S_\varepsilon} \frac{d\theta}{2\pi} \wedge (\psi_1|S_\varepsilon) = \int_{z_n=0} \psi_1.$$

But the restriction of ψ_1 to the set $z_n = 0$ (which is precisely V) is the same as the restriction of ψ to V. This proves the residue theorem.

The Spectral Theorem

The following is a set of notes from a seminar of Von Neumann around 1950.

APP., §1. HILBERT SPACE

Let E be a vector space over C (The real theory follows exactly the same pattern.) By an **inner product** on E we mean an R-bilinear pairing $\langle x, y \rangle \in C$ of $E \times E$ into C such that, for all complex numbers α, we have:

$$\langle \alpha x, y \rangle = \alpha \langle x, y \rangle, \qquad \langle x, y \rangle = \overline{\langle y, x \rangle},$$

$\langle x, x \rangle \geq 0$ and equals 0 if and only if $x = 0$.

We have the **Schwartz inequality**:

$$|\langle x, y \rangle|^2 \leq \langle x, x \rangle \langle y, y \rangle$$

whose proof is as follows. For all α, β complex,

$$0 \leq \langle \alpha x + \beta y, \alpha x + \beta y \rangle = \alpha \bar{\alpha} \langle x, x \rangle + \beta \bar{\alpha} \langle x, y \rangle + \alpha \bar{\beta} \langle x, y \rangle + \beta \bar{\beta} \langle y, y \rangle.$$

We let $\alpha = \langle y, y \rangle$ and $\beta = -\langle x, y \rangle$. The inequality drops out.

We define the **norm** of a vector x to be $\langle x, x \rangle^{1/2}$ and denote it by $|x|$. Using the Schwrtz inequality, one sees that $|x|$ defines a metric on E, the distance between x and y being $|x - y|$. The norm is continuous.

We write $x \perp y$ and say that x is **perpendicular** to y if $\langle x, y \rangle = 0$.

The following identities are useful and trivially proved.

Parallelogram Law. $|x + y|^2 + |x - y|^2 = 2|x|^2 + 2|y|^2$.

Pythagoras Theorem. *If* $x \perp y$, *then* $|x + y|^2 = |x|^2 + |y|^2$.

A **Hilbert space** is an inner product space which is complete under the induced metric. For the rest of this appendix, a **subspace** will always mean a closed subspace, with its structure of Hilbert space induced by that of **E**.

Lemma 1.1. *Let* **F** *be a subspace of* **E**, *let* $x \in$ **E**, *and let*

$$a = \inf |x - y|$$

the inf *taken over all* $y \in$ **F**. *Then there exists an element* $y_0 \in$ **F** *such that* $a = |x - y_0|$.

Proof. Let y_n be a sequence in **F** such that $|y_n - x|$ tends to a. We must show that y_n is Cauchy. By the parallelogram law,

$$|y_n - y_m|^2 = 2|y_n - x|^2 + 2|y_m - x|^2 - 4|\tfrac{1}{2}(y_n + y_m) - x|^2$$
$$\leqq 2|y_n - x|^2 + 2|y_m - x|^2 - 4a^2$$

which shows that y_n is Cauchy, converging to some vector y_0. The lemma follows by continuity.

Theorem 1.2. *If* **F** *is a subspace properly contained in* **E**, *then there exists a vector* z *in* **E** *which is perpendicular to* **F** (*and* $\neq 0$).

Proof. Let $x \in$ **E** and $x \notin$ **F**. Let y_0 be an element of **F** which is at minimal distance from x (use Lemma 1.1). Let a be this distance and let $z = y_0 - x$. After a translation, we may assume that $z = x$, so that $|x| = a$. For any complex number α and $y \in$ **F** we have $|x + \alpha y| \geqq a$, whence

$$\langle x + \alpha y, x + \alpha y \rangle = |x|^2 + \bar{\alpha}\langle x, y \rangle + \alpha\overline{\langle x, y \rangle} + \alpha\bar{\alpha}|y|^2$$
$$\geqq a^2.$$

Put $\alpha = t\overline{\langle x, y \rangle}$. We get a contradiction for small values of t.

APP., §2. FUNCTIONALS AND OPERATORS

A linear map A from a Hilbert space **E** to a Hilbert space **H** is **bounded** if there exists a positive real number α such that

$$|Ax| \leqq \alpha|x|$$

for all $x \in$ **E**. The norm of A, denoted by $|A|$ is the inf of all such α.

Proposition 2.1. *A linear map is bounded if and only if it maps the unit sphere on a bounded subset, if and only if it is continuous.*

Proof. Clear.

A **functional** is a continuous linear map into **C**. Functionals are bounded. We have the fundamental:

Representation Theorem. *A linear map $\lambda: \mathbf{E} \to \mathbf{C}$ is bounded if and only if there exists $y \in \mathbf{E}$ such that $\lambda(x) = \langle x, y \rangle$ for all $x \in \mathbf{E}$. If such a y exists, it is unique.*

Proof. If $\lambda(x) = \langle x, y \rangle$ then the Schwartz inequality shows that it is bounded, with bound $|y|$. It is obvious that y is unique.

Conversely, let λ be bounded. Let **F** be the kernel of λ. Then **F** is a subspace. If $\mathbf{E} = \mathbf{F}$ then everything is trivial. If $\mathbf{E} \neq \mathbf{F}$, then there exists $z \in \mathbf{F}$, $z \notin \mathbf{E}$ such that z is perpendicular to **F** by Theorem 1.2. We contend that some multiple $y = \alpha z$ does it. A necessary condition on α is that

$$\langle z, \alpha z \rangle = \bar{\alpha} |z|^2.$$

This is also sufficient. Namely, $x - \big(\lambda(x)|\lambda(z)\big)z$ lies in **F**. Put $\alpha = \overline{\lambda(z)}/|z|^2$. Then one sees at once that $\lambda(x) = \langle x, y \rangle$ as was to be shown.

By an **operator** we shall always mean a continuous linear map of a space into itself.

It is straightforward to show that operators form a Banach space, and in fact a normed ring. In other words, in addition to the Banach space property, we have

$$|AB| \leqq |A| |B|.$$

Proposition 2.2. *If A is an operator and $\langle Ax, x \rangle = 0$ for all x, then $A = O$.*

Proof. This follows from the polarization identity,

$$\langle A(x + y), (x + y) \rangle - \langle A(x - y), (x - y) \rangle = 2[\langle Ax, y \rangle + \langle Ay, x \rangle].$$

Replace x by ix. Then we get

$$\langle Ax, y \rangle + \langle Ay, x \rangle = 0,$$
$$i\langle Ax, y \rangle - i\langle Ay, x \rangle = 0,$$

for all x, y whence $\langle Ax, y \rangle = 0$ and $A = O$.

The above proposition is valid only in the complex case.

In the real case, we shall need it only when A is symmetric (see below), in which case it is equally clear. A similar remark applies to the next result.

Lemma 2.3. *Let A be an operator, and c a number such that*

$$|\langle Ax, x \rangle| \leq c|x|^2$$

for all $x \in \mathbf{E}$. Then for all x, y we have

$$|\langle Ax, y \rangle| + |\langle x, Ay \rangle| \leq 2c|x||y|.$$

Proof. By the polarization identity,

$$2|\langle Ax, y \rangle + \langle Ay, x \rangle| \leq c|x + y|^2 + c|x - y|^2 = 2c(|x|^2 + |y|^2).$$

Hence

$$|\langle Ax, y \rangle + \langle Ay, x \rangle| \leq c(|x|^2 + |y|^2).$$

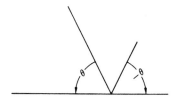

We multiply y by $e^{i\theta}$ and thus get on the left-hand side

$$|e^{-i\theta}\langle Ax, y \rangle + e^{i\theta}\langle Ay, x \rangle|.$$

The right-hand side remains unchanged, and for suitable θ, the left-hand side becomes

$$|\langle Ax, y \rangle| + |\langle Ay, x \rangle|.$$

(In other words, we are lining up two complex numbers by rotating one by θ and the other by $-\theta$.) Next we replace x by tx and y by y/t for t real and $t > 0$. Then the left-hand side remains unchanged, while the right-hand side becomes

$$g(t) = t^2|x|^2 + \frac{1}{t^2}|y|^2.$$

The point at which $g'(t) = 0$ is the unique minimum, and at this point t_0 we find that

$$g(t_0) = |x||y|.$$

This proves our lemma.

In our applications, we need the lemma only when A is self-adjoint (i.e. symmetric, see below), in which case it is even more trivial.

For fixed y, the function of x given by $\langle Ax, y \rangle$ is a functional (bounded because of the Schwartz inequality). Hence by the representation theorem, there exists an element y^* such that $\langle Ax, y \rangle = \langle x, y^* \rangle$ for all x. We define A^*, the **adjoint** of A, by letting $A^*y = y^*$. Since y^* is unique, we see that A^* is the unique operator such that

$$\langle Ax, y \rangle = \langle x, A^*y \rangle$$

for all x, y in **E**.

Theorem 2.4. *We have*:

$$(A + B)^* = A^* + B^*, \qquad A^{**} = A,$$

$$(\alpha A)^* = \bar{\alpha} A^*, \qquad |A^*| = |A|,$$

$$(AB)^* = B^* A^*, \qquad |AA^*| = |A|^2.$$

and the mapping $A \to A^*$ *is continuous.*

Proof. Exercise for the reader.

APP., §3. HERMITIAN OPERATORS

We shall say that an operator A is **symmetric** (or **hermitian**) if $A = A^*$.

Proposition 3.1. *A is hermitian if and only if $\langle Ax, x \rangle$ is real for all x.*

Proof. Let A be hermitian. Then $\overline{\langle Ax, x \rangle} = \overline{\langle x, Ax \rangle} = \langle Ax, x \rangle$. Conversely, $\langle Ax, x \rangle = \overline{\langle Ax, x \rangle} = \langle x, Ax \rangle = \langle A^*x, x \rangle$ implies that

$$\langle (A - A^*)x, x \rangle = 0$$

whence $A = A^*$ by polarization.

Proposition 3.2. *Let A be a hermitian operator. Then $|A|$ is the greatest lower bound of all values c such that*

$$|\langle Ax, x \rangle| \leqq c|x|^2$$

for all x, or equivalently, the sup of all values $|\langle Ax, x \rangle|$ taken for x on the unit sphere in E.

Proof. When A is hermitian we obtain

$$|\langle Ax, y \rangle| \leqq c|x||y|$$

for all x, $y \in E$, so that we get $|A| \leq c$ in Lemma 2.3. On the other hand, $c = |A|$ is certainly a possible value for c by the Schwartz inequality. This proves our proposition.

Proposition 3.2 allows us to define an ordering in the space of hermitian operators. If A is hermitian, we define $A \geq O$ and say that A is **semipositive** if $\langle Ax, x \rangle \geq 0$ for all $x \in E$. If A, B are hermitian we define $A \geq B$ if $A - B \geq O$. This is indeed an ordering; the usual rules hold: If $A_1 \geq B_1$ and $A_2 \geq B_2$, then

$$A_1 + A_2 \geq B_1 + B_2.$$

If c is a real number ≥ 0 and $A \geq O$, then $cA \geq O$. So far, however, we say nothing about a product of semipositive hermitian operators AB, even if $AB = BA$. We shall deal with this question later.

Let c be a bound for A. Then $|\langle Ax, x \rangle| \leq c|x|^2$ and consequently

$$-cI \leq A \leq cI.$$

For simplicity, if α is real, we sometimes write $\alpha \leq A$ instead of $\alpha I \leq A$, and similarly we write $A \leq \beta$ instead of $A \leq \beta I$. If we let

$$\alpha = \inf_{|x|=1} \langle Ax, x \rangle \qquad \text{and} \qquad \beta = \sup_{|x|=1} \langle Ax, x \rangle,$$

then we have

$$\alpha \leq A \leq \beta,$$

and from Proposition 3.1,

$$|A| = \max(|\alpha|, |\beta|).$$

Let p be a polynomial with real coefficients, and let A be a hermitian operator. Write

$$p(t) = a_n t^n + \cdots + a_0.$$

We define

$$p(A) = a_n A^n + \cdots + a_0 I.$$

We let $\mathbf{R}[A]$ be the algebra generated over $\mathbf{R}$ by A, that is the algebra of all operators $p(A)$, where $p(t) \in \mathbf{R}[t]$. We wish to investigate the closure of $\mathbf{R}[A]$ in the (real) Banach space of all operators. We shall show how to represent this closure as a ring of continuous functions on some compact subset of the reals. First, we observe that the hermitian operators form a closed subspace of $L(E, E)$, and that $\overline{\mathbf{R}[A]}$ is a closed subspace of the space of hermitian operators.

We can find real numbers α, β such that

$$\alpha I \leqq A \leqq \beta I.$$

We shall prove that if p is a real polynomial which takes on values $\geqq 0$ on the interval $[\alpha, \beta]$, then $p(A)$ is a semipositive operator.

The fundamental theorem is the following.

Theorem 3.3. *Let α, β be real and $\alpha I \leqq A \leqq \beta I$. Let p be a real polynomial, semipositive in the interval $\alpha \leqq t \leqq \beta$. Then $p(A)$ is a semipositive operator.*

Proof. We shall need the following obvious facts.

If A, B are hermitian, A commutes with B, and $A \geqq 0$, then AB^2 is semipositive.

If $p(t)$ is quadratic, of type $p(t) = t^2 + at + b$ and has imaginary roots, then

$$p(t) = \left(t + \frac{a}{2}\right)^2 + \left(b - \frac{a^2}{4}\right)$$

is a sum of squares.

A sum of squares times a sum of squares is a sum of squares (if they commute).

If $p(t)$ has a root γ in our interval, then the multiplicity of γ is even.

Our theorem now follows from the following purely algebraic statement.

Let $\alpha \leqq t \leqq \beta$ be a real interval, and $p(t)$ a real polynomial which is semipositive in this interval. Then $p(t)$ can be written:

$$p(t) = c\left[\sum Q_i^2 + \sum (t - \alpha)Q_j^2 + \sum (\beta - t)Q_k^2\right]$$

where Q^2 just denotes the square of some polynomial and c is a number $\geqq 0$.

In order to prove this, we split $p(t)$ over the real numbers into linear and quadratic factors. If a root γ is $\leqq \alpha$, then we write

$$(t - \gamma) = (t - \alpha) + (\alpha - \gamma)$$

and note that $(\alpha - \gamma)$ is a square. If a root γ is $\geqq \beta$, then we write

$$(\gamma - t) = (\gamma - \beta) + (\beta - t)$$

with $(\gamma - \beta)$ a square. We can then write, after expanding out the factorization of $p(t)$,

$$p(t) = c[\sum Q_i^2 + \sum (t - \alpha)Q_j^2 + \sum (\beta - t)Q_k^2 + \sum (t - \alpha)(\beta - t)Q_l^2]$$

with some constant c and Q_v^2 standing for the square of some polynomial. Note that c is ≥ 0 since $p(t)$ is semipositive on the interval. Our last step reduces the bad last term to the preceding ones by means of the identity

$$(t - \alpha)(\beta - t) = \frac{(t - \alpha)^2(\beta - t) + (t - \alpha)(\beta - t)^2}{\beta - \alpha}.$$

Corollary 3.4. *If $a \leq p(t) \leq b$ in the interval, then*

$$aI \leq p(A) \leq bI.$$

Suppose that $\alpha I \leq A \leq \beta I$. If $p(t)$ is a real polynomial, we define as usual

$$\|p\| = \sup |p(t)|$$

with t ranging over the interval.

Corollary 3.5. *Let $\alpha I \leq A \leq \beta I$. Let $p(t)$ be a real polynomial. Then $|p(A)| \leq \|p\|$.*

Proof. Let $q(t) = \|p\| \pm p(t)$. Then $q(t)$ is ≥ 0 on the interval. Hence $q(A) \geq O$ and our assertion follows at once.

As usual, we consider the continuous functions on the interval as a Banach space. If f is any continuous function on the interval, then by the Weierstrass approximation theorem, we can find a sequence of polynomials $\{p_n\}$ approaching f uniformly on this interval. We define $f(A)$ as the limit of $p_n(A)$. From Corollary 3.5 we deduce that $\{p_n(A)\}$ is a Cauchy sequences, and that its limit does not depend on the choice of the sequence $\{p_n\}$. Furthermore, by continuity, our corollary generalizes to continuous functions, so that $|f(A)| \leq \|f\|$.

We see that the map $f \mapsto f(A)$ is a continuous homomorphism from the Banach algebra of continuous functions on the interval into the closure of the subalgebra generated by A.

Proposition 3.6. *Let A be a semipositive operator. Then there exists an operator B in the closure of the algebra generated by A such that $B^2 = A$.*

Proof. The continuous function $t^{1/2}$ maps on $A^{1/2}$.

Corollary 3.7. *The product of two semipositive, commuting hermitian operators is again semipositive.*

Proof. Let A, C be hermitian and $AC = CA$. If B is as in Proposition 3.6, then

$$\langle ACx, x \rangle = \langle B^2 Cx, x \rangle = \langle BCx, Bx \rangle = \langle CBx, Bx \rangle \geqq 0.$$

The kernel of our homomorphism from the continuous functions to the operators is a closed ideal. Its zeros form a closed set called the **spectrum** of A and denoted by $\sigma(A)$.

Lemma 3.8. *Let X be a compact set, R the ring of continuous functions on X, and $\mathfrak{a}$ a closed ideal of R, $\mathfrak{a} \neq R$. Let C be the closed set of zeros of $\mathfrak{a}$. Then C is not empty and if a function $f \in R$ vanishes on C, then $f \in \mathfrak{a}$.*

Proof. Given ε, let U be the open set where $|f| < \varepsilon$. Then $X - U$ is closed. For each point $t \in X - U$ there exists a function $g \in \mathfrak{a}$ such that $g(t) \neq 0$ in a neighborhood of t. These neighborhoods cover $X - U$, and so does a finite number of them, with functions $g_1, \ldots, g_r$. Let $g = g_1^2 + \cdots + g_r^2$. Then $g \in \mathfrak{a}$. Our function g has a minimum on $X - U$ and for n large, the function

$$f \frac{ng}{1 + ng}$$

is close to f on $X - U$ and is $< \varepsilon$ on U, which proves what we wanted.

We now redefine the norm of a continuous function f to be

$$\|f\|_A = \sup_{t \in \sigma(A)} |f(t)|.$$

Theorem 3.9. *The map*

$$f(t) \mapsto f(A)$$

induces a Banach-isomorphism (i.e. norm-preserving) of the Banach algebra of continuous functions on $\sigma(A)$ onto the closure of the algebra generated by A.

Proof. We have already proved that our map is an agebraic isomorphism and that $|f(A)| \leqq \|f\|_A$. In order to get the reverse inequality, we shall prove:

If $f(A) \geq O$, then $f(t) \geq 0$ on the spectrum of A. Indeed, if $f(c) < 0$ for some $c \in \sigma(A)$, we let $g(t)$ be a function which is 0 outside a small neighborhood of c, is ≥ 0 everywhere, and is > 0 at c. Then $g(A)$ and $g(A)f(A)$ are both ≥ 0 by Corollary 3.7. But $-g(t)f(t) \geq 0$ gives $-g(A)f(A) \geq O$ whence $g(A)f(A) = O$. Since $g(t)f(t)$ is not 0 on the spectrum of A, we get a contradiction.

Let now $s = |f(A)|$. Then $sI - f(A) \geq O$ implies that $s - f(t) \geq 0$, which proves the theorem.

From now on, the norm on continuous functions will refer to the spectrum. All that remains to do is identify our spectrum with what can be called the **general spectrum**, that is those complex values ξ such that $A - \xi$ is not invertible. (By invertible, we mean having an inverse which is an operator.)

Theorem 3.10. *The general spectrum is compact, and in fact, if ξ is in it, then $|\xi| \leq |A|$. If A is hermitian, then the general spectrum is equal to $\sigma(A)$.*

Proof. The complement of the general spectrum is open, because if $A - \xi_0$ is invertible, and ξ is close to ξ_0, then $(A - \xi_0)^{-1}(A - \xi)$ is close to I, hence invertible, and hence $A - \xi$ is also invertible. Furthermore, if $\xi > |A|$, then $|A/\xi| < 1$ and hence $I - (A/\xi)$ is invertible (by the power series argument). So is $A - \xi$ and we are done. Finally, suppose that ξ is in the general spectrum. Then ξ is real. Otherwise, let

$$g(t) = (t - \xi)(t - \bar{\xi}).$$

Then $g(t) \neq 0$ on $\sigma(A)$ and $h(t) = 1/g(t)$ is its inverse. From this we see that $A - \xi$ is invertible.

Suppose ξ is not in the spectrum. Then $t - \xi$ is invertible and so is $A - \xi$.

Suppose ξ is in the spectrum. After a translation, we may suppose that 0 is in the spectrum. Consider the function $g(t)$ as follows:

$$g(t) = \begin{cases} 1/|t|, & |t| \geq 1/N, \\ N, & |t| \leq 1/N, \end{cases}$$

(g is positive and has a peak at 0.) If A is invertible, $BA = I$, then from $|tg(t)| \leq 1$ we get $|Ag(A)| \leq 1$ and hence $|g(A)| \leq |B|$. But $g(A)$ becomes arbitrarily large as we take N large. Contradiction.

Theorem 3.11. *Let S be a set of operators of the Hilbert space E, leaving no closed subspace invariant except 0 and $\mathbf{E}$ itself. Let A be a Hermitian operator such that $AB = BA$ for all $B \in S$. Then $A = \lambda I$ for some real number λ.*

Proof. It will suffice to prove that there is only one element in the spectrum of A. Suppose there are two, $\lambda_1 \neq \lambda_2$. There exist continuous functions f, g on the spectrum such that neither is 0 on the spectrum, but fg is 0 on the spectrum. For instance, one may take for f, g the functions whose graph is indicated on the next diagram.

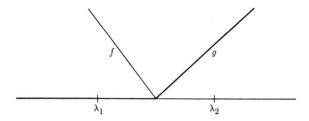

We have $f(A)B = Bf(A)$ for all $B \in S$ (because B commutes with real polynomials in A, hence with their limits). Hence $f(A)\mathbf{E}$ is invariant under S because

$$Bf(A)\mathbf{E} = f(A)B\mathbf{E} \subset f(A)\mathbf{E}.$$

Let $\mathbf{F}$ be the closure of $f(A)\mathbf{E}$. Then $\mathbf{F} \neq 0$ because $f(A) \neq O$. Furthermore, $\mathbf{F} \neq \mathbf{E}$ because $g(A)f(A)\mathbf{E} = 0$ and hence $g(A)\mathbf{F} = 0$. Since $\mathbf{F}$ is obviously invariant under S, we have a contradiction.

Corollary 3.12. *Let S be a set of operators of the Hilbert space $\mathbf{E}$, leaving no closed subspace invariant except 0 and $\mathbf{E}$ itself. Let A be an operator such that $AA^* = A^*A$, $AB = BA$, and $A^*B = BA^*$ for all $B \in S$. Then $A = \lambda I$ for some complex number λ.*

Proof. Write $A = A_1 + iA_2$ where A_1, A_2 are hermitian and commute $\left(\text{e.g. } A_1 = (A + A^*)/2\right)$. Apply the theorem to each one of A_1 and A_2 to get the result.

Bibliography

[Ab 62]   R. ABRAHAM, *Lectures of Smale on Differential Topology*, Columbia University, 1962

[AbM 78]   R. ABRAHAM and J. MARSDEN, *Foundations of Mechanics*, second edition, Benjamin-Cummings, 1978

[Am 56]   W. AMBROSE, Parallel translation of Riemannian curvature, *Ann. of Math.* **64** (1965) pp. 337–363

[Am 60]   W. AMBROSE, The Cartan structural equations in classical Riemannian geometry, *J. Indian Math. Soc.* **24** (1960) pp. 23–75

[Am 61]   W. AMBROSE, The index theorem in Riemannian geometry, *Ann. of Math.* **73** (1961) pp. 49–86

[APS 60]   W. AMBROSE, R.S. PALAIS, and I.M. SINGER, Sprays, *Acad. Brasileira de Ciencias* **32** (1960) pp. 163–178

[ABP 75]   M. ATIYAH, R. BOTT, and V.K. PATODI, On the heat equation and the index theorem, *Invent. Math.* **19** (1973) pp. 279–330; Errata, *Invent. Math.* **28** (1975) pp. 277–280

[AH 59]   M. ATIYAH and F. HIRZEBRUCH, Riemann–Roch theorems for differentiable manifolds, *Bull. Amer. Math. Soc.* **65** (1959) pp. 276–281

[APS 75]   M. ATIYAH, V.K. PATODI, and I. SINGER, Spectral asymmetry and Riemannian geometry: I. *Math. Proc. Camb. Philos. Soc.* **77** (1975) pp. 43–69

[Be 58]   M. BERGER, Sur certaines variétés Riemanniennes à courbure positive, *C.R. Acad. Sci. Paris* **247** (1958) pp. 1165–1168

[BGM 71]   M. BERGER, P. GAUDUCHON, and E. MAZET, *Le Spectre d'une Variété Riemannienne*, Lecture Notes in Mathematics **195**, Springer-Verlag, 1971

[BGV 92]   N. BERLINE, E. GETZLER, and M. VERGNE, *Heat Kernels and Dirac Operators*, Grundlehren der Math. Wiss. **298**, Springer-Verlag, 1992

[BiC 64]   R. BISHOP and R. CRITTENDEN, *Geometry of Manifolds*, Academic Press, 1964

[BoY 53]   S. BOCHNER and K. YANO, *Curvature and Betti Numbers*, Annals Studies **32**, Princeton University Press, 1953

[BoF 65]   R. BONIC and J. FRAMPTON, Differentiable functions on certain Banach spaces, *Bull. Amer. Math. Soc.* **71** (1965) pp. 393–395

[BoF 66]   R. BONIC and J. FRAMPTON, Smooth functions on Banach manifolds, *J. Math. Mech.* **15** No. 5 (1966) pp. 877–898

[Bo 60]   R. BOTT, *Morse Theory and its Applications to Homotopy Theory*, Lecture Notes by van de Ven, Bonn, 1960

[BoT 82]   R. BOTT and L. TU, *Differential Forms in Algebraic Topology*, Graduate Texts in Mathematics **82**, Springer-Verlag, 1982

[Bou 68]   N. BOURBAKI, *General Topology*, Addison-Wesley, 1968 (translated from the French, 1949)

[Bou 69]   N. BOURBAKI, *Fasicule de Résultats des Variétés*, Hermann, 1969

[Bou 81]   N. BOURBAKI, *Espaces Vectoriels Topologiques*, Masson, 1981

[Ca 28]   E. CARTAN, *Leçons sur la Géometrie des Espaces de Riemann*, Gauthier-Villars, 1928; second edition, 1946

[Cha 84]   I. CHAVEL, *Eigenvalues in Riemannian Geometry*, Academic Press, 1984

[ChE 75]   J. CHEEGER and D. EBIN, *Comparison Theorems in Riemannian Geometry*, North-Holland, 1975

[Che 55]   S. CHERN, On curvature and characteristic classes of a Riemann manifold, *Abh. Math. Sem. Hamburg* **20** (1958) pp. 117–126

[doC 92]   M.P. do CARMO, *Riemannian Geometry*, Birkhaüser, 1992

[Eb 70]   D. EBIN, The manifold of Riemannian metrics, *Proc. Symp. Pure Math. XV* AMS **26** (1970) pp. 11–40; and *Bull. Amer. Math. Soc.* **4** (1970) pp. 1002–1004

[EbM 70]   D.G. EBIN and J. MARSDEN, Groups of diffeomorphisms and the motion of an incompressible fluid, *Ann. of Math.* **92** (1970) pp. 102–163

[Ee 58]   J. EELLS, On the geometry of function spaces, *Symposium de Topologia Algebrica*, Mexico (1958) pp. 303–307

[Ee 59]   J. EELLS, On submanifolds of certain function spaces, *Proc. Nat. Acad. Sci. USA* **45** (1959) pp. 1520–1522

[Ee 61]   J. EELLS, Alexander–Pontrjagin duality in function spaces, *Proc. Symposia in Pure Math.* **3**, AMS (1961) pp. 109–129

[Ee 66]   J. EELLS, A setting for global analysis, *Bull. Amer. Math. Soc.* **72** (1966) pp. 751–807

[El 67]   H. ELIASSON, Geometry of manifolds of maps, *J. Diff. Geom.* **1** (1967) pp. 169–194

[GHL 87/93]   S. GALLOT, D. HULIN, and J. LAFONTAINE, *Riemannian Geometry*, second edition, Springer-Verlag, 1987, 1993 (corrected printing)

[Ga 87]   R. GARDINER, *Teichmuller Theory and Quadratic Differentials*, Wiley–Interscience, 1987

[God 58]   R. GODEMENT, *Topologie Algébrique et Théorie des Faisceaux*, Hermann, 1958

[Go 62]   S.I. GOLDBERG, *Curvature and Homology*, Academic Press, 1962

[Gr 61]   W. GRAEUB, Liesche Grupen und affin zusammenhangende Mannigfaltigkeiten, *Acta Math.* (1961) pp. 65–111

[Gray 73]   A. GRAY, The volume of small geodesic balls, *Mich. J. Math.* **20** (1973) pp. 329–344

[GrV 79]   A. GRAY and L. VANHECKE, Riemmanian geometry as determined by the volume of small geodesic balls, *Acta Math.* **142** (1979) pp. 157–198

[GriH 76]   P. GRIFFITHS and J. HARRIS, *Principles of Algebraic Geometry*, Wiley, 1976

[GrKM 75]   D. GROMOLL, W. KLINGENBERG, and W. MEYER, *Riemannsche Geometrie im Grössen*, Lecture Notes in Mathematics **55**, Springer-Verlag, 1975

[Gros 64]   N. GROSSMAN, *Geodesics on Hilbert Manifolds*, Princeton University Press Notes, 1964 (referred to in McAlpin's thesis)

[Gu 91]   P. GÜNTHER, Huygens' Principle and Hadamard's Conjecture, *Math. Intelligencer*, **13** No. 2 (1991) pp. 56–63

[Ha 1898]   J. HADAMARD, Les surfaces à courbures opposées, *J. Math. Pures Appl.* **4** (1898) pp. 27–73

[He 61]   S. HELGASON, Some remarks on the exponential mapping for an affine connections, *Math. Scand.* **9** (1961) pp. 129–146

[He 78]   S. HELGASON, *Differential Geometry, Lie Groups and Symmetric Spaces*, Academic Press, 1978 (later version of *Differential Geometry and Symmetric Spaces*, 1962)

[He 84]   S. HELGASON, *Wave Equations on Homogeneous Spaces*, Lecture Notes in Mathematics **1077**, Springer-Verlag (1984) pp. 254–287

[Hi 59]   N. HICKS, A theorem on affine connexions, *Ill. J. Math.* **3** (1959) pp. 242–254

[HoR 31]   H. HOPF and W. RINOW, Über den Begriff der vollständigen differentialgeometrischen Fläche, *Commentarii Math. Helv.* **3** (1931) pp. 209–225

[Ir 70]   M.C. IRWIN, On the stable manifold theorem, *Bull. London Math. Soc.* (1970) pp. 68–70

[Ke 55]   J. KELLEY, *General Topology*, Van Nostrand, 1955

[Ko 87]   S. KOBAYASHI, *Differential Geometry of Complex Vector Bundles*, Iwanami Shoten and Princeton University Press, 1987

[KoN 63]   S. KOBAYASHI and K. NOMIZU, *Foundations of Differential Geometry* I, Wiley, 1963 and 1969

[KoN 69]   S. KOBAYASHI and K. NOMIZU, *Foundations of Differential Geometry* II, Wiley, 1969

[Kos 57]   J.L. KOSZUL, *Variétés Kähleriennes*, Ist. Mat. Pura e Apl., Sao Paulo, 1957

[Kr 72]   N. KRIKORIAN, Differentiable structure on function spaces, *Trans. Amer. Math. Soc.* **171** (1972) pp. 67–82

[La 61]   S. LANG, Fonctions implicites et plongements Riemanniens, *Séminaire Bourbaki* 1961–62

358   BIBLIOGRAPHY

[La 62]   S. LANG, *Introduction to Differentiable Manifolds*, Addision-Wesley, 1962

[La 71]   S. LANG, *Differential Manifolds*, Addison-Wesley, 1971; Springer-Verlag, 1985

[La 75]   S. LANG, $SL_2(\mathbf{R})$, Addison-Wesley, 1975; reprinted by Springer-Verlag, 1985

[La 87]   S. LANG, *Introduction to Complex Hyperbolic Spaces*, Springer-Verlag, 1987

[La 93]   S. LANG, *Real and Functional Analysis*, third edition, Springer-Verlag, 1993

[Le 67]   J. LESLIE, On a differential structure for the group of diffeomorphisms, *Topology* **6** (1967) pp. 263–271

[LoS 68]   L. LOOMIS and S. STERNBERG, *Advanced Calculus*, Addison-Wesley, 1968

[Lo 69]   O. LOOS, *Symmetric Spaces*, I and II, Benjamin, 1969

[Mar 74]   J. MARSDEN, *Applications of Global Analysis to Mathematical Physics*, Publish or Perish, 1974; reprinted in "Berkeley Mathematics Department Lecture Note Series", available from the UCB Math. Dept.

[MaW 74]   J. MARSDEN and A. WEINSTEIN, Reduction of symplectic manifolds with symmetry, *Rep. Math. Phys.* **5** (1974) pp. 121–130

[Maz 61]   B. MAZUR, Stable equivalence of differentiable manifolds, *Bull. Amer. Math. Soc.* **67** (1961) pp. 377–384

[McA 65]   J. MCALPIN, *Infinite Dimensional Manifolds and Morse Theory*, thesis, Columbia University, 1965

[McK-S 67] H.P. MCKEAN and I. SINGER, Curvature and the eigenvalues of the Laplacian, *J. Diff. Geom.* **1** (1967) pp. 43–69

[Mi 58]   J. MILNOR, *Differential Topology*, Princeton University Press, 1968

[Mi 59]   J. MILNOR, *Der Ring der Vectorraumbundel eines topologischen Raümes*, Bonn, 1959

[Mi 61]   J. MILNOR, *Differentiable Structures*, Princeton University Press, 1961

[Mi 63]   J. MILNOR, *Morse Theory*, Ann. Math. Studies **53**, Princeton University Press, 1963

[Mi 64]   J. MILNOR, Eigenvalues of the Laplace operator on certain manifolds, *Proc. Nat. Acad. Sci. USA* **51** (1964) p. 542

[Min 49]   S. MINAKSHISUNDARAM, A generalization of Epstein zeta functions, *Can. J. Math.* **1** (1949) pp. 320–329

[MinP 49]   S. MINAKSHISUNDARAM and A. PLEIJEL, Some properties of the eigenfunctions of the Laplace operator on Riemannian manifolds, *Can. J. Math.* **1** (1949) pp. 242–256

[MokSY 93] N. MOK, Y-T. SIU, and S-K. YEUNG, Geometric superrigidity, *Invent. Math.* **113** (1993) pp. 57–83

[Mo 61]   J. MOSER, A new technique for the construction of solutions for nonlinear differential equations, *Proc. Nat. Acad. Sci. USA* **47** (1961) pp. 1824–1831

[Mo 65]   J. MOSER, On the volume element of a manifold, *Trans. Amer. Math. Soc.* **120** (1965) pp. 286–294

[Na 56]   J. NASH, The embedding problem for Riemannian manifolds, *Ann. of Math.* **63** (1956) pp. 20–63

[Om 70]   H. OMORI, On the group of diffeomorphisms of a compact manifold, *Proc. Symp. Pure Math. XV*, AMS (1970) pp. 167–184

[Pa 63]   R. PALAIS, Morse theory on Hilbert maifolds, *Topology* **2** (1963) pp. 299–340

[Pa 66]   R. PALAIS, Lusternik–Schnirelman theory on Banach manifolds, *Topology* **5** (1966) pp. 115–132

[Pa 68]   R. PALAIS, *Foundations of Global Analysis*, Benjamin, 1968

[Pa 69]   R. PALAIS, The Morse lemma on Banach spaces, *Bull. Amer. Math. Soc.* **75** (1969) pp. 968–971

[Pa 70]   R. PALAIS, Critical point theory and the minimax principle, *Proc. Symp. Pure Math. AMS* **XV** (1970) pp. 185–212

[PaS 64]   R. PALAIS and S. SMALE, A generalized Morse theory, *Bull. Amer. Math. Soc.* **70** (1964) pp. 165–172

[PaT 88]   R. PALAIS and C. TERNG, *Critical Point Theory and Submanifold Geometry*, Lecture Notes Mathematics **1353**, Springer-Verlag, 1988

[Pat 71a]   V.K. PATODI, Curvature and the eigenforms of the Laplace operator, *J. Diff. Geom.* **5** (1971) pp. 233–249

[Pat 71b]   V.K. PATODI, An analytic proof of the Riemann–Roch–Hirzebruch theorem, *J. Diff. Geom.* **5** (1971) pp. 251–283

[Po 62]   F. POHL, Differential geometry of higher order, *Topology* **1** (1962) pp. 169–211

[Pro 61]   *Proceedings of Symposia in Pure Mathematics III, Differential Geometry, AMS*, 1961

[Pro 70]   *Proceedings of the Conference on Global Analysis, Berkeley, Calif. 1968*, AMS, 1970

[Re 64]   G. RESTREPO, Differentiable norms in Banach spaces, *Bull. Amer. Math. Soc.* **70** (1964) pp. 413–414

[Ro 68]   J. ROBBIN, On the existence theorem for differential equations, *Proc. Amer. Math. Soc.* **19** (1968) pp. 1005–1006

[Sc 60]   J. SCHWARZ, On Nash's implicit functional theorem, *Commun. Pure Appl. Math.* **13** (1960) pp. 509–530

[Sm 61]   S. SMALE, Generalized Poincaré's conjecture in dimensions greater than four, *Ann. of Math.* **74** (1964) pp. 391–406

[Sm 63]   S. SMALE, Stable manifolds for differential equations and diffeomorphism, *Ann. Scuola Normale Sup. Pisa* **III** Vol. XVII (1963) pp. 97–116

[Sm 64]   S. SMALE, Morse theory and a non-linear generalization of the Dirichlet problem, *Ann. of Math.* **80** No. 2 (1964) pp. 382–396

[Sm 67]   S. SMALE, Differentiable dynamical systems, *Bull. Amer. Math. Soc.* **73**, No. 6 (1967) pp. 747–817

[Sp 70/79]   M. SPIVAK, *Differential Geometry* (5 volumes), Publish or Perish, 1970–1979

[Ste 51]   N. STEENROD, *The Topology of Fiber Fundles*, Princeton University Press, 1951

[Str 64]   S. STERNBERG, *Lectures on Differential Geometry*, Prentice-Hall, 1964

[Sy 36]   J.L. SYNGE, On the connectivity of spaces of positive curvature, *Quart. J. Math. Oxford Series* 7 (1936) pp. 316–320

[Vi 73]   L. VIRTANEN, *Quasi Conformal Mappings in the Plane*, Springer-Verlag, 1973

[We 67]   A. WEINSTEIN, A fixed point theorem for positively curved manifolds, *J. Math. Mech.* 18 (1968) pp. 149–153

[Wel 80]   R. WELLS, *Differential Analysis on Complex Manifolds*, Graduate Texts in Mathematics 65, Springer-Verlag, 1980

[Wh 32]   J.H.C. WHITEHEAD, Convex regions in the geometry of paths, *Quaterly J. Math.* (*Oxford*) 3 (1932) pp. 33–42

Index

A

adjoint of d 330, 332, 333
admit partitions of unity   32
almost compact support   312
alternating   142
alternating product   126, 129
analytic   336
approximate solution   71
arc length   187
atlas   20

B

Banachable   4
base space   41
Bianchi identity   227
bilinear map associated with spray   99, 194
bilinear tensor   141, 214
bilinear tensor field   142
block   288
boundary   38, 315
bounded functional   300
bounded operator   4
bracket of vector fields   115
bundle along the fiber   54

C

canonical 1-form and 2-form   147
canonical lifting   95
canonical spray   189

Cartan–Hadamard theorem   246
category   2
Cauchy theorem   337
C_c-functional   300
change of variables formula for integration   295
change of variables formula for sprays   100
chart   21
Christoffel symbols   208
class C^p 7, 56
closed form   135, 281
closed graph theorem   4
closed submanifold   24
cocycle condition   42
cokernel   512
commuting vector fields   120
compactly exact   322
compatible   21
complete   218
complex analytic   335
compressible   110, 180
connection   101, 102
constant curvature   238, 247
contraction   137
contraction lemma   13
convex neighborhood   214
corners   39
cotangent bundle   58, 146
covariant derivative   191, 195, 265
critical point   92, 182
curvature operator   233
curvature tensor   227

curve 87
cut-off function 196

D

d* 277
Darboux theorem 150
decomposable 127, 128
degenerate block 288
density 304
dependence on parameters 70, 74
de Rham cohomology 322
derivation 115
derivative 6
differentiable 6
differential equations 65, 158
differential form 58, 122
differential of exponential map 240
direct product 49
direct sum 59
divergence 263
divergence theorem 329
divisor 339
domain of definition 78, 84, 90, 105
dual bundle 55, 58, 146
duality of higher degree forms 273
duality of vector fields and
    1-forms 141

E

embedding 25
energy 184, 189, 251
exact form 135, 281, 321, 322
exact sequence 50, 52
exponential map 105, 109, 209, 240,
    259
exterior derivative 125, 130

F

factor bundle 51
fiber 41
fiber bundle 101
fiber product 29
finite type 62
first variation 253
flow 84, 89
forms 3
frame 43
Frobenius theorem 155
frontier 315
function 31
functional 36, 187, 300, 302, 345

functor 2, 57
functor of class C^p 56

G

Gauss lemma 214, 241
Gauss theorem 330
g-distance 185
geodesic 95
geodesic flow 107
geodesically complete 218
global smoothness of flow 85
gradient 144
Green's formula 332
group manifold 163

H

half plane 36
Hamiltonian 145
harmonic function 332
HB-morphism 177
hermitian operator 347
Hilbert bundle 176
Hilbert group 173, 177
Hilbert space 344
Hilbert trivialization 177
Hodge conditions 279
Hodge decomposition 279
Hodge star 273
Hodge theorem 273
homomorphism 165
Hopf–Rinow theorem 219
horizontal subbundle 103
hyperplane 36
hypersurface 212

I

immersion 24
implicit mapping theorem 18
index form 250
initial condition 65, 88
inner product 343
integrable vector bundle 154,
    166
integral 11
integral curve 65
integral manifold 160
integration of density 305
integration of forms 302
interior 38
isometry 187
isomorphism 83

isotopic   111
isotopy of tubular neighborhoods   111

J

Jacobi differential equation   233
Jacobi field or lift   233, 269
Jacobi identity   116
Jacobian of exponential map   268

K

kernel   52
kinetic energy   145, 189
Koszul's formalism   273

L

Laplace operator   264, 267, 278
Laut   3
left invariant   164
length   184
level hypersurface   212
Levi-Civita derivative   204
lie above   199
Lie algebra   164
Lie derivative   121, 138, 140
Lie group   163
Lie subalgebra   165
Lie subgroup   165
lifting   95, 199
linear differential equation   74
Lipschitz condition   66, 284
Lipschitz constant   66
Lis   3
local coordinates   21
local flow   65
local isomorphism   13, 109
local projection   17
local representation   44, 65, 87, 96, 98,
     122, 147, 189, 194, 202, 205
local smoothness   76, 78
locally closed   23
locally convex   3
locally finite   31

M

manifold   21
manifolds of maps   23
manifold with boundary   36, 297
manifold without boundary   37
mean valuer theorem   11
measure associated with a form   302

measure 0   285
metric   170
metric derivative   204
metric isomorphism   187
metric spray   207
minimal geodesic   217
modeled   21, 41, 142
momentum   148
morphism   2, 8, 22
Morse–Palais lemma   182
Moser's theorem   328
multilinear tensor field   59

N

natural transformation   2
negligible   315
non-degenerate   182
non-singular   141, 142, 149, 182
norm   343
norm of operator   4
normal bundle   54, 109
normal chart   211
normal neighborhood   211

O

one-parameter subgroup   168
operation on vector field   115
operations on vector bundles   56
operator   141, 174
orientation   261, 297
oriented chart   262
oriented volume   289
orthonormal frame   262, 266

P

paracompact   131
parallel   100, 202
parameter   70, 158
parametrized by arc length   187
partial derivative   8
partition of unity   32
path lifting   222
perpendicular   343
Poincaré lemma   135
Poisson bracket   146
polar coordinates   216, 269
positive definite   170
projection   17
proper domain of isotopy   111
pseudo Riemannian derivative   204
pseudo Riemannian manifold   171

pseudo Riemannian metric 170
pull back 30

R

reduction to Hilbert group 177
refinement 31
regular 315
related vector fields 117
representation, local, *see* local
     representation
residue theorem 340
Ricci tensor 232
Riemann tensor 227
Riemannian density 306
Riemannian manifold 171
Riemannian metric 170
Riemannian volume form 262
rule mapping 10

S

Sard theorem 173, 286
scalar curvature 232
scaling 230
scalloped 34
second-order differential equation 96
second-order vector field 95
second variation 252
section 3
sectional curvature 230
self dual 141
semi Riemannian 173
seminegative curvature 245
semipositive operator 348
shrinking lemma 13, 67
singular 199, 315
skew symmetric 173
spectral theorem 351
spectrum 352
sphere 212
split (injection) 16
split subspace 4
split vector bundle 62
spray 97, 104, 189, 194, 208
standard 2-form 150
star operator 273
step mapping 10
Stokes' theorem for rectangles 308
Stokes' theorem on a manifold 310
Stokes' theorem with singularities 316
subbundle 50, 54
submanifold 24
submersion 25

support 31, 300, 321
symmetric 142, 173
symmetric bilinear form on vector
     bundle 171
symplectic manifold 145

T

tangent bundle 49
tangent curves 88
tangent space 26
tangent subbundle 153
tangent to 0 6
tangent vector 26
Taylor expansions 257
Taylor formula 11
tensor bundle 59
tensor field 59
time dependent 65, 69
toplinear isomorphism 3
topological vector space 3
total space 41
total tubular neighborhood 108
transition map 42
transpose of $d\exp_x$ 243
transversal 28
trivial vector bundle 43
trivializing covering 41
trivializable 62
tube 108
tubular map 108
tubular neighborhood 108, 178

U

uniqueness theorem 68

V

variation formula 252
variation of a curve 236
variation at end points 242
variation through geodesics 237
VB (vector bundle) equivalent 41
VB chart 44
VB morphism 44
vector field 87, 115
vector field along curve 201
vector subbundle 103
volume form 262, 265, 303

W

wedge product 125
Whitney sum 59